# Water Security in a New World

There is a greatly heightened sense of awareness amongst politicians, policymakers, researchers and the general public that water security is a new and emerging threat. Just in the past few years, a number of high-level meetings involving the world's leaders and thinkers have focused on water security. With water security now commanding global attention, specific questions are posed on the likelihood of armed conflict and war over shared water resources, on the continuing availability of water resources to produce sufficient food for 9 or 10 billion people, on the probability of providing safe drinking water to every man, woman and child, and on the impact of climate change to create extreme water events – such as typhoons, floods and droughts – for which we are not prepared. By bringing together inputs from the world's leading thinkers, experts, practitioners and researchers, the Water Security in a New World series aims to provide evidence-based and policy-relevant responses to these and many other questions related to water security. The volumes in this series will provide in-depth analysis of the various dimensions of water security and are meant to be used by researchers, policymakers and practitioners alike.

**Editorial Board:**

Zafar Adeel, Pacific Water Research Centre, Simon Fraser University, Canada (Editor-in-Chief)
David Devlaeminck, Xiamen University, China
Dustin Garrick, University of Oxford, UK
Michael Glantz, University of Colorado, Boulder, USA
Joyeeta Gupta, University of Amsterdam, The Netherlands
Uma Lele, Independent Scholar, Formerly Senior Advisor, The World Bank, USA
Bob Sandford, University of British Columbia, Canada.

Elena López Gunn • Peter van der Keur
Nora Van Cauwenbergh • Philippe Le Coent
Raffaele Giordano

Editors

# Greening Water Risks

Natural Assurance Schemes

 Springer

*Editors*
Elena López Gunn
ICATALIST S.L.
Las Rozas, Madrid, Spain

Nora Van Cauwenbergh
IHE Delft Institute for Water Education
Delft, The Netherlands

Raffaele Giordano
CNR IRSA
Bari, Italy

Peter van der Keur
Geological Survey of Denmark and
Greenland (GEUS)
Copenhagen, Denmark

Philippe Le Coent
French Geological Survey
Montpellier, France

ISSN 2367-4008          ISSN 2367-4016    (electronic)
Water Security in a New World
ISBN 978-3-031-25310-2          ISBN 978-3-031-25308-9    (eBook)
https://doi.org/10.1007/978-3-031-25308-9

This work was supported by EU Horizon 2020 NAIAD Project (Grant No. 730497).

This Springer imprint is published by the registered company Springer Nature Switzerland AG
The registered company address is: Gewerbestrasse 11, 6330 Cham, Switzerland

# Contents

**1 Concepts in Water Security, Natural Assurance Schemes and Nature-Based Solutions** .................................... 1
Peter van der Keur, Nora Van Cauwenbergh, Elena López Gunn,
Jonatan Godinez Madrigal, Philippe Le Coent,
and Raffaele Giordano

**2 A Reader's Guide to Natural Assurance Schemes** ............... 19
Elena López Gunn, Laura Vay, and Carlos Marcos

**3 Insurance and the Natural Assurance Value (of Ecosystems) in Risk Prevention and Reduction** ........................... 35
Roxane Marchal, David Moncoulon, Elena López Gunn,
Josh Weinberg, Kanika Thakar, Mónica Altamirano,
and Guillaume Piton

**4 Methodologies to Assess and Map the Biophysical Effectiveness of Nature Based Solutions** ...................... 51
Mark Mulligan, Sophia Burke, Caitlin Douglas,
and Arnout van Soesbergen

**5 Participatory Modelling for NBS Co-design and Implementation** ... 67
Raffaele Giordano and Alessandro Pagano

**6 Economic Assessment of Nature-Based Solutions for Water-Related Risks** ..................................... 91
Philippe Le Coent, Cécile Hérivaux, Javier Calatrava,
Roxane Marchal, David Moncoulon, Camilo Benítez Ávila,
Mónica Altamirano, Amandine Gnonlonfin, Ali Douai,
Guillaume Piton, Kieran Dartée, Thomas Biffin, Nabila Arfaoui,
and Nina Graveline

7    **Designing Natural Assurance Schemes with Integrated
     Decision Support and Adaptive Planning** . . . . . . . . . . . . . . . . . . . . . . .   113
     Laura Basco-Carrera, Nora Van Cauwenbergh,
     Eskedar T. Gebremedhin, Guillaume Piton, Jean-Marc Tacnet,
     Mónica A. Altamirano, and Camilo A. Benítez Ávila

8    **NAS Canvas: Identifying Business Models to Support
     Implementation of Natural Assurance Schemes** . . . . . . . . . . . . . . . .   135
     Beatriz Mayor, Elena López Gunn, Pedro Zorrilla-Miras,
     Kieran Dartée, Thomas Biffin, and Karina Peña

9    **Closing the Implementation Gap of NBS
     for Water Security: Developing an Implementation Strategy
     for Natural Assurance Schemes** . . . . . . . . . . . . . . . . . . . . . . . . . . . . . .   149
     Mónica A. Altamirano, Hugo de Rijke, Begoña Arellano,
     Florentina Nanu, Marice Angulo, Camilo Benítez Ávila,
     Kieran Dartée, Karina Peña, Beatriz Mayor, Polona Pengal,
     and Albert Scrieciu

10   **Reducing Water Related Risks in the Lower Danube
     Through Nature Based Solution Design: A Stakeholder
     Participatory Process** . . . . . . . . . . . . . . . . . . . . . . . . . . . . . . . . . . . . . .   171
     Albert Scrieciu, Sabin Rotaru, Bogdan Alexandrescu,
     Irina Catianis, Florentina Nanu, Roxane Marchal,
     Alessandro Pagano, and Raffaele Giordano

11   **Multidisciplinary Assessment of Nature-Based Strategies
     to Address Groundwater Overexploitation and Drought
     Risk in Medina Del Campo Groundwater Body** . . . . . . . . . . . . . . . .   201
     Beatriz Mayor, África de la Hera-Portillo, Miguel Llorente,
     Javier Heredia, Javier Calatrava, David Martínez, Marisol Manzano,
     María del Mar García-Alcaraz, Virginia Robles-Arenas,
     Gosia Borowiecka, Rosa Mediavilla, José Antonio de la Orden,
     Julio López-Gutiérrez, Héctor Aguilera-Alonso,
     Laura Basco-Carrera, Marta Faneca, Patricia Trambauer,
     Tiaravani Hermawan, Raffaele Giordano, Eulalia Gómez,
     Pedro Zorrilla-Miras, Marta Rica, Laura Vay, Félix Rubio,
     Carlos Marín-Lechado, Ana Ruíz-Constán,
     Fernando Bohoyo-Muñoz, Carlos Marcos, and Elena López Gunn

12   **Natural Flood Management in the Thames Basin:
     Building Evidence for What Will and Will Not Work** . . . . . . . . . . . .   223
     Mark Mulligan, Arnout van Soesbergen, Caitlin Douglas,
     and Sophia Burke

13 Giving Room to the River: A Nature-Based
Solution for Flash Flood Hazards? The Brague River
Case Study (France) ........................................ 247
Guillaume Piton, Nabila Arfaoui, Amandine Gnonlonfin,
Roxane Marchal, David Moncoulon, Ali Douai,
and Jean-Marc Tacnet

14 Can NBS Address the Challenges of an Urbanized
Mediterranean Catchment? The Lez Case Study ................ 269
Philippe Le Coent, Roxane Marchal, Cécile Hérivaux,
Jean-Christophe Maréchal, Bernard Ladouche, David Moncoulon,
George Farina, Ingrid Forey, Wao Zi-Xiang, and Nina Graveline

15 Glinščica for All: Exploring the Potential
of NBS in Slovenia: Barriers and Opportunities ................ 297
Polona Pengal, Alessandro Pagano, Guillaume Piton,
Zdravko Kozinc, Blaž Cokan, Zarja Šinkovec,
and Raffaele Giordano

16 The Opportunities and Challenges for Urban
NBS: Lessons from Implementing the Urban Waterbuffer
in Rotterdam ............................................. 325
Kieran Wilhelmus Jacobus Dartée, Thomas Biffin,
and Karina Peña

17 Urban River Restoration, a Scenario for Copenhagen ........... 347
Morten Ejsing Jørgensen, Jacob Kidmose, Peter van der Keur,
Eulalia Gómez, Raffaele Giordano, and Hans Jørgen Henriksen

18 Enabling Effective Engagement, Investment
and Implementation of Natural Assurance Systems
for Water and Climate Security.............................. 367
Josh Weinberg, Kanika Thakar, Roxane Marchal, Florentina Nanu,
Beatriz Mayor, Elena López Gunn, Guillaume Piton, Polona Pengal,
and David Moncoulon

19 The Natural Assurance Schemes Methodological
Approach – From Assessment to Implementation................ 385
Nora Van Cauwenbergh, Raffaele Giordano, Philippe Le Coent,
Elena López Gunn, Beatriz Mayor, and Peter van der Keur

20 Looking into the Future: Natural Assurance Schemes
for Resilience ............................................ 407
Elena López Gunn, Nina Graveline, Raffaele Giordano, Nora Van
Cauwenbergh, Philippe Le Coent, Peter van der Keur,
Roxane Marchal, Beatriz Mayor, and Laura Vay

# Chapter 1
# Concepts in Water Security, Natural Assurance Schemes and Nature-Based Solutions

**Peter van der Keur, Nora Van Cauwenbergh, Elena López Gunn, Jonatan Godinez Madrigal, Philippe Le Coent, and Raffaele Giordano**

**Highlights** This chapter illustrates how nature-based solutions, operationalized in natural assurance schemes can increase water security using the readiness level concept to address barriers to implementation

- The concept of water security strategies in the context of water related hazards and mitigated by Nature-based Solutions is analyzed and conceptualized in Natural Assurance Schemes
- Operationalization of Natural Assurance Schemes are tailored to the specific regulatory context of the insurance sector and its stakeholders
- Readiness levels with respect to technology, institutions and investment are developed to address and overcome barriers to implement Nature-based Solutions and Natural Assurance Schemes.

P. van der Keur (✉)
Geological Survey of Denmark and Greenland (GEUS), Copenhagen, Denmark
e-mail: pke@geus.dk

N. Van Cauwenbergh · J. Godinez Madrigal
IHE Delft Institute for Water Education, Delft, The Netherlands

E. López Gunn
ICATALIST S.L., Las Rozas, Madrid, Spain

P. Le Coent
G-EAU, BRGM, Université de Montpellier, Montpellier, France

R. Giordano
CNR-IRSA, Bari, Italy

© The Author(s) 2023
E. López Gunn et al. (eds.), *Greening Water Risks*, Water Security in a New World, https://doi.org/10.1007/978-3-031-25308-9_1

## 1.1 Introduction

In the face of the looming water crisis, the concept of water security has been positioned by many nations and international organizations as a major societal objective in recent years. But given the importance of water security, how can we define this concept? What does water security mean in practice, and how does it relate to ecosystems and nature-based solutions? In a way, water security seems to be negatively defined as the avoidance or absence of water crises, conflicts or even wars (The World Climate and Security Report 2020). However, the scientific community has proposed more elaborate definitions of this concept. Table 1.1 offers four different definitions based on four approaches to water security we have identified in the literature review.

This chapter will first introduce a brief discussion on the concept and framing of water security, then introduce the concept of assurance schemes to present the readiness approach (Fig. 1.1) adopted in our case studies and finally conclude by presenting the main questions that will be addressed in this edited book.

**Table 1.1** Approaches and definitions of water security

| Approach | Reference | Definition |
|---|---|---|
| Water availability and reduced risks | Grey and Sadoff (2007) | Water security is the availability of an acceptable quantity and quality of water for health, livelihoods, ecosystems and production, coupled with an acceptable level of water-related risks to people, environments and economies" |
| Sufficient, safe and efficient supply of water | Falkenmark (2013) | Water security is, therefore, essential for a society's survival, health, and prosperity. Scarcity of water or difficulty to safeguard access, is consequently an obstacle and functions as a bottleneck in socioeconomic development. |
| | Rijsberman (2006) | Water security is sufficiency of water supply for humans. It exists when access is secured to sufficiently safe and affordable water to satisfy individual needs for drinking, washing, and livelihood. |
| Social equity and link to other types of security | Zeitoun (2011) | Sustainable water security is interpreted as a function of the degree of equitability and balance between the six related security areas (human-community security, national security, climate security, energy security, food security, water resources security), as this plays out within a web of socioeconomic and political forces at multiple spatial levels. |
| Governance for equitable access of people and ecosystems | Boelens and Seemann (2014) | Water security refers to people's and ecosystems' secure, sustainable access to water, including equitable distribution of advantages/disadvantages related to water use, safeguarding against water-based threats, and ways of *sharing decision-making power in water governance.* |

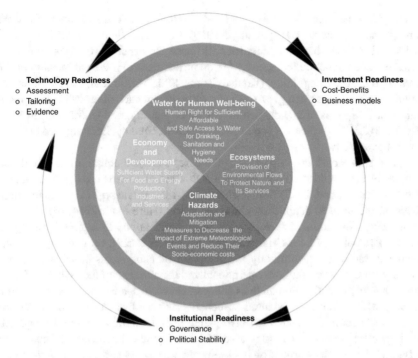

**Fig. 1.1** Water security concept (modified from UN-water.org, 2013) Ecosystems-based adaptation (EbA) and Ecosystem Disaster Risk Reduction (ecoDRR) have enormous potential for reducing losses and damage from natural hazards and therefore contributes to water security

## 1.2    The Evolution of the Concept of Water Security

The concept of water security has evolved over the last twenty years to incorporate different aspects of, and also as an answer to, certain criticisms and shortcomings from earlier conceptualisations. Table 1.1 summarises this evolution which we discuss now to frame natural assurance schemes as part of water security and place special emphasis on environmental and social dimensions.

A first approach by Grey and Sadoff (2007) defines the concept of water security in terms of 'acceptable quantity and quality´, such approaches promote large grey infrastructure such as dams and water transfers as a vital strategy to address water scarcity, but as is demonstrated in e.g. the case study of Rotterdam (Chap. 16, this volume), nature-based solutions can complement rainwater harvesting to mitigate the effects of cloudburst generated flooding with water storage to anticipate water scarcity. A similar definition of water security is provided by UN-water.org (2013) (Fig. 1.1). Therefore, water insecurity is not a matter of scarcity of water resources, but rather the absence of storage, including green infrastructure, to increase water supply and provide water availability at certain times of the year when it is most needed, e.g. for irrigation.

Despite the criticisms of the social and environmental costs of large infrastructure (Garandeau et al. 2014; Jeuland 2010; Molle and Floch 2008; Wang et al. 2014), it is argued that some countries still need the benefits from this built infrastructure, and cannot be replaced by 'green infrastructure' if they are to support the needs of a growing population (Koutsoyiannis 2011; Muller et al. 2015), but hybrid solutions where green and grey solutions complement each other have large potential. Four categories of key challenges for water security are identified in Investing in Nature (2019): surface water quality, groundwater quality, flooding, and droughts and water scarcity.

These approaches are important in the cases studies included in this book, but can easily be extended to e.g. the region of Central America and the Caribbean (Mysiak and Calliari 2013) or elsewhere in disaster-prone regions or increasingly focus on the need to adapt throughout the world (Hare et al. 2014; Van der Keur et al. 2016). Institutional factors, including combinations of international agreements, national regulations and planning, as well as local level capacity development can facilitate substantially the adoption of ecosystems-based approaches. Notably nature-based solutions operationalized in natural assurance schemes (NAS) have value in protecting human lives and infrastructure against the effects of natural hazards while offering substantial co-benefits like biodiversity, carbon sequestration, better health. Urban and regional planning and natural resource management are important areas that can play a central role in the enhancement of ecosystem services. The positive correlation between enhanced ecosystems and poverty reduction warrants more attention as the poor are frequently the most vulnerable to the effects of disasters (Hare et al. 2014).

Ecosystems are constituting components of the natural and semi-natural environment, and a source of vital services, benefits and goods to humankind. Ecosystem services (ES) inhabit provisioning, cultural, supporting and regulating properties and embody the benefits people obtain from ecosystems that are eventually translated into valuable goods. The ES regulatory services thus include natural hazard mitigation and contribute effectively to tackling the drivers of social and economic vulnerability to natural hazards. Ecosystem services delivered by NBS are often, as illustrated in case studies in this volume, a more cost-effective way of dealing with climate extremes than 'hard infrastructure engineering solution (grey) alone and where nature-based solutions can supplement grey as hybrid solutions (e.g. Kazmierczak et al. 2020; Browder et al. 2019; Mysiak and Calliari 2013).

A second approach emphasizes that water insecurity is the result of context-specific increasing rates of population and economic growth and their relation to water utilization, consumption and availability. This approach considers a more dynamic and complex conceptualization of water insecurity by adding water crowding (number of people per million cubic meter per year) as a key indicator and linking the concept to food security. Falkenmark (2013) argues that there is an inexorable link between water scarcity and human and food security with projections to 2050 estimating a "carrying capacity overshoot in water-short countries with continuing population growth and therefore, meaning that there will be a massive dependence on food imports". Vörösmarty et al. (2010) estimated that nearly 80% of the world's population suffer high levels of water insecurity, specifically regions of intensive agriculture and dense settlements. In practice, this approach promotes two key strategies to bring about water security: vital large infrastructure as the main approach, but adding improved efficiency of water use as a key strategy, especially for agriculture.

A third approach to water security is related to the second one by emphasizing the importance of agriculture in the quantity and quality of available water. Allan (2013) criticises the apparent invisibility of the food-water relation and the pressure that is put on farmers, which in his view are 'society's water managers', to decrease their water use in favour of more productive uses often localized in cities (Molle and Berkoff 2006). However, this approach departs from the previous one by also incorporating other dimensions, such as household water security, urban water security, environmental water security, resilience to water-related disasters and economic water security (Van Beek and Arriens 2014). This approach acknowledges water as a highly complex natural resource that reaches out to other important issues and dimensions, whereby urban water security, often conceived as the main target, is just a fraction amidst the great challenges of "water for food, water for nature, sustainable use of water resources, closing water and nutrient cycles [...]" (Savenije 2002).

In the context of this book, important aspects of water security related to water related hazards and mitigation by means of nature-based solutions, are analysed, conceptualised and operationalised in natural assurance schemes. Following this idea, Zeitoun (2011) argues for the need to go beyond water, and conceive water as a web, in which food security, energy security, climate security, human/community security and national security are also included. In Chap. 7 (Basco-Carrera et al., this volume) IWRM is, in line with the third approach, elaborated for providing the guiding principles to achieve water security for all by means of strategic planning. Water security is here defined as the capacity of a population to safeguard sustainable access to adequate quantities of acceptable quality water for sustaining livelihoods, human well-being, and socio-economic development, for ensuring protection against water-borne pollution and water-related disasters, and for preserving ecosystems in a climate of peace and political stability (UN-Water 2013).

The fourth approach we identified, is a social perspective on the concept of water security. Bakker and Morinville (2013) analysed the role of social power, or power relations, as another cause of water insecurity besides "poor management decisions, suboptimal processes, insufficient science, and evolving environmental pressures", and conclude that vulnerability and uneven water security emerges from the exclusion of stakeholders from decision-making processes. The approach criticizes strategies that advocate for seemingly uncontroversial desirable objectives such as increasing water efficiency by promoting controversial measures such as water utility privatizations, which have been widely promoted by international institutions like the World Bank. Some of the unintended consequences of these measures is the worsening conditions of large sections of the population once the private water utility takes over (Bakker 2013: 215). This approach brings back the political dimension of water security, when other approaches intend to depoliticize agendas that include large grey infrastructure as a vital component of water security without considering alternatives to large grey infrastructure nor the uneven distribution of costs and benefits of large grey infrastructure across different sectors of society and natural systems (Boelens and Seemann 2014; Godinez-Madrigal et al. 2020). The focus of this book is to compare such grey infrastructures to smaller green intervention measures and the protective function of green infrastructure functions and its services.

Despite the multiple socio-ecological costs associated with grey infrastructure, the discursive power of the State to 'securitize' certain agendas may close social debates and the decision space, thereby depoliticizing water problems and what to do about them (Molle 2009). This political reality is important when discussing nature-based solutions, because when decision makers consider which kind of strategies are best suited to bring about water security, they may be biased towards tried-and-true large grey infrastructure given the risk aversion of many water managers and utility managers to explore different strategies (Marlow et al. 2013). However, when considering water security as a complex multi-dimensional concept, one-size fits-all, and one function only solutions may entail multiple pitfalls in related dimensions of water security.

Figure 1.2 shows this increasing level of complexity and new dimensions added with the evolution of the concept. New transdisciplinary knowledge has lately been developed that sheds light on the importance of considering a more complex approach to water security. In natural sciences, recent research has contributed to the understanding of long-term unintended consequences of large grey infrastructure in societies such as reservoirs, drainage networks and levees. The 'reservoir effect' emerges when societies over-rely on large water transfers for their water supply, an unintended consequence of the long-term dependency of societies to reservoirs and their increasing vulnerability to hydro-climatic events (Di Baldassarre et al. 2018). The 'levee effect' emerges when levees are built to protect societies from flooding, but in turn diminishes social memory over larger periods of time by

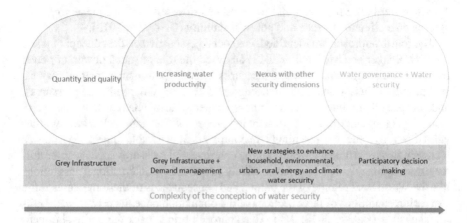

**Fig. 1.2** Evolution of conceptions of water security and related strategies. (Source: authors' own)

giving a sense of unfounded sense of security that can backfire with an extreme hydro-climatic event (Di Baldassarre et al. 2013). In the social sciences, recent research shows that large infrastructure can trigger intractable social conflicts that forestalls the development of any kind of solution, while urban water systems continue to deplete groundwater with increasing socio-ecological costs (Godinez-Madrigal et al. 2020).

This knowledge may force decision makers to reconsider alternatives to large grey infrastructure. Raymond et al. (2017) argues that nature-based solutions not only have the potential for bringing about the benefits purported by grey infrastructure, but also bring about socio-ecological co-benefits along. Therefore, in the pursuit of resilient societies and urban and rural water systems, the decision space needs to include nature-based solutions as necessary components to bring about water security, understood as a complex, multi-dimensional and transdisciplinary process.

## *1.2.1   The Assurance and Insurance Value of Ecosystems*

Ecosystems have a value based on their sustained capacity to maintain their functioning and production of benefits despite any disturbance by reducing risks to human society caused by e.g. climate change related excess precipitation, temperature or by natural disasters. In the context of NAS, the value of ecosystems is understood as the extent to which nature-based solutions operationalised in natural assurance schemes reduce water related risks from e.g. extreme events while at the same time provide co-benefits including increased health (physical and mental), increased attractiveness of living areas, especially in cities, and also benefits as improved water quality and quantity. The latter can be exemplified by e.g. re-infiltration and recharge to groundwater contributing to water quality and quantity respectively.

"Natural Assurance Schemes" (NAS) build on the potential that ecosystems have in reducing damage costs by mitigating water related risks, notably floods and drought under climate change, and increasing resilience of society. Natural assurance schemes are thus framed under the ecosystem services concept, i.e. the benefits people obtain from ecosystems, implemented as nature-based solutions for risk reduction and accompanying co-benefits. In general, ecosystem services include provisioning services such as food, water, timber and fiber; regulating services that affect climate, floods, disease, wastes, and water quality; cultural services that provide recreational, aesthetic, and spiritual benefits; and supporting services such as soil formation, photosynthesis, and nutrient cycling. Burkhard et al. (2012), define ecosystem services as the flow of materials, energy, and information from natural capital stocks, which combined with manufactured and human capital services to produce human welfare, which gives centrality to the concept of natural capital to human well-being.

### 1.2.2 The Concept of Natural Assurance Schemes

The underpinning concept for this edited volume is that of a natural assurance scheme or NAS. The NAS concept is itself based on the concept of natural assurance value, previously defined as the reduction of risks that natural systems can produce and associated benefits. The protective value of NAS must be economically and financially viable and include the multifunctional aspects of nature-based solutions, i.e. the primary function of mitigation and avoiding damages as well as accompanying co-benefits.

Assurance and insurance refer to a risk transfer mechanism, where a premium payment by a household, company or community to an insurer in return for having to reimburse their clients after a disaster occurs. Assurance generally applies to persistent coverage over extended periods of time or until death, whereas insurance refers to coverage over a limited amount of time. The insurance companies deal with natural hazards by modelling and pricing risks and therefore a strong knowledge on natural hazards is required by insurers to be able to face shocks and change induced by natural hazards under climate change. That compensation role is crucial for the national economy resilience. The insurance industry has a key role in protecting society from natural disasters throughout the insurance coverage providing financial support to society to reestablish itself as quick as possible, limiting damaging economic domino effects.

Natural Assurance Schemes denote a range of institutional, technological and financial mechanisms to operationalize the value of green infrastructure in the green/grey mix of mitigation against water risks. Natural assurance schemes are devised, with stakeholders, based on locally relevant but EU-wide physical, socio-cultural and economic valuation and could constitute a mechanism to finance the mitigation of risks through targeted investment of a proportion of insurance revenues.

Operationalization of these natural assurance schemes for the primary benefits of risk prevention and reduction and associated co-benefits must be tailored to the specific context and regulatory settings. The insurance industry is a heavily regulated industry at national level, and marked by high competition and range from the fully solidarity based, more hybrid systems to fully private (Fig. 1.3).

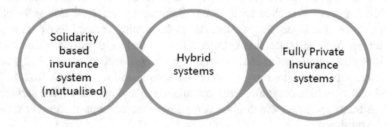

**Fig. 1.3** Insurance models from fully public to fully private. (Source: authors' own)

The design of a natural assurance scheme entails a series of different steps: (i) to undertake a robust physical assessment of the hazard and exposure and subsequently identify preferred and socially acceptable nature based strategies for stakeholders including an effective and a viable business model and financing scheme; (ii) to consider the potential role of notably the insurance industry in these natural assurance schemes which may include investors, data providers and other information facilitators to focus on potentially increased risks of natural hazards as a consequence of increased exposure and vulnerability compounded by climate change.

Improved understanding of natural assurance schemes can help develop Climate Resilient Investments. A climate-resilient investment is an investment that results from a process where governments, planners and developers integrate climate change in project planning and design, including climate finance (UNFCCC 2018) from relevant sources such as the Green Climate Fund, green bonds and now under the EU Green Deal. The proper integration of climate change in the planning and design of infrastructure investments including green infrastructure and hybrid infrastructure, may considerably reduce the risk of damage to national assets.

## 1.3 Readiness Level Concepts to Overcome Barriers and Implement NBS and NAS

When operationalizing the value of ecosystems in natural assurance schemes, a number of barriers have to be overcome. Those barriers relate to the technical, social and economic dimensions. Nesshöver et al. (2017) identified the following key elements and drivers needed to operationalize NBS and ensure its integration in climate adaptation plans: (i) dealing with uncertainty and complexity (e.g. by adaptive management); (ii) involvement of multiple stakeholders; (iii) use of multi- and transdisciplinary knowledge; (iv) common understanding of multifunctional solutions, trade-offs and natural adaptation; and (v) evaluate and monitor for mutual learning. Moreover, in a multiple case study analysis of NBS implementation in European cities by (Frantzeskaki 2019) highlights that NBS require multiple disciplines for their design, diversity (of settings) for co-creation and recognition of the place-based transformative potential of NBS as 'superior' to grey infrastructure. In this book, we build on the concept of NBS readiness by Van Cauwenbergh et al. (2020) and frame the operationalizing of insurance value of ecosystems as a process of increasing readiness for the implementation of NBS, rather than framing it as a process of overcoming barriers, and management of uncertainty, that may hamper integration in climate adaptation plans.

Three types of readiness are considered: (1) Technology and Knowledge readiness – linked to barriers on knowledge and performance (generation of evidence) + inclusion of certain benefits such as aesthetic appeal in the design– related to setting up an appropriate level of experimentation in a context of trust; (2) Socio-Institutional readiness – linked to barriers on acceptance, trust, handling uncertainty and ambiguity, multi-functional solutions and coordination, as well as innovative regulatory frameworks to deal with the inherent uncertainty of NBS and potential liabilities; and (3) <u>Investment readiness</u> – linked to capturing multiple values and valorizing the multiple benefits in public-private-people partnerships.

### 1.3.1 Technology and Knowledge Readiness

Technology readiness levels (TRL) were developed by NASA in the 1970s, as a common way was needed to describe the maturity and state of flight readiness of technology projects for which a 9-level description in a thermometer analogy was invented. TRL is widely used in industrial sectors that want to gauge the development and prospective market value of innovative developments as well as by potential investors or users of the technology as it gives an indication of utility of reliability (Webster and Gardner 2019).

TRL are also used by funding agencies, as a guideline for researchers, developers and innovators to target technology toward higher TRL. In its Horizon 2020 and Horizon Europe research and innovation framework, the EC foresees the use of TRL for non technological development. Figure 1.4 shows the different TRL levels, based on NASA and including some additional explanation to understand the definitions.

In the context of this book technology is defined not only as bio-physical components of the NBS and interaction with the natural environment, notably the hydrological cycle and related risks. Technology is also considered as a body of knowledge and general perception of the multi-functional performance of NBS by diverse stakeholder groups, in analogy with (Arthur 2010).

### 1.3.2 Institutional Readiness

The concept of Institutional Readiness Level (IRL) follows Webster and Gardner (2019) as a combination of 8 categories that have to be fulfilled for readiness to be achieved: (1) demand for technology, (2) strategic focus, (3) relative need and benefit of the new technology, (4) (e)valuation processes in place, (5) IRL enacted through specific enablers within and outside of the organization, (6) receptivity, (7) adaptive capacity and (8) sustainability (see Table 1.2 below). Originally developed in the field of philanthropic studies to understand which features and characteristics are more likely to improve the 'success' of an organisation, Barnes and Brayley (2006) and Webster and Gardner (2019) applied the concept to the field of regenerative medicine. In this book it is applied to mainstream NBS as a novel technology/approach in climate adaptation.

Contrary to the TRL which uses a numerical scale, the IRL categories in Table 1.2 differ in levels of maturity expressed qualitatively. Each of the categories needs to

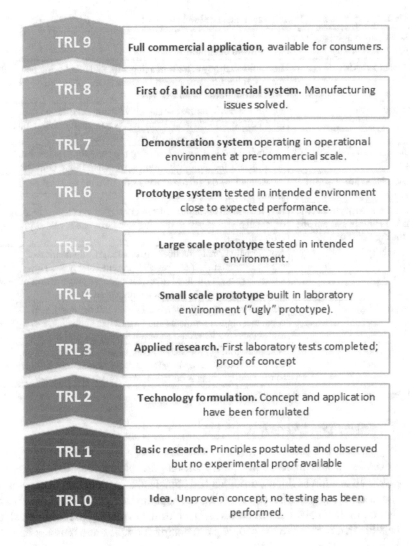

| TRL 9 | Full commercial application, available for consumers. |
| TRL 8 | First of a kind commercial system. Manufacturing issues solved. |
| TRL 7 | Demonstration system operating in operational environment at pre-commercial scale. |
| TRL 6 | Prototype system tested in intended environment close to expected performance. |
| TRL 5 | Large scale prototype tested in intended environment. |
| TRL 4 | Small scale prototype built in laboratory environment ("ugly" prototype). |
| TRL 3 | Applied research. First laboratory tests completed; proof of concept |
| TRL 2 | Technology formulation. Concept and application have been formulated |
| TRL 1 | Basic research. Principles postulated and observed but no experimental proof available |
| TRL 0 | Idea. Unproven concept, no testing has been performed. |

**Fig. 1.4** TRL levels

be at a sufficient level of maturity or readiness for socio-institutional readiness to be achieved. At the core of IR is the existence of an effective communication between stakeholders as well as pluri- and transdisciplinary cooperation to capture the multi-functional character of NBS and translate it into a fair and sustainable distribution of its multiple benefits. This means that in order to successfully integrate NBS in water related risk strategies, substantial coordination across municipal or sectoral organizations is needed as discussed by Van Cauwenbergh et al. (2021).

**Table 1.2** Key aspects of institutional readiness levels for NBS

| Institutional readiness levels (IRL) category | Operational definition |
| --- | --- |
| Demand for NBS | Institution has key actors engaging with and identifying new NBS that meet field/organizational needs in CCA, DRR and Water Resources Planning (WRP) |
| Strategic focus | Institution has identified potential NBS and determined their relationship with existing technologies and grey infrastructure to achieve water-related security and resilience |
| Relative need and benefit of NBS | Institution has key actors assessing the capacity to take-on and develop new technologies within current and future contexts |
| (E)valuation processes in place | Assessment of the (diverse) values of NBS are undertaken and shared |
| IR enacted through specific enablers within and outside of the organization | Key individuals/groups are formally tasked to enable adoption especially with regard to meeting standards and regulatory requirements |
| Receptivity | Novel institutional structures are created, in anticipation of expected challenges / affordances presented by NBS. These structures reflect the need to retrain staff, the construction of new innovation spaces and new technology platforms. |
| Adoptive capacity | NBS aligns with institutional priorities and organisational capacities. Initial problems and unanticipated challenges/affordances are identified and seen to be manageable |
| Sustainability | NBS is routinely produced/used/assessed within institutions. Current institutional arrangements and resources are sufficient for routine and ongoing production, assessment and deployment |

Adapted from Webster and Gardner (2019)

### 1.3.3 Investment Readiness

While academia and increasingly policy makers are promoting NBS as a cost-effective way to address floods, droughts and climate change resulting in economic, social and environmental benefits, investment appetite in NBS is still low. This is largely due to the unclear return on investment, as the capturing of multiple values and benefits in a public-private-people partnership is complex and requires innovative business models.

In this book we discuss business models specifically developed for NBS and NAS and link it to the concept of investment readiness. We use the definition by Blank (2014) for investment readiness or IVRL. In analogy to TRL, eight levels of investment readiness are indicated as a simple and visual way to share a common understanding of investment readiness status (Fig. 1.5). The development of the business canvas (Chap. 8, this volume) is at the core of generating investment readiness in where the canvas is representing how lower levels of IVRL can be overcome by various ways to mobilize the funds and finance. The modified business canvas for NBS-based Natural Assurance Schemes by (Mayor et al. 2017; Chap. 8, this volume) underlies the considered IVRL and addresses the growing interest of leaders

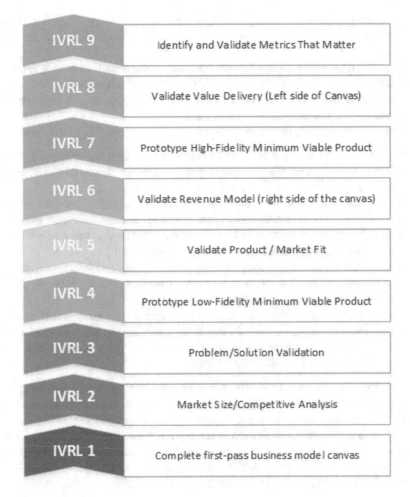

**Fig. 1.5** IVRL levels

from philanthropy, development and finance to mobilise capital to effectively solve social and environmental issues (Höchstädter and Scheck 2015) (Fig. 1.5).

## 1.4   Main Questions to be Addressed by the Book

Nature-based solutions, as operationalised in Natural Assurance Schemes, are important to ensure water security in various ways. The leading thread in this book is to increase our understanding of NBS through a multidisciplinary approach and investigate how NBS can contribute to water security by helping mitigate water related hazards while at the same time contributing to maintaining water resources of sufficient quality and quantity. Strengthening the knowledge on NBS can help scaling up implementation of NBS in Natural Assurance Schemes in cities and basins.

The main questions addressed by the book follow the overall chapter structure of this work and help to understand how to improve readiness at the level of Technology (TRL), Institutions (IRL) and Business (IVRL) for developing and implementing NBS and NAS while increasing water security.

Essential overarching questions include:

- how to develop and apply methodologies to assess the effectiveness of NBS for different water related natural hazards, physical environments and spatial scales; how do they add to water security; (contribute to TRL)
- how can understanding and mapping stakeholder participation processes and risk perception help NAS development in the planning process; contribute to IRL
- what is the economic value of NBS and how can NBS be assessed through a cost-benefit analysis framework; (contribute to IRL and IVRL)
- What decision process can support analysing, selecting and implementing NBS with the view to reach a robust strategy that contributes to the different dimensions of water security; (contribute to TRL, IRL and IVRL)
- how to operationalise NBS and identify suitable business models and enabling environment in order to build effective NAS; (contribute to IRL and IVRL) what business models emerge from capturing the assurance value of ecosystems? Can this be insured? how would these be financed?

The leading questions are addressed and illustrated in a range of contrasting case studies in Europe both with respect to (1) varying environmental, physical conditions, spatial scale and vulnerability to water related natural hazards that require diverse NAS approaches, but also to (2) varying readiness level of technology, institutionality and investment for implementation of nature-based solutions in NAS. Illustration of how developed methodologies and strategies are applied in the case studies serve important lessons learned in the final chapters of this book.

# References

Allan JA (2013) Food-water security: beyond water resources and the water sector. In: Lankford B, Bakker, Zeitoun M, Conway D (eds) Water security: principles, perspectives and practices. Routledge

Arthur WB (2010) What is technology and how does it evolve? (Book Excerpt). The New York Academy of Sciences Magazine

Bakker K (2013) Neoliberal Versus Postneoliberal Water: Geographies of Privatization and Resistance. Ann Assoc Am Geogr 103(2):253–260. https://doi.org/10.1080/0004560 8.2013.756246

Bakker K, Morinville C (2013) The governance dimensions of water security: a review. Philos Trans R Soc A Math Phys Eng Sci 371(2002):20130116–20130116. https://doi.org/10.1098/rsta.2013.0116

Barnes ML, Brayley RE (2006) Institutional readiness and grant success among public recreation agencies. Manag Leis 11(3):139–150. https://doi.org/10.1080/13606710600720739

Blank S (2014) How investors make better decisions. The investment readiness level. Consulted May 2021

Boelens R, Seemann M (2014) Forced engagements: water security and local rights formalization in Yanque, Colca Valley, Peru. Hum Organ 73(1):1–12. https://doi.org/10.17730/humo.73.1.d44776822845k515

Browder R, Ozment S, Rehberger Bescos I, Gartner T, Lange GM (2019) Integrating green and gray, creating next generation infrastructure, Worldbank/WRI report, p 7

Burkhard B, Kroll F, Nedkov S, Müller F (2012) Mapping ecosystem service supply, demand and budgets. Ecol Indic 21. https://doi.org/10.1016/j.ecolind.2011.06.019

Di Baldassarre G, Kooy M, Kemerink JS, Brandimarte L (2013) Towards understanding the dynamic behaviour of floodplains as human-water systems. Hydrol Earth Syst Sci Discuss 10(3)

Di Baldassarre G, Wanders N, AghaKouchak A, Kuil L, Rangecroft S, Veldkamp TI et al (2018) Water shortages worsened by reservoir effects. Nat Sustain 1(11):617–622

Falkenmark M (2013) The multiform water security dimension. In: Lankford B, Bakker, Zeitoun M, Conway D (eds) Water security: principles, perspectives and practices. Routledge

Frantzeskaki N (2019) Seven lessons for planning nature-based solutions in cities. Environ Sci Policy. https://doi.org/10.1016/j.envsci.2018.12.033

Garandeau R, Edwards S, Maslin M (2014) Biophysical, socioeconomic and geopolitical impacts assessments of large dams: an overview. University College London. Retrieved from https://www.ucl.ac.uk/hazardcentre/documents/Mega-dam-overview

Godinez-Madrigal J, Van Cauwenbergh N, Van der Zaag P (2020) Unravelling intractable water conflicts: the entanglement of science and politics in decision-making on a large hydraulic infrastructure project. Hydrol Earth Syst Sci Discuss. https://doi.org/10.5194/hess-2020-86, in review, 2020

Grey D, Sadoff CW (2007) Sink or Swim? Water security for growth and development. Water Policy 9(6):545–571. https://doi.org/10.2166/wp.2007.021

Hare MP, van Bers C, van der Keur P, Henriksen H-J, Luther J, Kuhlicke C, Jaspers F, Terwisscha van Scheltinga C, Mysiak J, Calliari E, Warner K, Daniel H, Coppola J, McGrath PF (2014) CATALYST – a multi-regional stakeholder Think Tank for fostering capacity development in disaster risk reduction and climate change adaptation Nat. Hazard Earth Syst Sci 14:2157–2163. https://doi.org/10.5194/nhess-14-2157-2014

Höchstädter AK, Scheck B (2015) What's in a name: an analysis of impact investing understandings by academics and practitioners. J Bus Ethics. https://doi.org/10.1007/s10551-014-2327-0

Investing in Nature (2019) Private finance for nature-based resilience. The Nature Conservancy. Environmental Finance, report, 2019. Retrieved from: www.nature.org

Jeuland M (2010) Social discounting of large dams with climate change uncertainty. Water Altern 3(2):185–206

Kazmierczak et al. (2020). Urban adaptation in Europe: how cities and towns respond to climate change. EEA Report, No.12/2020

Koutsoyiannis D (2011) Scale of water resources development and sustainability: small is beautiful, large is great. Hydrol Sci J 56(January 2015):553–575. https://doi.org/10.1080/02626667.2011.579076

Marlow DR, Moglia M, Cook S, Beale DJ (2013) Towards sustainable urban water management: a critical reassessment. Water Res 47(20):7150–7161

Mayor B et al (2017) Guidelines for the application of the NAS canvas to NBS strategies. EU Horizon 2020 NAIAD Project, Grant Agreement N°730497

Molle F (2009) Water, politics and river basin governance: repoliticizing approaches to river basin management. Water Int 34(1):62–70. https://doi.org/10.1080/02508060802677846

Molle F, Berkoff J (2006) Cities versus Agriculture: revisiting intersectoral water transfers, potential gains and conflicts. Assessment. Colombo, Sri Lanka: Comprehensive Assessment Secretariat: Comprehensive Assessment Research Report 10

Molle F, Floch P (2008) Megaprojects and social and environmental changes: the case of the Thai "water grid". Ambio 37(3):199–204. https://doi.org/10.1579/0044-7447(2008)37[199:MASA EC]2.0.CO;2

Muller M, Biswas A, Martin-hurtado R, Tortajada C (2015) Built infrastructure is essential. Science 349(6248):585–586

Mysiak J, Calliari E (2013) Dwindling land: small island developing states. In: Hare M, van Bers C, Mysiak J (eds) A best practices notebook for disaster risk reduction and climate change adaptation: guidance and insights for policy and practice from the CATALYST Project. TWAS

Nesshöver C, Assmuth T, Irvin KN, Rusch GM, Waylen KA, Delbaere B, Haase D, Jones-Walters L, Keune H, al., et. (2017) The science, policy and practive of nature-based solutions: An inter-disciplinary perspective. Sci Total Environ 579:1215–1227

Raymond CM, Frantzeskaki N, Kabisch N, Berry P, Breil M, Nita MR et al (2017) A framework for assessing and implementing the co-benefits of nature-based solutions in urban areas. Environ Sci Pol 77:15–24

Rijsberman FR (2006) Water scarcity: fact or fiction? Agric Water Manag 80(1–3 SPEC. ISS):5–22. https://doi.org/10.1016/j.agwat.2005.07.001

Savenije HHG (2002) Why water is not an ordinary economic good, or why the girl is special. Phys Chem Earth 27(11–22):741–744. https://doi.org/10.1016/S1474-7065(02)00060-8

The World Climate and Security Report (2020) Product of the Expert Group of the International Military Council on Climate and Security. Authors: Steve Brock (CCS), Bastien Alex (IRIS), Oliver-Leighton Barrett (CCS), Francesco Femia (CCS), Shiloh Fetzek (CCS), Sherri Goodman (CCS), Deborah Loomis (CCS), Tom Middendorp (Clingendael), Michel Rademaker (HCSS), Louise van Schaik (Clingendael), Julia Tasse (IRIS), Caitlin Werrell (CCS). Edited by Francesco Femia & Caitlin Werrell. Published by the Center for Climate and Security, an institute of the Council on Strategic Risks. Feb 2020

UNFCCC (2018) UNFCCC Standing Committee on Finance 2018 Biennial assessment and overview of climate finance flows technical report

UN-Water (2013) https://www.unwater.org/publications/water-security-infographic

Van Beek E, Arriens WL (2014) Water security: putting the concept into practice. Global Water Partnership, Stockholm

Van Cauwenbergh N, Dourojeanni P, Mayor B, Altamirano M, Dartee K, Basco-Carrera L, Piton G, Tacnet JM, Manez M, Lopez-Gunn E (2020) Guidelines for the definition of implementation and investment plans for adaptation. EU Horizon 2020 NAIAD Project, Grant Agreement N°730497

Van Cauwenbergh N, Dourojeanni P, Van Der Zaag P, Brugnach M, Dartee K, Giordano R, Lopez-Gunn E (2021) Beyond TRL – Understanding institutional readiness for implementation of nature-based solutions. Environmental Science and Policy – in review

Van der Keur P, van Bers C, Henriksen HJ, Nibanupudi HK, Yadav S, Wijaya R, Subiyono A, Mukerjee N, Hausmann H-J, Hare M, Terwisscha van Scheltinga C, Pearn G, Jaspers F (2016) Identification and analysis of uncertainty in disaster risk reduction and climate change adaptation in South and Southeast Asia. Int J Disaster Risk Reduct 16(2016):208–214. https://doi.org/10.1016/j.ijdrr.2016.03.002

Vörösmarty CJ, McIntyre PB, Gessner MO, Dudgeon D, Prusevich A, Green P et al (2010) Global threats to human water security and river biodiversity. Nature 467(7315):555–561. https://doi.org/10.1038/nature09549

Wang P, Dong S, Lassoie J (2014) The large dam dilemma. An exploration of the impacts of hydro projects on people and the environment in China. Springer, Dordrecht. Retrieved from http://link.springer.com/content/pdf/10.1007/978-94-007-7630-2.pdf

Webster A, Gardner J (2019) Aligning technology and institutional readiness: the adoption of innovation. Tech Anal Strat Manag 31(10):1229–1241. https://doi.org/10.1080/09537325.2019.1601694

Zeitoun M (2011) The Global Web of National Water Security. Global Pol 2(3):286–296. https://doi.org/10.1111/j.1758-5899.2011.00097.x

# Chapter 2
# A Reader's Guide to Natural Assurance Schemes

**Elena López Gunn, Laura Vay, and Carlos Marcos**

**Highlights**

- Natural assurance schemes emerge from a structured methodological approach with a number of sequential steps.
- The main aim of a natural assurance scheme is to mitigate the impact from water related risks (avoided costs and damages) and additional co-benefits.
- Natural assurance schemes can be implemented at any scale (micro, meso and large) to cover water related risks like floods and droughts.

## 2.1 Introduction

This Reader's Guide presents the overall framing for this book, introducing and explaining the logic for the structure in the main sections of the publication, based on the main conceptual framework around natural assurance schemes (or NAS for short), underpinning the book. It looks at the main methodological components, the integration of these components and their testing in specific real-life conditions in nine case studies.

The aims of this chapter are fourfold:

1. First to provide a Reader's guide for different potential users and readers for this book to help navigate the content of the book, and the sections that might be more relevant depending on the specific aspect sought.

E. López Gunn (✉) · L. Vay
ICATALIST S.L., Las Rozas, Madrid, Spain
e-mail: elopezgunn@icatalist.eu

C. Marcos
Confederación Hidrográfica del Duero, Valladolid, Spain

© The Author(s) 2023        19
E. López Gunn et al. (eds.), *Greening Water Risks*, Water Security in a New
World, https://doi.org/10.1007/978-3-031-25308-9_2

2. Second to present the tools and methods (or "NAS toolbox and Methodological assessment frame") co-developed to assess the physical impacts of NAS and to value NAS in monetary and non-monetary terms. That is how these NAS can be turned into strategies and bankable projects to be fully developed and implemented. In short, to present the Natural assurance scheme assessment frame, tools and methods developed under different disciplines, by looking in detail at each of its components, as well as the sequence of analysis.
3. Third, by looking at concrete examples of how these methodologies, tools and methods have been applied and tested in nine case studies across Europe. In this in depth look at the case studies, we show the advantages and limitations of the NAS approach. It will be seen how the NAS assessment framework is a modular and scalable (flexible) approach, where some or all components can be applied to assess the role and value of nature-based solutions (NBS) and of nature-based strategies for mitigating the effects of water related natural hazards at the urban, peri-urban and catchment scale and linked co-benefits.
4. Fourth, to provide some preliminary thoughts on transferability to other contexts and location.

## 2.2 A Technical Expert and Researcher's Guide to Natural Assurance Schemes: The Assessment Frame

A modular methodological assessment was developed to help design natural assurance schemes. This modular approach has several elements, which cover a biophysical, social, and economic assessment of the specific area where the scheme could be potentially implemented. The assessment frame can be applied at different scales from large basins like the lower Danube to small scales (in our case a football stadium in Rotterdam). What changes between scales is not so much the approach as the range of tools and methods to be deployed. The aim of this (modular) robust assessment framework is to provide a structured and replicable methodology for the testing, data collection and operationalization of the assurance value of nature-based solutions (NBS) as strategic investments for risk reduction, mitigation, and the valorisation of co-benefits. Also, for the monitoring and evaluation frame to be able to collect the evidence in a systematic way on their effectiveness to reduce or prevent risks and facilitate their replication.

In the case of biophysical assessments for the large scale, as described in Burke et al. (this volume, Chap. 4), the use of Eco:Actuary, a web based spatial policy support system - developed with the insurance industry and end users, to map and understand the biophysical basis to value natural capital for different stakeholders and events, as well as the impacts of land use and climate change upon it.

For other scales, like e.g., the city of Copenhagen, other tools are more suitable like the hydrological model MIKESHE or Bayesian Networks (BN), also known as belief networks (or Bayes nets for short). Hence, BNs combine principles from graph theory, probability theory, computer science, and statistic (Ben-Gal 2008). The tools, in many ways, must match the relevant decision support and planning scale and phase (as discussed in Basco et al. 2023 – this volume Chap. 7). Therefore, the right biophysical assessment tools and frameworks can support the eventual design and implementation of these nature-based strategies, capable of delivering the right (science/evidence- based) key performance indicators.

In terms of social assessment, the 'social man/woman' performs a number of functions like the establishment of social institutions that match social norms and rules, the forming of social organisations, formulating laws, principles and policies that turn social norms into the formal rules of the game, often crystallized into contracts to safeguard the existence, interest and social welfare of the community (in healthy social systems and institutions), or for the benefit of a smaller groups, while preventing or avoiding captured or clientelistic systems (like extractive institutions) (Singh 2006; Acemoglu and Robinson 2012). Its relevance to the natural assurance schemes is based on the level of collective action that must be facilitated to align the different interests and incentives of the different agents. For this, an understanding of their risk perception and their interests, as well as the current state of play (rules in norm and rules in use) is key. A range of tools and methods have been used to undertake the social assessment, mainly social network analysis, fuzzy cognitive maps, and ambiguity analysis (see Giordano et al. 2023 – this volume Chap. 5).

In terms of economic assessment, the project adopted a cost benefit analysis. In particular, the process of quantifying the costs and benefits of an NBS over a certain period, and those of its alternatives within the same period, in order to have a single scale of comparison and a robust and unbiased evaluation (Atkinson and Mourato 2015). The NAS economic frame gave specific attention to the economic benefits in terms of the economic advantages of designing and implementing a set of NBS (or more comprehensive nature-based strategies) over a certain period, quantifiable in terms of monetised costs and benefits, including generated cash flows. Also, the economic cost, i.e., the cost of designing and implementing a NAS over a certain period. It may include acquisition, management, transaction, damage, and opportunity costs (Naidoo et al. 2006). The cost benefit analysis specific to a NAS is one of the most important foundations and innovation that has been developed to construct a natural assurance scheme. This consists of several elements to estimate the costs with the use of life cycle costs of nature-based solutions vis-à-vis normal infrastructure and the opportunity costs, which as discussed in Le Coent et al. (2023, – this volume in Chap. 6) often refer to land use. In terms of benefits the focus was on combining the benefit from avoided damages, as well as other co-benefits, which is central to the definition of a NAS. An important element in the Natural Assurance frame is the link between the elicitation of pluralistic values through biophysical, social, and economic value assessments to a multicriteria assessment frame that can generate a set of key performance indicators. This can eventually be linked to the achievement of specific policy goals (or levels of service, as described by Altamirano et al., 2023 – this volume Chap. 9) and thus e.g., to potential impact investments.

## 2.3 A Planners, Business and Financial Guide: Integration of the Assessment Frame into Real Cases

The methodologies and tools developed were piloted in "DEMO Living Solution Labs" which in this book we call case studies (Dell'Era and Landoni 2014). These case studies span across diverse hazards, risks, scales, environmental and NBS contexts, to provide locally nuanced co-developed models and integrated analytical frames. The modular assessment frame of biophysical, social, and economic analysis was tested in nine different case studies with the main aim to integrate knowledge generated in real environments. Our Demonstration Living Labs (see case studies in Sect. 2.4) are innovation ecosystems, where research organizations collaborate with users and early adopters to create participative strategies to co-define, co-design, co-develop, and validate new products, services, and business models, in our case the development of Natural Assurance Schemes. For this kind of innovation cluster to succeed, effective practices must be implemented. The capturing of the full value of these nature-based strategies was integrated in several ways.

First, through its strong framing under adaptive planning as introduced by Basco et al. (2023 – this volume, Chap. 7) and analysed and discussed by Van Cauwenbergh et al. (2023 – this volume, Chap. 19). Adaptive planning is a structured, iterative process of robust yet flexible decision making in the face of uncertainty, with the aim to manage uncertainty over time through system monitoring and learning from what is experienced as the future unfolds. Using some of the models developed specifically for many of our case studies, it is possible to potentially develop Dynamic Adaptive Policy Pathways (DAPP), which is an iterative policy analysis process for adaptive planning that allows to adjust future action when events, that are presently unknown, unfold in the future. The DAPP approach combines "Adaptive Policymaking" with "Adaptation Pathways", and the developed plans include a strategic vision of the future, commit to short-term actions, and establish a framework to guide future actions. This was not implemented in our case studies, but it could be integrated into the current method.

Second, through the natural assessment business canvas that is explained in Mayor et al. (2023 – this volume, Chap. 8) and Mayor et al. 2021), the value proposition is elicited collaboratively. A business model is a conceptual tool containing a set of concepts and their relationship to each other, to fully develop the value proposition of a specific product or service. It allows for a simplified description and representation of what value is provided to customers, how this is captured, with which funding sources and its financial elements (Osterwalder et al. 2010; Osterwalder and Pigneur 2010; Burkhard et al. 2012; Raymond et al. 2017; Jarzabkowski et al. 2019). The NAS Canvas is different on two accounts; first, because it is structured based on a logic of supply and demand of ecosystem services, and because it is based on a pluralistic understanding of value (Jacobs et al. 2016) and relational values (Mouraca and Himes 2018). These are part of the IPBES Framework and defined as "… imbedded in desirable relationships (sought after), including those between nature and people" (Díaz et al. 2015). Therefore, the natural assurance business canvas captures not just the fully private values, but also the collective and public values, preparing the ground for the collective alignment of a number of interested parties and their collective co-benefits, and willingness to pay for different services provided by multifunctional solutions like nature based strategies (Fig. 2.1). These NBS often deliver simultaneously a bundle of services (collective benefits), i.e., the various benefits that can be provided by a NBS simultaneously over a certain period (Jiang et al. 2016).

Third, the financing framework for water security as described by Altamirano et al. (2023, – this volume Chap. 9), further develops and tests the "Better Business Case approach" (Smith and Flanagan 2001). This includes 5 elements of analysis (a) the "strategic case" to demonstrate that the proposed nature based solutions (or strategies) are strategically aligned and is supported by a compelling case for change, (b) the "economic case" to ensure that a wide range of investment options (in our case also comparing green, hybrid and grey options) have been evaluated and that the preferred option optimises value and benefits, (c) the "commercial case" to facilitate that any proposed procurement is commercially attractive and viable, which in relation to nature based solutions offers specific challenges, (d) the "financial case" to demonstrate that the preferred solution is affordable and can be funded, (e) the "management case" to provide a guarantee where processes and capabilities are in place to ensure that the preferred solution can be successfully delivered. In our case – as will be seen shown – quite often this is spearheaded by public authorities since these are often the problem owners and most exposed directly (or indirectly through their citizens and businesses) to natural hazards (Fig. 2.2).

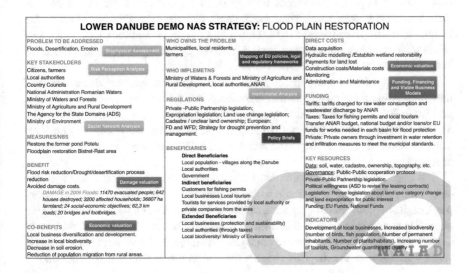

**Fig. 2.1** Example of the NAS Business Canvas applied to the Lower Danube Natural Assurance Scheme. (Source: authors' own)

**Fig. 2.2** Stakeholder
engagement process to
move from risk perception
to socially acceptable
natural assurance schemes.
(Source: authors' own)

## 2.4   A Practitioner's Guide: Applied Case Studies

Our case study Living Labs or case studies span nine locations in eight countries across the European Union and the UK (Dell'Era and Landoni 2014). The methodological tools and methods described earlier are used and integrated in real place-based locations to operationalize the assurance value of ecosystems to reduce the human and economic cost of water-related natural hazards and water related risks like floods and droughts. Our nine case studies have different geographical spatial scales: micro, meso and large scale. These scales for example range from relatively large scales (>5000 km$^2$), mesoscale (200–2000 km$^2$) and some at microscale (<20 km). The spatial boundaries used to delineate our cases studies cover both rural and urban, with small river catchments like the Glinščica, the Lez or the Brague to entire river basins (16,000 km$^2$) and one large aquifer (5000 km$^2$). The urban scale in some cases ranges from the city of Copenhagen to a neighbourhood in the smallest case study with 4 hectares in Rotterdam.

An embedded case study methodology was adopted which provides a means of integrating quantitative and qualitative methods into a single research study (Scholz and Tietje 2002). This embedded approach and identification of sub-units allows for a more detailed level of inquiry (Yin 2003). This opens the possibility of considering EU level data, like e.g., the current SEEAW initiative in natural capital accounting.

Our case studies address different natural (water related) hazards. Most of our cases focused on floods as the main problem identified by the stakeholders. For example, Lez, Rotterdam and Brague are developing Natural Assurance Schemes that give particular attention to flash floods (pluvial floods). Other case studies like the city of Copenhagen are focused on how to manage cloud bursts and how to manage groundwater/waterlogging floods. This renders the soil unproductive and infertile due to excessive moisture and due to the creation of anaerobic conditions.[1] Meanwhile our case studies of Glinscisca, Medina and Lower Danube are dealing with river floods, one of the most common forms of natural disaster when a river fills with water beyond its capacity, and the surplus water overflows the banks and runs into adjoining low-lying lands, causing loss of human life and the damage of property.[2]

*Nora Taylor[3] from Live Science describes floods as follows: "Water from floods can take time to build up, allowing the population in an area time to be warned in advance. But sometimes flooding occurs quickly. Flash floods gather steam within six hours of the events that spawned them. They are characterized by a rapid rise of fast-moving water. Fast-moving water is extremely dangerous — water moving at 10 miles an hour can exert the same pressures as wind gusts of 270 mph (434 kph), according to a 2005 article in USA Today. Water moving at 9 feet per second (2.7 meters per second), a common speed for flash floods, can move rocks weighing almost a hundred pounds (aprox. 45 kg). Flash floods carry debris that elevate their potential to damage structures and injure people".*

All the case studies relied on a stakeholder engagement protocol which structured the process of interaction between the different stakeholders (public bodies, NGOs, SMEs, universities, cities, citizens), with the direct involvement of the insurance industry, end users and implementers as far as possible. In other words, these theoretical approaches and disciplinary assessments have been translated into a case study roadmap as an important step of the operationalization of NAS and the inter-disciplinarity approach with inputs from a range of scientific disciplines (including social sciences). Stakeholders were defined as "individuals and organizations that have an interest in or are affected by your evaluation and/or its results". Another definition by the *Accountability 1000 Stakeholder Engagement Standard* defines stakeholders as "… those groups who affect and/or could be affected by an organisation's activities, products or services and associated performance". Stakeholders will each have distinct types and levels of involvement, and often with diverse and sometimes conflicting interests and concerns. This is relevant because one of the main objectives of the social assessment was precisely to undertake ambiguity analysis, seeing these potential divergences of opinions as a key area of research and knowledge gathering that can open opportunities for collaboration and collective action for mutual protection. The stakeholder engagement process is defined as "… the process used by an organisation to engage relevant stakeholders for a purpose to achieve accepted outcomes" in our case to develop a NAS.

---

[1] http://www.yourarticlelibrary.com/water/waterlogging/waterlogging-definition-causes-effects-with-statistics/61000/

[2] http://www.ehow.com/about_6310709_river-flood_.html

[3] http://www.livescience.com/23913-flood-facts.html

**Fig. 2.3** Stakeholder workshops (co-design process). (Source: authors' own)

Through a co-design process, our stakeholders provided a reality check on the appropriateness and feasibility of proposed nature-based solutions, offering insights on the potential barriers and drivers to NBS, providing relevant feedback and recommendations to help Natural Assurance Schemes become actionable (Fig. 2.3).

Within this range, there is also a social and technical gradient of demos, from those where NBS have been already implemented (like Rotterdam- see Chap. 16) to those were the stakeholders had low awareness of the NBS options (Fig. 2.4). Through a process of co-design and the use of different tools and methods, an assessment was made of the water-related natural hazards in each demo. Therefore, through this social engagement process, the vulnerability aspects were addressed, i.e., the characteristics and circumstances of a community, system or asset that make it susceptible to the damaging effects of a hazard (UNISDR 2009). In the case of the Thames a new tool (Eco:actuary) has been developed which has analysed risk portfolios with consistent multi-hazard analysis and data, focused on process-based and spatially specific information to evaluate the role of NBS for natural flood management (Mulligan et al. 2023 – this volume Chap. 12).

For the specific cases of Medina, Glinščica and the Lower Danube, the social acceptance of NBS was also studied. What social barriers exist towards NBS acceptance and implementation, and an analysis of the institutional settings that will hamper or accelerate the setting up and adoption of a NAS.

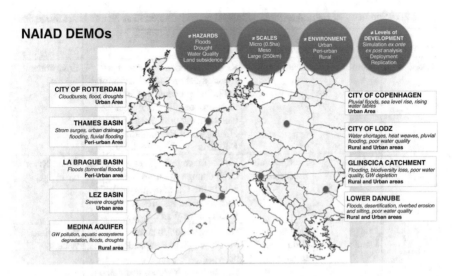

**Fig. 2.4** Summary of case study Living Labs. (Source: authors' own)

Other case studies, in particular the Lez and the Brague (Le Coent et al. 2023, this volume Chap. 14, and Piton et al. 2023, this volume Chap. 13) were able to undertake a full economic assessment to evaluate the economic costs and benefits of green (NBS) and grey solutions, along their life cycle, considering implementation costs, opportunity costs, assurance value (diminished risks costs or avoided damages from natural functions) and co-benefits (productive market values and environmental values).

The integration of these different modular components in the case studies, has allowed to develop several decision supports tools for stakeholders and a common integrated and the holistic evaluation framework of Natural Assurance Schemes. Furthermore, the case studies – once their nature-based solutions and strategies were identified – have developed a set of business models for their nature-based strategies as natural assurance schemes.

Through capacity building activities, including their contributions to the preparation of a MOOC, case studies have been supported to identify how these natural assurance schemes could be funded and financed in the future, to identify the most relevant sectors and actors, funding streams and financial options.

This has produced a toolkit of plausible business cases that will facilitate the implementation of NBS for increasing the resilience towards natural hazards, including an online Handbook on Financing (Altamirano et al. 2021). The approach has a fundamental orientation towards co-design, developing a continuous stakeholder and end user engagement process in each case study, with interviews and workshops. The approach aims to also facilitate policy dialogues in the political arena on key topics through a set of policy roundtables and dialogues as summarised in Table 2.1.

**Table 2.1** Summary of Natural hazards addressed and NBS selected as Nature bases strategies

| Scale | Case study | Hazard(s) addressed | NBS selected | Lead Agency |
|---|---|---|---|---|
| Large | Thames | River floods | Retention ponds landscaped as recreational areas; Conservation agriculture- changes in soil tillage to improve infiltration; Leaky dams, Forest protection and afforestation; Restoration of wetlands | Environment Agency |
| | Medina | Drought and River floods | MAR (managed aquifer recharge); Crop change (to crops more adapted to CC and droughts); Soil conservation; Reforestation; Small dams; Water re- use | Duero River basin agency |
| | Lower Danube | Floods and Droughts, erosion | Building retention areas Forest windbreak expanded network. Creating buffer zones dedicated to flood prevention. Smart sediment management Reconnecting former wetlands | Rumanian Waters |
| Medium catchment | Brague | Flash floods | Riparian woodland and large dead wood integrated management Giving room to the river; Large retention areas Small natural retention areas, cumulated area ~200 ha; Widening of the Brague river bed (~10–40 m); Wetland restoration (11 ha); Riparian forest restoration (13 ha) | CASA: Brague Basin Agency |
| | Lez | Pluvial floods | Green infrastructure: bioswales, open vegetated retention basins, green roofs; city deproofing; conservation of agriculture and natural land through urbanization strategies; Karst active management | City of Montpellier |
| | Glinscica | River floods | Re-meandering & Re-vegetation; Opening natural floodplains; Small multi-functional dry retention areas; Roof rain water tanks; Remove crosswise barriers/ dams | City of Ljubljana |

(continued)

**Table 2.1** (continued)

| Scale | Case study | Hazard(s) addressed | NBS selected | Lead Agency |
|---|---|---|---|---|
| Small | Rotterdam | Pluvial floods | Biofiltration (Constructed wetlands); Buffer / retention; Aquifer Storage & Recovery (ASR) | City of Rotterdam |
| | Lodz | Pluvial floods, droughts, heat waves | Blue-green network; Green ring around the city; Woonerfs; Pocket parks; Green backyards; River rehabilitation; Reservoirs and biofilters in the city | City of Lodz |
| | Coppenhagen | Pluvial floods, groundwater floods, cloudbursts | Green infrastructure (parks, green beds etc). Retention areas where water is kept and evaporates. Retention areas in which stormwater is infiltrated to groundwater; Blue infrastructure (surface channels) (Retention areas from which stormwater is routed to open waters (e.g. harbour); LAR (SUDS) solutions (Stormwater from roof areas is collected in local urban drainage systems and locally infiltrated | City of Copenhagen |

Source: Authors'own

Therefore, our chapters (from Chaps. 10, 11, 12, 13, 14, 15, 16, 17) summarise the experience of our pilot methodologies in nine case studies across diverse risk and NBS contexts to provide locally nuanced co-developed and tested examples.

## 2.5   A Policy Maker's Guide: Policy Uptake of Natural Assurance Schemes

One of the main aims of this publication is to make the results of our Natural Assurance schemes methods, tools, testing and implementation accessible and useful to different stakeholders like policy makers, insurers, water users, etc. Frameworks and tools are used to support of NBS planning and implementation (gathering evidence of the effectiveness of the measures implemented).

On a higher level this publication aims to help identify and address specific barriers and opportunities for the uptake of NBS and natural assurance schemes and how to strengthen or develop policy instruments, business models and innovations in this area to prevent and reduce risks to increase water security.

One of the main areas, based on the interaction with the insurance sector is the different ways in which insurance companies, re-insurance, and public authorities (particularly cities and regional governments/basins) can incorporate nature into risk reduction and the awareness of their co-benefits. In the context of the EU Green Deal, the Sendai Framework (UNISDR 2015), the Paris Agreement and the SDGS, together with the EU adaptation strategy (EC 2021), These can help us to identify robust opportunities based on evidence-based policy making and science-based impact investments and to provide the knowledge base to increase funding for NAS implementation.

In relation to funding and finance, it is important to consider NBS and nature-based strategies in relation to potential budget reallocations in view of the implementation of new ambitious EU strategies under the EU Green Deal umbrella through the multiannual financing framework. In terms of policy the book wants to contribute to the area of ecosystem-based adaptation and eco DRR, i.e., how Natural Assurance schemes can contribute to adaptation, understood as the adjustment in ecological, social or economic systems in response to observed or expected changes in climatic stimuli and their effects and impacts in order to alleviate adverse impacts of change or take advantage of new opportunities (Adger et al. 2005). For example, green adaptation is an application of eco-engineering and aims at adaptation to the pressures of climate change, population growth and economic development, making use of ecosystem services. Ecosystems can adapt to changing circumstances and therefore, might be more robust in the light of climate change under lower temperature increase scenarios. The approach has a strong connection to protection and support of local communities and their livelihoods. One area where this contribution aims to add is to facilitate adequate training and capacity building, which this book hopes to support and promote.

## 2.6 Conclusions

This Reader's Guide has outlined the main structure of this publication in relation to main concept(s) analysed, the methodological assessment of key elements (biophysical, social, and economic), which are then integrated through a series of adaptive planning, business models and financing, which are tested and developed in nine case studies.

In terms of conclusions, key lessons, and recommendations we can summarise these as follows:

- First, the development of so-called Natural Assurance Schemes, which are characterized by the incorporation of the quantified avoided damages and the qualified co-benefits which create a value proposition for *ex ante* investment in nature-based solutions for risk reduction and well-being. The focus is mainly on how to increase loss prevention knowledge (and the tools designed to do so), raising awareness on the potential of NBS for risk reduction and co-benefits.
- Second, as will be seen and described in other chapters, a set of methodologies have been developed to assess NBS. This set of tools and methods can be used as

modular components to develop Natural Assurance Schemes. The tools used are scalable, and the methods have to be matched with the scale and the hazard to be addressed in a process of co-design with the local stakeholders through a structured stakeholder engagement protocol. The focus of this methodology development has been the concept of the assurance value of ecosystems. An EGuide is now available for others to further develop the development and implementation of natural assurance schemes.

- Third, the different key elements need to be integrated to provide the full perspective on nature-based solutions and all their multifunction, following an adaptive planning approach that would make it easier for the lead agencies (like e.g. cities, regions or basins) to incorporate these solutions as part of their DRR and CCA portfolio. The business canvas facilitates the elicitation of plural values spanning the fully private benefits, collective and public good elements. The financing framework, with attention to the five "better business case" for natural assurance schemes that can then help take the natural assurance schemes from design to full implementation (including their funding and finance).

- Fourth, our nine case studies served as real life spaces for the validation of methodologies by feeding and retrofitting our methodologies. Insights from theory to practice and which has led to a better alignment between science and practitioners, offering tools to better understand, assess and implement NBS. By testing methodologies in different contexts, institutional settings, scales, climatic regions and risk types, etc. created a baseline for future actions and for how to use ecosystem services to mitigate water risks (i.e the assurance value). There is impact on methods used by the case studies to finally quantify processes (risks, vulnerability, potential, etc.). Furthermore, due to the macro, meso and microscale of our case studies, there is a focus on NBS across scales & the importance of scale in relation to e.g., effectiveness.

- Fifth, natural assurance schemes are policy relevant because one of the main aims is to shift earlier in the risk management cycle to prevention and mitigation in line with Sendai's risk paradigm. It also helps to support the implementation of the EU Water Framework Directive and the EU Floods Directive by incorporating green infrastructure and e.g., natural flood management and natural water retention measures as part of the risk management portfolio. Finally, it also supports other clear policy objectives on biodiversity and climate change commitments, as outlined in the new EU Biodiversity Strategy and EU Adaptation strategy, as well as facilitate training material with methods accessible to decision makers and technical experts that allow for the greening risk reduction with NBS as part as the curriculum of water managers and decision makers.

We hope that this book provides a useful reference for those aiming to produce their own natural assurance schemes to reduce risks while putting value in nature's protection.

# References

Acemoglu D, Robinson JA (2012) Why nations fail: the origins of power, prosperity and poverty. Profile Books, London

Adger W, Arnell N, Tompkins E (2005) Successful adaptation to climate change across scales. gec. 15:77. https://doi.org/10.1016/j.gloenvcha.2004.12.005

Altamirano MA, de Rijke H, Basco Carrera L, Arellano Jaimerena B (2021) Handbook for the Implementation of Nature-based Solutions for Water Security: guidelines for designing an implementation and financing arrangement, DELIVERABLE 7.3: EU Horizon 2020 NAIAD Project, Grant Agreement N°730497 Dissemination. Can be download: http://naiad2020.eu/wp-content/uploads/2021/03/D7.3REV.pdf

Atkinson G, Mourato S (2015) Cost-benefit analysis and the environment, OECD Environment Working Papers No. 97. OECD Publishing, Paris. https://doi.org/10.1787/5jrp6w76tstg-en

Ben-Gal, I. (2008). Bayesian Networks. In Encyclopedia of Statistics in Quality and Reliability (eds F. Ruggeri, R.S. Kenett and F.W. Faltin). https://doi.org/10.1002/9780470061572.eqr089

Burkhard B, Kroll F, Nedkov S, Müller F (2012) Mapping ecosystem service supply, demand and budgets. Ecol Indic 21:17–29. https://doi.org/10.1016/j.ecolind.2011.06.019

Dell'Era C, Landoni P (2014) Living Lab: a methodology between user-centred design and participatory design. Creat Innov Manag 23. https://doi.org/10.1111/caim.12061

Díaz S, Demissew S, Carabias J, Joly C, Lonsdale M, Ash N, Larigauderie A, Adhikari JR, Arico S, Báldi A, Bartuska A, Baste IA, Bilgin A, Brondizio E, KMA C, Figueroa VE, Duraiappah A, Fischer M, Hill R, Koetz T, Leadley P, Lyver P, Mace GM, Martin-Lopez B, Okumura M, Pacheco D, Pascual U, Pérez ES, Reyers B, Roth E, Saito O, Scholes RJ, Sharma N, Tallis H, Thaman R, Watson R, Yahara T, Hamid ZA, Akosim C, Al-Hafedh Y, Allahverdiyev R, Amankwah E, Asah ST, Asfaw Z, Bartus G, Brooks LA, Caillaux J, Dalle G, Darnaedi D, Driver A, Erpul G, Escobar-Eyzaguirre P, Failler P, Fouda AMM, Fu B, Gundimeda H, Hashimoto S, Homer F, Lavorel S, Lichtenstein G, Mala WA, Mandivenyi W, Matczak P, Mbizvo C, Mehrdadi M, Metzger JP, Mikissa JB, Moller H, Mooney HA, Mumby P, Nagendra H, Nesshover C, Oteng-Yeboah AA, Pataki G, Roué M, Rubis J, Schultz M, Smith P, Sumaila R, Takeuchi K, Thomas S, Verma M, Yeo-Chang Y, Zlatanova D (2015) The IPBES Conceptual Framework — connecting nature and people. Curr Opin Environ Sustain 14:1–16, ISSN 1877-3435. https://doi.org/10.1016/j.cosust.2014.11.002

European Commission (2021) Forging a climate-resilient Europe – the new EU Strategy on Adaptation to Climate Change, COM/2021/82 final, Brussels, 24.2.2021

Jacobs S, Dendoncker N, Martín-López B, Barton DN, Gomez-Baggethun E, Boeraeve F, McGrath FL, Vierikko K, Geneletti D, Sevecke KJ, Pipart N, Primmer E, Mederly P, Schmidt S, Aragão A, Baral H, Bark RH, Briceno T, Brogna D, Cabral P, De Vreese R, Liquete C, Mueller H, Peh KS-H, Phelan A, Rincón AR, Rogers SH, Turkelboom F, Van Reeth W, van Zanten BT, Wam HK, Washbourne C-L (2016) A new valuation school: Integrating diverse values of nature in resource and land use decisions. Ecosyst Serv 22(Part B):213–220, ISSN 2212-0416. https://doi.org/10.1016/j.ecoser.2016.11.007

Jarzabkowski P, Chalkias K, Clarke D, Iyahen E, Stadtmueller D, Zwick A (2019) Insurance for climate adaptation: opportunities and limitations. Rotterdam/Washington, DC. Available online at www.gca.org

Jiang P, Xu B, Dong W, Chen Y, Xue B (2016) Assessing the environmental sustainability with a co-benefits approach: a study of industrial sector in Baoshan District in Shanghai. J Clean Prod 114:114–123. https://doi.org/10.1016/j.jclepro.2015.07.159

Mayor B, Zorrilla-Miras P, Coent PL, Biffin T, Dartée K, Peña K, Graveline N, Marchal R, Nanu F, Scrieu A et al (2021) Natural assurance schemes Canvas: a framework to develop business models for nature-based solutions aimed at disaster risk reduction. Sustainability 13:1291. https://doi.org/10.3390/su13031291

Mouraca B, Himes A (2018) Relational values: the key to pluralistic valuation of ecosystem services. Curr Opin Environ Sustain 35:1–7. https://doi.org/10.1016/j.cosust.2018.09.005

Naidoo R, Balmford A, Ferraro PJ, Polasky S, Ricketts TH, Rouget M (2006) Integrating economic costs into conservation planning. Trends Ecol Evol 21:681–687. https://doi.org/10.1016/j.tree.2006.10.003

Osterwalder A, Pigneur Y (2010) Business model generation: A handbook for visionaries, game changers, and challengers. Wiley

Osterwalder A, Pigneur Y, Tucci C (2010) Clarifying business models: origins, present, and future of the concept. Commun AIS 16:10.17705/1CAIS.01601

Raymond CM, Berry P, Breil M, Nita MR, Kabisch N, de Bel M, Enzi V, Frantzeskaki N, Geneletti D, Cardinaletti M, Lovinger L, Basnou C, Monteiro A, Robrecht H, Sgrigna G, Munari L, Calfapietra C (2017) An impact evaluation framework to support planning and evaluation of nature-based Solutions Projects. Report prepared by the EKLIPSE Expert Working Group on Nature-based Solutions to Promote Climate Resilience in Urban Areas. Centre for Ecology & Hydrology, Wallingford

Scholz RW, Tietje O (2002) Embedded case study methods: Integrating quantitative and qualitative knowledge. Sage Publications, London. ISBN 0-7619-1946-5

Singh YK (2006) Environmental Science Faculty of Arts, Mahatma Gandhi Chitrakoot Rural University Chitrakoot, Satna, Madhya Pradesh

Smith C, Flanagan J (2001) Making sense of public sector investments: The five case model in decision making paperback: 232 pages Publisher: Radcliffe Publishing; New edition (1 Jun. 2001) ISBN-10: 1857754328 http://www.projectresults.co.nz/bbc-overview.html

UNISDR (2009) Terminology on Disaster Risk reduction (English) ISDR Strategy for Disaster Risk Reduction https://www.unisdr.org/files/7817_UNISDRTerminologyEnglish.pdf

UNISDR (2015) Working background text on Disaster Risk Reduction Terminology_23 October_2015

Yin RK (2003) Case study research, design and methods, 3rd edn. Sage Publications, Newbury Park. ISBN 0-7619-2553-8

# Chapter 3
# Insurance and the Natural Assurance Value (of Ecosystems) in Risk Prevention and Reduction

**Roxane Marchal, David Moncoulon, Elena López Gunn, Josh Weinberg, Kanika Thakar, Mónica Altamirano, and Guillaume Piton**

**Highlights**

- The (re)insurance sector is found to play five roles in natural disasters loss prevention.
- The five roles lead to one objective: reducing exposure to risks using preventive measures.
- The impact of such roles could be fostered through partnerships.
- Further research is needed on the effectiveness of NBS on hazard reduction.
- Using challenging climate change to improve knowledge on natural disasters and NBS to ensure insurability of risks.

R. Marchal · D. Moncoulon
Caisse Centrale de Réassurance (CCR), Paris, France
e-mail: rmarchal@ccr.fr

E. López Gunn
ICATALIST S.L., Las Rozas, Madrid, Spain

J. Weinberg · K. Thakar
Stockholm International Water Institute (SIWI), Stockholm, Sweden

M. Altamirano
Deltares, Delft, The Netherlands

Faculty of Technology, Policy and Management, Delft University of Technology,
Delft, The Netherlands

G. Piton (✉)
Université Grenoble Alpes, INRAE, CNRS, IRD, Grenoble INP, IGE, Grenoble, France
e-mail: guillaume.piton@inrae.fr

E. López Gunn et al. (eds.), *Greening Water Risks*, Water Security in a New World, https://doi.org/10.1007/978-3-031-25308-9_3

## 3.1 Introduction

Flood events have huge impacts worldwide. In Europe, numerous examples can be found from the past decade that caused extensive damages (e.g., cloudburst in Copenhagen, Elbe floods in 2002, 2013, Danube floods in 2006, Alpes Maritimes floods in 2015, Lez floods in 2014, Seine floods in 2016 and 2018, etc.). Around 90% of natural hazards are water-related and these are likely to become more frequent and more severe as a result of climate change. Climate change is projected to increase damage up to 50% by 2050 in France (Moncoulon et al. 2018). Caisse Centrale de Réassurance (CCR) estimates that mean annual insured of flood hazards will be up to 38%, respectively 50% related to runoff hazard and 34% to river flooding.

Climate change is already resulting in rising levels of risk posed by natural disasters and the related costs these create (Lawrynuik 2019). The total reported losses caused by natural disasters over the period 1980–2014 reached approximatively 453€ billion, with only 45% of these economic losses insured in Europe.[1] This is why the (re)insurance industry is a critical actor to engage and to understand its current and potential contributions to the assessment of the Natural Assurance Schemes (NAS) (Lopez-Gunn et al. 2022, Chap. 2 of this book).

The impact of climate change on the frequency and the intensity of natural hazards can create a myriad of challenges for the insurance industry' business model (Fédération Française de L'Assurance 2015a; Surminski and Kool 2016; Moncoulon et al. 2018). General insurers may face on the one hand, increasing physical risks which increases their underwriting liability and the number and costs of claims made and on the other hand, transition risks i.e. when there is increased market demand and high premiums could lead to limitations on coverage. Responding the climate change impacts by increasing premiums poses challenges to make both insurance and reinsurance coverage available and affordable. Higher premiums can make insurance coverage unaffordable and then lead to a greater protection gap i.e. when less people are covered by an insurance contract. Nowadays, the industry is slowly moving towards ex-ante actions (Cardone 2018; Marchal et al. 2019). It means that the industry is not only involved after disasters strike with compensation, but can also participate in loss prevention assessments and dialogues.

The rising awareness of the insurance sector can be seen in its engagements in multiple European and global policy frameworks that support climate change adaptation (CCA) and eco-DRR (e.g., European Union's Floods Directive, the Sendai Framework for Disaster Risk Reduction (DRR, 2015–2030), the Sustainable Development Goals and the COP Paris Agreement (2015)) (Warner et al. 2009; Surminski et al. 2015, 2016; Nussbaum et al. 2017; Cremades et al. 2018). EU Flood Directives are seen as bottom-up approaches making countries aware of their risk exposure. The frame of the Directives could be used to foster the implementation of

---

[1] Economic losses from climate-related extremes in Europe, European Environment Agency, https://www.eea.europa.eu/data-and-maps/indicators/direct-losses-from-weather-disasters-3/assessment-2

NBS. The on-going developments on the sustainable finance taxonomy, the European Green Deal to a low carbon future, offer a significant window of opportunities for the NAS concept to be used to promote effective green investments in resilience.

While there is growing evidence of NBS benefits, the assessment of the Natural Assurance Value (NAV), the reduction of risks that NBS can produce is still an emerging process within the insurance industry (Narayan et al. 2016, 2017; Colgan et al. 2017). The chapter aims to fill the knowledge gap by demonstrating the beneficial uses of a catastrophe modelling approach, developed by the industry, to quantitatively estimate the avoided damage thanks to the implementation of NBS on insured damages. Catastrophe modelling offers a powerful framework to estimate the economic damage caused to property from natural disasters. Such models are routinely used in the insurance industry globally to help insurers and reinsurers price and manage catastrophe risk. Case studies on NAS (e.g., Chaps. 6, 14, and 13) show that the (re)insurance industry could effectively use catastrophe modelling, in combination to scientific partners knowledge on co-benefits assessment, to assess the effectiveness of NBS.

In this Chap. 3, we first conceptualize the assurance value of nature and its current implementation within the (re)insurance industry. Next, we present the different roles of insurers in loss prevention using the on-going awareness of the sector in assessing preventive measures to focus their interest on NBS.

## 3.2  Evaluation of the Natural Assurance Value (NAV) Integration within the (Re)Insurance Industry for Now and for the Future

### 3.2.1  Methodology

Traditionally, the main focus of researchers has been on the consequences of climate change on the insurance business rather than insurers role in responding to climate impacts (Surminski et al. 2015). This chapter considers how insurance can support loss prevention and action for Climate Change Adaptation (CCA).

It is based upon findings from several research activities. First, an in-depth literature review on the different European natural hazards insurance business models has been performed (Marchal et al. 2019). Then a "country fact sheet – do you know European natural hazard insurance system well?" for 11 countries were designed. In addition, the literature provides more references linking climate change, disaster risk reduction, NBS and the insurance industry (European Commission 2017; Francis et al. 2016; Narayan et al. 2016; WBCSD 2017; World Wildlife Fund 2017; Weingärtner et al. 2017; Tipper and Francis 2017). Interviews were performed with (re)insurers (Marchal et al. 2019) to assess current status and potential opportunities for the sector to target loss prevention actions. Taken as a whole, this broad set of information has the advantage of providing a robust picture of the current integration of NAV within the (re)insurance industry.

### 3.2.2 Fit for Today?

The engagement in loss prevention by the insurance industry began with insurers specialized in industrial risks coverage. This branch of the industry has had a strong sensitivity to risk culture, particularly for fire and theft prevention. Increasing experiences with losses in these areas, the sector expanded its work with prevention measures and we observed its diffusion into general home insurance for fire and thefts. It was not the case for natural hazards prevention. Indeed, the industry considered that it was not its role to be involved in natural hazard prevention (Fédération Française de L'Assurance 2015b).

One reason is that the industry considers thefts/fire prevention as individual measures that can be financed by insurers for the reduction of damages (e.g., alarms, remote monitoring etc.). These individual measures reduce individual exposure with a direct effect on the insurance premium value. On the contrary, natural hazards prevention measures are often considered as collective measures (e.g., dikes, NBS). Since these are considered as collective prevention, these measures are differently integrated or challenge their integration within the insurance business. For example, in some countries (e.g. those with mandatory insurance coverage), it is not possible to adjust premiums according to the presence of collective protective measures, as the premiums are flat-based and not risk-based (Le Den et al. 2017). On the contrary, voluntary insurance schemes can manage premiums according to the risk exposure. There are however examples of natural hazard insurance schemes that can link the insurance industry to loss prevention. This is the case in France, for example, through the Barnier Fund and article L-121-1 of the Insurance Code which considers the integration of loss prevention. The current commitment of the insurance industry to support NBS development depends on the national natural hazard insurance scheme and the degree of loss prevention considered within. The main limit is the delay of awareness by the insurance industry to consider natural hazard prevention as an area were these could intervene. On the contrary, individual measures at the building level against natural hazards can be financed by insurers (e.g., moving electrical sockets higher, sand bags) (Association of British Insurers 2016). It is related to the on-going involvement of the industry to encourage land-use policies, building codes and build-back-better measures as resilient tools against disasters (Association of British Insurers 2016; Michell 2016; Le Quesne et al. 2017; Nussbaum et al. 2017; Leymarie 2019).

### 3.2.3 Fit for the Future?

Affordability, availability and solvability are the main challenges for the natural hazards insurance schemes, depending on their specificities, to cope with growing insured damage costs (European Commission 2013a; IPCC 2014; Munich Re 2017). Extreme events are going to be more frequent with climate change (physical risk) and without any adaptive measures, insured damage costs of natural disasters will largely increase, as well as premiums (underwriting risk) (see Chaps. 13 and 14, (Moncoulon et al. 2018)). Thus, with the expected more frequent and intense events, the industry' risk solvability is challenged to maintain damage caused by natural events to tolerable limits. If the risks are not well assessed and anticipated, the premiums will increase making insurance contracts unaffordable notably at high risk areas (market and reputational risks). This may lead to an increase in the insurance coverage gaps and create risk exposure through spatial segregation. It could also lead to the bankruptcy of the system and the burden could then fall on corporates or states the impacts of which include lower future investment and high credit exposure. Improved knowledge about future expected risks is necessary to avoid unanticipated losses and to understand loss prevention. Natural hazard risk prevention measures can therefore become necessary to ensure both affordability and insurability to avoid future coverage gaps.

The more that physical risks occur, the higher the interest on prevention. Exploring this insurance's interest throughout NBS fits the objective of maintaining risk exposure and premiums to an affordable level in a climate change context. Potentially, the risk reduction benefits will be integrated into premium calculations, notably for schemes pricing premiums depending on historical claims data and not on current risk exposure. Currently, the pricing of individual risks is based on a statistical view and not on a risk-by-risk analysis. It is still too early to be able to model and to price every risk. For example, in France it runs counter to the principle of solidarity based on flat premiums. In other words, the same amount of premium (env. €20/year/household) is paid for natural hazards insurance. That challenge takes on its full meaning with the European policy objective to raise insurance coverage rate in some countries where the rates are currently low (European Commission 2013b; CRO Forum 2019).

Nevertheless, currently most of the collective preventive measures are not considered for premiums calculations. It is rarely done in some countries with risk-based scheme or even in risk maps. The integration of prevention measures (grey or NBS) relies on the availability of internal catastrophe models and research and development tools that can model the consequences of prevention on damages.

The specific demand for studies on the role of NBS has largely emerged in the past years and is in constant development (see Chaps. 13 and 14, or Lloyd's Tercentenary Research). A high-performing analysis of NBS requires specialized capacities accompanied by the assessment of co-benefits through the intervention of experts in the area. Insurers can help evaluate avoided damage related to NBS by using catastrophe models integrating NBS effects on hazard reduction and related impact on insured losses. The science behind the NBS is under development and while the physical effects of natural disasters can be modelled, the full extent of nature's role on hazards is complex to model. However, enough is known to state the importance of nature in risk reduction to integrate NBS into the catastrophe models, even while considering the uncertainties.

Overall, information disclosed about risk exposure and avoided damages through NBS will be useful for decision making processes. It is a way for achieving tandem efforts with local authorities. Risk informed decision making encourage forward thinking, policy development and investments in longer term projects as exemplified by NBS.

### 3.2.3.1 The French Example to Mainstream Insurers' Involvement into Loss Prevention: The Barnier Fund and the GEMAPI Policy

In practice, protection measures, including NBS, are already funded by insurance companies in France through the Barnier Fund mechanism (FPRNM, Fonds de Prévention des Risques Naturels Majeurs). The mechanism was created by the Law n°95-101 of the 2nd February 1995. It is funded by a 12% levy on the additional premium linked to the compulsory extended natural catastrophe coverage on all property damage insurance contracts (Law 82-600 of the 13rd July 1982, Fig. 3.1). The Barnier Fund is funded by taxpayers.

This fund is dedicated to reducing vulnerability of the assets (local communities and homeowners) exposed to natural hazards. It is used for different risk reduction measures, both DRR and vulnerability reduction measures: studies and works (up to 50% of funding), structural protective measures and different measures such as amiable land acquisition and targeted communication to raise risk awareness. Nevertheless, it is not dedicated for maintenance or reconstruction.

The repartition of the funding allowed by the Barnier fund are: 70% dedicated to dike rings, 20% towards slowdown water engineering and 10% for adaptive measures. The related repartition is relatively stable over the periods 2003–2009, 2011–2016 and 2016–2017.

This fund is mono-specific this means that it will preferentially fund a prevention project that is dedicated to a single hazard risk reduction. The special case of NBS with co-benefits and multi-hazard effects is challenging its use to fund NBS implementation (see discussion in Chap. 14 on the Brague case study).

In France, Water Agencies can also finance NBS and local departments can support up to 20% of financing of protective measures for residential and professional areas, particularly in rural areas. In addition, the GEMAPI policy (Integrated Management of Aquatic Environments and Flood Prevention) aims to create an integrated water cycle governance at intercommunal authority level. Water agencies support local and regional authorities to implement GEMAPI jurisdiction. The objectives are a combination of river management and flood prevention based on natural measures. GEMAPI uses NBS to better protect catchment communities in an upstream/downstream relationships.

We can briefly exemplify one example of NBS implementation in France through the Barnier Fund (Fig. 3.1). The Isère-Amont project for the protection of Grenoble city in the frame of the 2012 French action plans for flood protection (PAPI). Four partners are involved to fund the project: the Isère Department, the Water Agency Rhône Méditerranée Corse, the two local intercommunalities and the State through the Barnier Fund and PAPI funds.

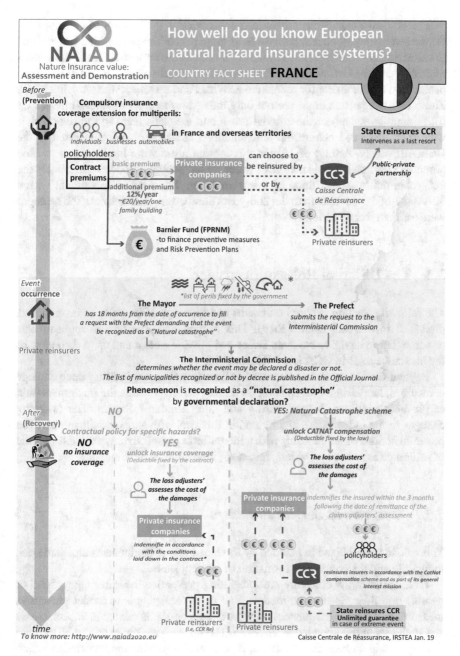

**Fig. 3.1** Country Fact Sheet, explanation of the French natural hazards insurance scheme. Extract from a series of 11 country fact sheets. (Source: NAIAD H2020, authors' own)

The objective is to protect urbanized areas through the concept of giving-room-for the river using floodable field areas. A large part of the programme is dedicated to valorizing co-benefits related to the NBS such as the protection of the local natural heritage of the Grésivaudan valley and to mainstream appropriation of the riverbanks by the inhabitants for recreative activities and urban movement using bikes. These flood expansion areas are not only located on agricultural areas but also on forest areas to reconnect the alluvial forest and wetlands areas to the river. Dikes were then displaced behind the alluvial forest areas. The project is fitted to the GEMAPI policy. A total of 16 flood expansion areas were implemented to hold around 35 million of cubic meters of water for the potential flood event, namely the 200-years return period flood of the Isère River. The NBS is operational with a 30-years flood event and up to the 200-years flood event, in which case the damages were estimated at €1 billion.

## 3.3   The Different Roles of Insurance

The following section is based on results and typology of the roles of insurance as discussed in Marchal et al. (2019). This section does not aim to replicate this research, but to navigate the main issues and provide a clear and up to date view of the insurance roles in CCA and DRR for NBS.

Beyond the claims management role of the (re)insurance industry once damage is incurred, increasingly focus is on increasing efforts towards the implementation of preventive actions, before, during and after a natural disaster. The on-going changes in the industry are highlighted in this part with an exemplification of the potential actions of the industry with a longer-term perspective. We discuss the role of insurance as service providers, investors, innovators and partners.

### 3.3.1   Insurers as Service Providers

This role in materialized in practice by forward-looking components alongside risk modelling and the continuing formation of loss adjusters.

First, some of the company are integrating preventive measures (grey, NBS or hybrid) into their catastrophe models. The main challenge is to model preventive measures and their consequences on hazard reduction, it is related to access to data and to the scientific knowledge available. The insurers are not expert in NBS functioning, and in co-benefit assessment, it is necessary to combine knowledge of scientis/ts with insurers' expertise on damage. A lot has to be performed to find a frame to increase that kind of collaboration. Nevertheless, in the literature there are some relevant examples of the use of catastrophe modelling to value the hazard reduction benefits from NBS (Beck and Lange 2016; Maynard et al. 2017; Reguero et al. 2018; Caisse Centrale de Réassurance 2018).

Second, the sector is encouraging better formation of loss adjusters, actuaries and catastrophe models teams. It supports also early warning alerts to insured properties with instructions to be followed to contain and/or avoid damages. Thanks to their special relationship with insured people, the sector participates in the awareness-raising of the population, which can be residential, commercial, and industrial agents, through the diffusion of educational information such as:

- Sharing maps dedicated to assessing hazard exposure;
- Vulnerability assessments;
- Sharing information on regulatory and opportunity to finance loss prevention projects;
- Sharing information of good practices before, during and after a natural event.

The sector is also developing specific location exercises on crisis management for natural disasters, which constitute a non-negligible interest in sharing risk culture.

### 3.3.2  Insurers as Investors

Insurance companies are changing how they invest in light of climate change. In the context of integration of Environmental, Social and Corporate governance criteria (ESGs) the investment of the industry in NBS projects could take multiple forms. The set of policies previously presented (ESGs, Green Deal, Sendai Framework, etc.) frame decision on strategic sustainable investments to adequately support society financial resilience.

While the sector is slowly moving towards long-term risk investments (multiannual financial framework), it is traditionally involved in short-term investments. Long-term investments require high-quality data to clearly understand the risk exposure and appropriate investments vehicles. To date, the role of the industry in financing is more in the allocation in alternative projects such like green projects, especially for energy production. A large disinvestment in the coal industry is currently underway due to the assessment of transition risks related to carbon sectors.

These investments take the form of the bonds as the primary financial tools. Catastrophe Bond, Climate Bond and Green Bond markets are not new (1992), but the novelty is that corporate sector (banks, insurers etc.) and local collectivities – since the signature of the Paris Agreement (2015) – are issuing an increasing number of bonds.

A new type of bond, Resilience Bonds is emerging. Their potential to incentivize investment in prevention, and therefore hopefully in NBS, is large since these are designed to fund risk reduction projects via a resilience rebate that turns avoided losses in to a revenue stream. The Climate Bonds Initiative and the Green Bond Principles have been developed to assess and to standardize bonds issuance. Nevertheless, these still only represent a small portion of the amount of bonds issued. In addition, as the frame is not specifically targeted and related to insurance as users.

A key issue is to promote actions to not only make bonds available, but to push for the insurance industry to catalyze 'green' investments in risk reduction measures with these bonds, like the sector already does for green energy. One example is the case of a low-carbon investment strategy framed as responsible investments (UN Environment Inquiry and MATTM 2016). This could be linked to the NBS role as part of CCA strategies. Ecosystems (mangroves, peatlands) or NBS that sequester greenhouse gases (Tuffnell and Bignon 2019) could be evaluated as an asset with values incorporating the amount of carbon these can capture. Based on it, the insurance sector could be interested to finance this kind of green projects, as disinvestment from carbon energy. Nevertheless, it is important to focus on getting investments in NBS as resilience investments for risk reduction where carbon reduction is seen as a co-benefit.

It is also worth noting that NBS along with co-benefits can also be eligible to multiple funding sources from multiple scales. The water quality improvement potential NBS can also trigger special funding from, for example, water agencies. NBS could benefit from multiple funding sources that could be pooling options with co-financiers. Eligibility from multiple funding sources is a financial opportunity to absorb risk exposure among partners (public authority, water agencies, national funds and the private sector). The complexity of exploiting these multiple sources is however often encountered in the way the current national investment planning systems and even private investment planning practices follow a sectoral and siloed approach. Nevertheless, no individual insurance company will finance a preventive measure alone at the watershed scale. Indeed, each insurer in the area will have only a sample portfolio of the global at-risk buildings in the exposed areas. This discussion is related to the business model and to the NAS chapter of this book. The bond mechanism could be one of the solutions, as well as sponsorship programs, where the insurance industry may have a role in financing NBS in this kind of scheme.

### 3.3.3   Insurers as Innovators

There are opportunities for insurers to push innovation across multiple areas of their competences to mitigate natural disaster risks. The research and development departments of insurance companies can have a tremendous impact on scientific innovations. The industry is now, for example, developing specialized skillsets building upon a catastrophe loss modelling framework. Efforts are needed, however, to take greater advantage of technical innovations and progress made throughout the sector. This could help harmonize the level of knowledge and practices among companies to speed up the uptake of the state-of-the-art methods and practices.

Innovations in insurance products used in the market are also emerging. There is a growing demand, for example, in specialized weather-related insurance products, including parametric products or micro-insurance (see CCRIF example[2]). Other innovations in insurance product development can be used to spread risk and allow cities to pool their insurance protection. For example, a coastal municipality that faces a primary risk of coastal flooding can create a common coastal insurance product to cover that risk. Another innovation related to NBS is the possibility to ensure a natural environment, such as the case of the parametric insurance policy to cover Mexico coral reef.

The insurance industry can then become as an innovator into two areas. First, at a global level with the development of financial and insurance products. Second, from a territorial perspective, companies rooted at a catchment or city scales can foster the promotion or sponsorship of activities for loss prevention.

### 3.3.4   Insurers as Partners

Globally, the insurance industry serves as an important institutional partner for sharing risk knowledge and providing assessment of the countries exposure by working with the OECD, the World Bank, the European Commission, the Disaster Risk Management Knowledge Centre (DRMKC), the World Forum of Catastrophe Programmes, etc. This enables important opportunities for the exchange of knowledge, practice, and expertise among the insurance industry and with scientists that can consider the diversity of insurance schemes. These exchanges in these kind of areas can unlock partnerships to provide studies on the consequences of climate change on insured damages.

Partnerships at the national and local level may be even more critical. At the national level for example, the industry often relays information from the State and delegated ministries regarding risk preventive policies to its customers. In France, for example, insurers share information on on-going actions performed by the Directorate-General for Risk Prevention (DGPR in French). It is perceived as a social obligation to improve risk knowledge globally. In practice, the (re)insurance sector is interested to reduce the impacts of climate change by not only providing financial resilience but also sharing knowledge on risk management.

Partnerships between insurers with national and local authorities often also support the development and implementation of building codes, land-use planning, flood hazard zone regulations, resilience engineering and other measures. One area that can be expanded upon in future partnerships are efforts to promote build back better measures, as well as collective NBS prevention. This requires that the insurance industry participates in decision-making processes by providing simplified evaluations of water risks and information on loss data. In relation to this, the ongoing discussions about sharing insurance data with communities or scientists is a key issue. A notable European example can be found in Norwegian cities, where

---

[2] The Caribbean Catastrophe Risk Insurance Facility (CCRIF) www.ccrif.org

data sharing from the insurance companies has been applied to improve municipal authorities knowledge on the specific locations of their exposures to risks that can support planning measures taken (Danish Insurance Association et al. 2013; Berg et al. 2014). Nevertheless there are some limits. The first limit for sharing data is to ensure protection of personal data since the insured claims and policies are address-based and geo-localized. The second limit is linked to the market and commercial information available in that kind data. In that context, the industry is willing to be considered as a partner in risk assessment and not as a data provider. As for example in some countries, the involvement of the insurance sector to support planning in the form of public-private partnerships to assess to effectiveness of protective measures through cost-benefit analysis. The NAIAD project, the Oasis platform, research performed by The Nature Conservancy and Lloyd's or French examples of research performed by CCR for local communities etc. Or again, it could take, for example, the form of investments capital towards research companies.

Figure 3.2 summarizes the four roles of the insurance industry in loss prevention and questions raised for insuring NBS. Currently only a few schemes and protective measures are insured. In the case of NBS, the time return to normal situation and temporality of rebuilding is longer than grey measures. In the frame of an insurance contract for NBS, the compensation by insurance could be done only if a natural disaster strikes. The definition of an NBS value could be the insurance value

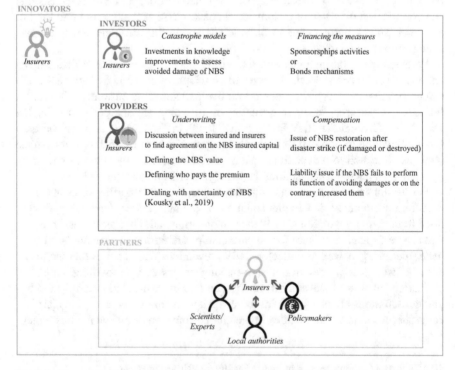

**Fig. 3.2** Exemplifying insurers' roles and challenges for insuring NBS. (Source: authors' own)

(avoided damages and co-benefits). There are other aspects that require careful consideration and planning. For example, if one links an NBS to environmental liability insurance, it is complex to assess and to provide values to an environmental area and to replace that area to the same status if it is damaged by pollution or destroyed by a natural disasters. The rebuild and replacement costs are important and complicated issues. There are some interesting ways to address this that have been proposed. One potential method is to target the premium buyer as the global beneficiaries of the NBS, such as cooperative of farmers, cities, hotels, tourisms association as did e.g., (Francis et al. 2016; Tipper and Francis 2017).

## 3.4 Conclusion

This chapter highlights the (re)insurance industry's role in NAS and initial progress to integrate the concept in the sector. Analyzing evidence of the effects of NBS to cope with climate change adaptation and disaster risk reduction is critical in this process, which is why the analysis done within case studies should of great interest to many in the insurance industry (see case studies chapters) for more detailed analysis.

Thus the main key lessons and recommendations can be summarized as follows:

- **Maintaining sustainability of the insurance system under climate change is a critical challenge for societies worldwide.** The biggest threats to the industry arise from physical and transition risks. Weather catastrophes threaten the entire insurance business, regardless the specificities of the national insurance schemes. If change towards sustainable and resilient economic development is not taken, both economic and human losses from the physical impacts will be far more severe. The transition risks will be manageable for companies that are well-prepared, and opportunities will arise from changes in society and technology used to support low emissions and natural carbon capture. The maintenance of affordable insurance coverage is crucial to fill the low diffusion rate gaps (i.e. the protection gap). Insurers support people and companies to tackle risks. The development of strong risk modelling knowledge and risk awareness can make a big difference to supports systems to enable risk transfer. It offers unique financial capacity to support society's recovery and to cope with uncertainty. It would have a major impact in achieving the SDG2030 agenda to insure vulnerable poorest group physical and financial resilience perspective.
- **Insurers' have an important role to engage in and support society in CCA and DRR, including NBS, and efforts in these areas should expand further.** It no longer appears to be a question of whether NBS have an impact on natural hazards, but how much value they present through the Natural Assurance Schemes (avoided damage and co-benefits). The role of NBS in raising systemic resilience and making at-risk areas insurable again is strategic; therefore, it contributes to close the protection gap issue. It offers the opportunity for the sector

to provide affordable insurance in new areas and thus grow their markets while developing risk knowledge. A coordinated assessment is required to assess as much as possible their effectiveness.

- **The insurance sector can contribute to develop knowledge on natural hazards in a changing climate and in the systematic examination of NBS in risk industry models.** The number of studies on the effectiveness of NBS in loss prevention are growing but this field is still under development. A continuous improvement of tailored catastrophe models is needed to ensure a correct vision of the natural hazards and of NBS. This includes innovating with new products, managing investment portfolios in a way that mainstream NBS in the protection of assets. It also includes support through providing other benefits, t partnerships with local authorities or scientists to increase knowledge on risk and loss and how to manage them. It is a long-term process for the insurance sector to move towards prevention. In the future, systematic examination of NBS by using catastrophe modelling should become a new standard practice.
- **The development of policies in the area of NBS will have to consider the diversity of insurance schemes between countries.** It is important for the sector to be onboard during the policy discussions and requires clear policy recommendations to link sectors working on the same questions for a continued dialogue on CCA and DRR. Thus it is particularly the case for the on-going discussions on data sharing. Considering the efficacy of EU Directives to frame the opportunity for low and high penetration gap to support knowledge improvement on NBS (relevant example of the Floods Directive).
- **Broad multi-stakeholder collaborations will be needed to take full advantage of NAS through NBS projects.** Commitment, engagement and investment of time, knowledge and resources are needed from many actors, ranging from urban planners, architects, insurers, land managers, and management authorities to design and construct NBS. These actors can also support projects linking research institutions to the public institutions and private companies involved.

# References

Association of British Insurers (2016) ABI guide to resistant and resilient repair after a flood. Association of British Insurers, London

Beck MW, Lange GM (2016) Managing coasts with natural solutions, guidelines for measuring and valuing the coastal protection Services of Mangroves and Coral Reefs. World Bank Group, Washington, DC

Berg H, Ebeltoft M, Nielsen J (2014) Flood damage survey after a major flood in Norway 2013: cooperation between the insurance business and a government agency. In: University of the West of England Wessex Institute of Technology (ed) Flood recovery, innovation and response IV. Norwegian Water Resources and Energy Directorate, Norway, pp 227–235

Caisse Centrale de Réassurance (2018) Retour sur les inondations de janvier et février 2018, modélisation des dommages et évaluation des actions de prévention. Caisse Centrale de Réassurance, Paris

Cardone A (2018) Experts urge more prevention efforts, financial resilience to manage disasters. In: Xinhuanet. Available via http://www.xinhuanet.com/english/2018-11/23/c_137625274. htm. Accessed 25 Feb 2020

Colgan CS, Beck MW, Narayan S (2017) Financing natural infrastructure for coastal flood damage reduction. Lloyd's Tercentenary Research Foundation, London, p 44. https://doi.org/10.7291/ V9PN93H3

Cremades R, Surminski S, Máññez Costal M et al (2018) Using the adaptive cycle in climate-risk insurance to design resilient futures. Nat Clim Chang 8:4–7. https://doi.org/10.1038/ s41558-017-0044-2

CRO Forum (2019) The heat is on insurability and resilience in a changing climate emerging risk initiative – position paper. CRO Forum, Amterdam

Danish Insurance Association, Finance Norway, The Federation of Finnish Financial Services, Insurance Sweden (2013) Weather related damage in the Nordic countries – from an insurance perspective. Nordic Insurance Associations, Finance Norway, Insurance Sweden and The Federation of Finnish Financial Services

European Commission (2013a) Greenpaper on the insurance of natural and man-made disasters. European Commission, Brussels

European Commission (2013b) An EU strategy on adaptation to climate change. Communication from the Commission to the European Parliament, the Council, the European Economic and Social Committee and the Committee of the Regions, Brussels, Belgium

European Commission (2017) Horizon 2020 work Programme 2016–2017 in the area of climate action, environment, resource efficiency and raw materials. European Commission, Brussels

Fédération Française de L'Assurance (2015a) Impact du changement climatique sur l'assurance à horizon 2040. FFA, Paris

Fédération Française de L'Assurance (2015b) Livre blanc pour une meilleure prévention et protection contre les aléas naturels. FFA, Paris

Francis A, Amstrong Brown S, Tipper AW, Wheeler N (2016) New markets for a land and nature. How natural infrastructure schemes could pay for a better environment. Green Alliance, London

IPCC (2014) Climate change 2014: impacts, adaptation and vulnerability. Part B: regional aspects. Contribution of working group II to the fifth assessment report of the intergovernmental panel on climate change. Cambridge University Press, Cambridge/New York

Lawrynuik S (2019) Insurance companies push cities to take climate action. In: The Sprawl. Available via https://www.sprawlcalgary.com/insurance-companies-push-climate-action. Accessed 26 Feb 2020

Le Den X, Persson M, Benoist A et al (2017) Insurance of weather and climate-related disaster risk: Inventory and analysis of mechanisms to support damage prevention in the EU. EU Publications, p 196. https://doi.org/10.2834/40222

Le Quesne F, Tollmann J, Range M et al (2017) The role of insurance in integrated disaster & climate risk management: evidence and lessons learned. MCII, United Nations University and GIZ, Bonn

Leymarie J (2019) Climat: le coût des dommages pourrait augmenter de 50% d'ici 2050 selon le réassureur CCR. Available via France Info https://www.francetvinfo.fr/replay-radio/l--interview-eco/climat-le-cout-des-dommages-pourrait-augmenter-de-50-d-ici-2050-selon-le-reassureur-ccr_3212029.html. Accessed on 29 Mar 2019

Marchal R, Piton G, Lopez-Gunn E et al (2019) The (Re)insurance industry's roles in the integration of nature-based solutions for prevention in disaster risk reduction—insights from a European survey. Sustainability 11:6212. https://doi.org/10.3390/su11226212

Maynard T, Stanbrough L, Stratton-Short S, Belinda H (2017) Future cities: building infrastructure resilience. Lloyd's of London/Arup, London

Michell N (2016) The new role of insurers in resilience planning. In: Andrews J (ed) Cities Today. Cities Today, London, pp 26–30

Moncoulon D, Desarthe J, Naulin JP et al (2018) Conséquences du changement climatique sur le coût des catastrophes naturelles en France à l'horizon 2050. Caisse Centrale de Réassurance, Météo-France, Paris

Munich Re (2017) Natural catastrophes 2016 analyses, assessment positions, 2017 issue. Munich Re, Germany

Narayan S, Beck MW, Wilson P et al (2016) Coastal Wetlands and Flood Damage Reduction: Using Risk Industry-based Models to Assess Natural Defenses in the Northeastern USA Lloyd's Tercentenary Research Foundation 25. https://doi.org/10.7291/V93X84KH

Narayan S, Beck MW, Wilson P et al (2017) The value of coastal wetlands for flood damage reduction in the Northeastern USA. Sci Rep 7:9463. https://doi.org/10.1038/s41598-017-09269-z

Nussbaum R, Schwarze R, Surminski S (2017) 20 – Economic resilience, Total loss control and risk transfer. In: Vinet F (ed) Floods. Elsevier, pp 297–320

Reguero BG, Beck MW, Bresch DN et al (2018) Comparing the cost effectiveness of nature-based and coastal adaptation: a case study from the Gulf Coast of the United States. PLoS One 13:e0192132. https://doi.org/10.1371/journal.pone.0192132

Surminski S, Kool D (2016) Taking a risk on the weather. In: Financial World (ed) Financial world, pp 49–53

Surminski S, Aerts JCJH, Botzen WJW et al (2015) Reflections on the current debate on how to link flood insurance and disaster risk reduction in the European Union. Nat Hazards 79:1451–1479. https://doi.org/10.1007/s11069-015-1832-5

Surminski S, Bouwer LM, Linnerooth-Bayer J (2016) How insurance can support climate resilience. Nat Clim Chang 6:333–334. https://doi.org/10.1038/nclimate2979

Tipper AW, Francis A (2017) Natural infrastructure schemes in practice: how to create new markers for ecosystem services from land. Green Alliance, London

Tuffnell F, Bignon J (2019) Terres d'eau, terres d'avenir "faire de nos zones humides des territoires pionniers de la transition écologique". Assemblée Nationale, Sénat, CGEDD, Paris

UN Environment Inquiry, MATTM (2016) Report of the Italian National Dialogue on Sustainable Finance. Inquiry: Design of a Sustainable Financial System, Geneva Switzerland

Warner K, Ranger N, Surminski S et al (2009) Adaptation to climate change: linking disaster risk reduction and insurance. United Nations Office for Disaster Risk Reduction (UNISDR), Geneva

WBCSD (2017) Incentives for natural infrastructure, review of existing policies, incentives and barriers related to permitting, finance and insurance of natural infrastructure. World Business Council for Sustainable Development (WBCSD), Geneva

Weingärtner L, Simonet C, Caravani A (2017) Disaster risk insurance and the triple dividend of resilience. Overseas Development Institute (ODI), London

World Wildlife Fund (2017) Natural and nature-based flood management: a green guide. World Wildlife Fund (WWF), Washington, DC

# Chapter 4
# Methodologies to Assess and Map the Biophysical Effectiveness of Nature Based Solutions

**Mark Mulligan, Sophia Burke, Caitlin Douglas, and Arnout van Soesbergen**

**Highlights**

- It is vital that Natural Flood Management interventions are carefully designed and fully tested for their effectiveness
- We outline newly developed methods for assessing the effectiveness of NFM with a focus on lowcost and open-access solutions.
- We show how modelling is best suited to large scale strategic assessment of the optimal type, magnitude and locations of interventions
- Assessment of the effectiveness of specific interventions is best achieved through a field measurement approach
- Where modelling or measurement are not possible, space-for-time substitution with comparable sites for which the intervention has already been applied and tested should be considered

## 4.1 Introduction: Understanding the Biophysical Effectiveness of Nature Based Solutions

This chapter defines what is meant by effectiveness of nature based solutions (NBS) for flood mitigation, highlights the opportunities and challenges provided by a range of methods for assessing the effectiveness of NBS and introduces new methods for

M. Mulligan (✉) · C. Douglas · A. van Soesbergen
Department of Geography, King's College London, London, UK
e-mail: mark.mulligan@kcl.ac.uk

S. Burke
AmbioTEK Community Interest Company, Essex, UK

© The Author(s) 2023
E. López Gunn et al. (eds.), *Greening Water Risks*, Water Security in a New World, https://doi.org/10.1007/978-3-031-25308-9_4

low-cost assessment of NBS effectiveness in a range of flood mitigation contexts. The methods discussed include monitoring, modelling, and space for time substitution. In the latter an intervention in a different location but similar context can be used in substitution for the potential impact of an intervention not yet built, see Pickett (1989). We discuss methods that can be used to assess the effectiveness of different flood-focused NBS in different biophysical contexts (i.e to understand what type of NBS works where), and to understand the scale required for effectiveness in flood mitigation, including the generation of co-benefits which, for nature based solutions, often by far outweigh the direct benefits of flood risk reduction (Chap. 6).

The methods outlined here can be used to indicate the biophysical effectiveness of the investment made in terms of reducing flood risk and, where NBS are shown to be effective, to advocate for greater investment. To measure NBS effectiveness in mitigation of floods one needs to know:

- the size or capacity of the NBS, for example the volume of upstream flood storage captured in relation to that needed for downstream damage loss mitigation,
- where in the catchment the NBS should be located to be most effective in reducing flows to downstream assets at risk,
- how effective the NBS are for different events, ie to what extent does the intervention mitigate the hazard for different storm sizes, durations and antecedent conditions,
- the magnitude of losses mitigated by NBS in relation to other (grey infrastructure) mitigation methods.

Here we describe new methods for assessing the effectiveness of the NBS before discussing challenges that still exist for assessing the effectiveness of NBS. We focus on no or low-data situations, which is the norm for most NBS investments, in contract to grey infrastructure investments where detailed feasibility studies are required ahead of any major investment. Our focus is a subset of NBS: natural flood management (NFM) and it is the low-data situations associated with NFM investments for which we have developed the Eco:Actuary toolkit, a suite of tools to support risk analysis, NFM investment planning and NFM effectiveness monitoring for installed interventions. These methods have been implemented in the Thames case study (Chap. 12). For NAIAD case studies that cover data-rich situations where extensive and detailed background hydrological data are available, high resolution 'engineering' models can potentially be adapted and used to assess NBS, for example, the MIKE-SHE/HYDRO model application in Copenhagen (Chap. 17), and HEC-HMS and HEC-RAS applications in Slovenia (Chap. 15) and the Lower Danube (Chap. 10).

## 4.2   The Eco:Actuary Toolkit

### 4.2.1   Introduction

The Eco:Actuary toolkit was designed to assess different types of NFM at scales from small to very large, and is designed for locations where there is little or no pre-existing data available to support assessment, which is often the case for non-engineered solutions, like NFM (Dadson et al. 2017). The toolkit comprises three components:

1. The //Smart:River and //Smart:Soil systems for in-field monitoring and web-based analysis of the effectiveness of specific interventions made. //Smart: is intended as a turnkey system for monitoring the effectiveness of specific in-place NFM investments accurately, at low cost and with ease of construction, deployment and analysis. It combines self-build open source hardware (FreeStations, www.freestation.org) alongside custom on-device firmware for data collection and pre-processing and server-side software (serverware) to provide near real-time visualisation and analysis of NFM performance. FreeStation internet connecting environmental monitoring designs are around 1/30th the cost of commercial devices, make use of open source technology, but must be self built according to the instructions provided. //Smart: includes configurations for use on point source in-line storage, such as leaky woody debris dams (Thomas and Nisbet 2012), and retention ponds (Wilkinson et al. 2019) and non-point source land management options such as conservation (regenerative) agriculture (Boardman and Vandaele 2020). Example live data are available at www.policy-support.org/smart

2. The Eco:Actuary web-based Investment planner, which is an empirically based spreadsheet model capable of connection to data APIs (Application Programming Interfaces) such as the UK Environment Agency Real Time Flood Monitoring API, for understanding the total magnitude and type of upstream interventions necessary to make a significant difference to observed flood peaks at a monitored location. This tool is further described in Chap. 12.

3. The Eco:Actuary web-based spatial policy support system (PSS) – which uses a physically based raster spatial model and globally available data to map flood risk, asset exposure, damage, mitigation and avoided loss by current green infrastructure and under scenarios for climate change and NFM investments. This coupled physically-based, probabilistic model can be used to generate a large population of rainfall total, spatial distribution and intensity events that are applied to a mapped human and physical landscape. As with all policysupport.org tools, all data are provided for application of the EA PSS anywhere, globally.

## 4.2.2 The //Smart: Tools for Monitoring NFM Effectiveness

### 4.2.2.1 The //Smart:River System

The //Smart:River System is for monitoring the effectiveness of NFM measures used in small rivers and streams such as an in-line storage intervention (eg leaky woody debris dams, Fig. 4.1). There are three methods for monitoring effectiveness using //Smart:. The **first** focuses on a single intervention and involves a FreeStation water level (stage) and soil moisture device either side of the intervention (Fig. 4.1) alongside local measurement of rainfall, temperature, humidity and air pressure using a FreeStation weather station. The data can then be analysed using the //Smart: visualisation and analysis interface (Fig. 4.2) to assess the differences in soil moisture and stage either side of the intervention which to calculate the operation of the intervention. A typical visualisation for FreeStage is shown in Fig. 4.2 with measured stage shown alongside live rainfall for the nearest station as captured from the Environment Agency Rainfall API in this case.

The **second** method for monitoring the effectiveness of in-line storage interventions calculates the storage behind the dam (or in the retention pond) either as a volume or as a proportion of measured flow downstream at the asset at risk from another FreeStation stage gauge. To calculate the storage, the stage data are analysed by the //Smart: serverware in relation to the cross sectional area of the store (and any downstream asset at risk station location). Cross sectional area is measured

**Fig. 4.1** Live FreeStations monitoring water level and soil moisture upstream and downstream of a Leaky Dam (*left*) and close-up of a water level FreeStation (*right*). Each has a pole-mounted aerial to improve mobile connectivity

**Fig. 4.2** The //Smart: visualisation and control panel interface of hourly measured river depth and live hourly rainfall from the Environment Agency Rainfall API

using a low-cost hand-held FreeStation LIDAR (the 'FreeDAR'). The storage is calculated using the //Smart: serverware, by calculating the change in storage behind the dam in periods when water is rising ("the rising limb") as an absolute volume stored or as a proportion of a downstream flow. Volumes are calculated from surface areas assuming a unit velocity (1 m/s) to provide unit discharges. This assumption holds for small streams and rivers in similar geomorphological settings as the assets at risk or comparator dam, as is the case for most leaky debris dam and retention pond NFM since they address small scale flooding. When the water level is receding, the dam is draining and these "falling limbs" represent drainage of the stores. For example, Fig. 4.3 shows that during a week-long period some ~3 m$^3$ in total are captured during the rising limbs by a dam on this small experimental stream. This can be compared in the system to the corresponding discharge over the same period at the asset at risk downstream.

The **third** method is usually applied to networks of multiple interventions and involves analysis of unit discharge upstream of the interventions in comparison to downstream. Differences in the magnitude of flow peaks and the duration of recession limbs reflect the impact of the combined interventions. Figure 4.4 shows how

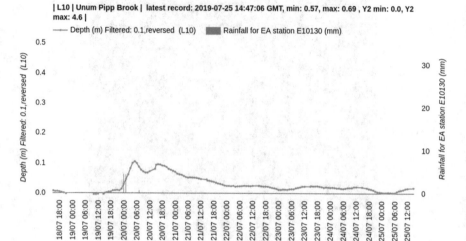

**Fig. 4.3** Cumulative inline water storage behind a leaky dam calculated by //Smart:River

a network of 31 leaky debris dams decreases the flow peaks downstream (red) compared with upstream (blue) and extends the recession of flow.

We find that, usually the total storage behind a single leaky dam is a very small fraction of the flow downstream, and so – unless the asset at risk is very close downstream or the dam raises flood waters to spill over onto a low asset value river floodplain – many hundreds or even thousands of these leaky dam interventions are needed in order to make a significant impact on downstream flood peaks for anything other than the smallest rivers. Chapter 12 discusses this further and demonstrates how this has been applied in the Thames.

### 4.2.2.2 The //Smart:Soil System

//SmartSoil: continually assesses how much water enters and is stored in the soil, in absolute volumes or in proportion to downstream flow. It combines FreeStation soil moisture sensors with specific serverware for analysis of storage by soil over a given area (as absolute volume or relative to downstream flow) and can also be used to measure the impact of soil management interventions aimed at NFM (such as low-till or other regenerative agriculture methods) relative to a counterfactual in a neighbouring field (eg conventional tillage). When used to assess the effectiveness of soil

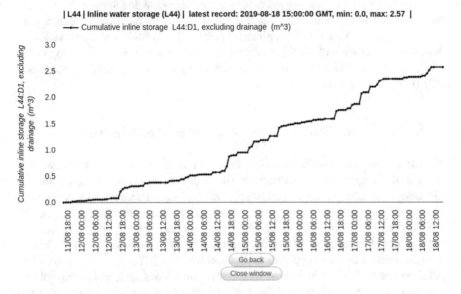

**Fig. 4.4** Flow peaks downstream (red) compared with upstream (blue) of 31 leaky dams as monitored by FreeStation //Smart

management interventions, one sensor is deployed within the land use where the soil management intervention is practiced (eg the regenerative agriculture field) and another in the counterfactual (land use without intervention, eg under conventional tillage, but all other hydrological factors the same, such as slope, soil type). Each device measures soil moisture at one or two depths (e.g. 30 and 60 cm). With information of the field areas, soil bulk density and soil depth, //Smart: can calculate the difference in soil water storage and drainage between the intervention and the counterfactual. There is also a local measurement of rainfall, temperature, humidity and air pressure. The result can be calculated as an absolute volume of water retained or percentage of downstream flow (with downstream flow volume data).

These methods have been applied in the Thames demo during the NAIAD project and are discussed in Chap. 12. //Smart:Soil was used to assess the contribution of regenerative agriculture to increased flood storage, by increasing infiltration and thus reducing runoff over farmland. Since these techniques can be applied over large areas of agricultural land, they can be effective NFM for evening large rivers. The benefit of stream based NFM over farm based NFM as that the former, whilst requiring an initial investment and maintenance, are not dependent on the landholder changing their land use/management practice. Farm based NFM require buy-in and/or incentivisation of the landowner.

### 4.2.3 The Eco:Actuary Investment Planner

The Eco:Actuary Investment Planner (EIP) is designed as a simple, online spreadsheet-based tool that allows assessment of the scale and approximate cost of upstream interventions of different types that would be required to reduce peak flows at a monitored station, by a given proportion. It is useful to investors considering the type and scale of NBS investment that will be effective at mitigating flooding for a river for which long term measured river flow data is available, and thus is only relevant for gauged catchments.

#### 4.2.3.1 Using the EIP to Assess the Effectiveness of NBS Investment

The EIP is available at www.policysupport.org/ecoactuary. The user enters the location and the proportion of the discharge (flood peak) they wish to mitigate and the tool calculates the volume of flood storage that NFM must hold back to achieve that level of flow mitigation. For applications in England, this tool uses data supplied by Eco:Actuary (EA) through DEFRA Opendata (data.gov.uk) in particular the UK Environment Agency Real Time Flood Monitoring API and connects to this directly through a real time API (application protocol interface). The non-UK spreadsheet tab allows users to paste in their own discharge time-series data, which the tool will then use. Users can update the default per-unit volume construction cost for different Natural Flood Management (NFM) techniques (leaky debris dam, retention pond, regenerative agriculture) and the tool will estimate the budget required to mitigate flow at the level required. The user sets the flow mitigation goal as a percentage of flow. An example application of this tool is given in Chap. 12. EIP takes no account of how the placement of NBS in specific catchment positions can make a difference: a spatial model like Eco:Actuary is necessary for that (see below).

### 4.2.4 The Eco:Actuary Spatial Policy Support System (PSS)

#### 4.2.4.1 Purpose of Eco:Actuary

Eco:Actuary (E:A) is a spatial Policy Support System (PSS) focused on assessing fluvial flood risk, exposure of multiple asset types and their values, estimation of baseline damage losses and better understanding the mitigation of potential damage losses by existing and proposed green infrastructure and NFM interventions. As well as assessing baseline risk and mitigation E:A can also be used to understand the impact of scenarios of climate, land use, asset value and asset adaptation (damage function modification) change. Like all policysupport.org tools, Eco:Actuary is applicable anywhere globally based on datasets provided with the model. As a coupled physically based, probabilistic model the EcoActuary PSS can be used to assess risk and mitigation in unmonitored catchments. A detailed user guide to Eco:Actuary and the PSS itself are available at www.policysupport.org/ecoactuary.

### 4.2.4.2 Using Eco:Actuary to Assess NBS Effectiveness under Different Scenarios

The impact of climate change or proposed land management interventions can be achieved by comparing modelled outputs for flood damage losses under the applied scenario compared with those in the baseline simulation (representing climate and land use at 2015). Within Eco:Actuary the following types of scenarios can be assessed:

- *Climate change:* Eco:Actuary provides access to downscaled monthly CMIP5 (Emori et al. 2016) and AR4 (IPCC 2007) climate change scenarios for 17 climate models (GCMs) and a range of Intergovernmental panel for climate change (IPCC) scenarios which, combined with user-specified changes to the rainfall intensity and storm size distributions can be used to examine impacts of likely climate changes, spatially.
- *Impact of extreme events:* Changes to rainfall extremes can be simulated by changing the maximum intensity and maximum monthly rainfall parameters for the distribution of rainfall event intensities and event volumes generated in the "Hazard Ensemble" of the catastrophe model. This is in addition to any changes in monthly rainfall volume under climate change, which in turn determine the number of events.
- *Land Use and Cover Change:* Eco:Actuary allows users to generate large-scale land use and cover change scenarios to represent landscape scale NFM interventions such as afforestation or rewilding. To understand impacts of observed deforestation or afforestation, users can also use the land use modeller QUICKLUC (Mulligan 2015) within Eco:Actuary. QUICKLUC projects recent rates of afforestation/deforestation forward on the basis of past observations and other relevant factors.
- *Asset adaptation:* scenarios with different flood damage curves (representing damage losses for different depths of flood water) can be used in Eco:Actuary to represent asset adaptation (such as lifting assets from the ground).
- *Natural flood mitigation infrastructure:* Eco:Actuary allows users to assess the impact of existing natural flood, mitigating infrastructure (wetlands, canopies, soils, floodplains, water bodies) on reducing flood risk and can also change the volume and distribution of these infrastructures.

We have outlined the Eco:Actuary suite of tools to assess NFM. These tools can be used as part of the NFM planning and preparation stage to assess the current location and value of assets at risk and of mitigation by natural infrastructure (Eco:Actuary PSS), or for exploratory Cost Benefit Analysis for proposed intervention types and magnitudes using available discharge records (EIP). The toolkit can also be used post-investment to assess the effectiveness of NFM. A detailed review of the application of the toolkit in the UK Thames Demo is given in Chap. 12, whilst the following section demonstrates an application of the Eco:Actuary PSS to the Lower Danube.

## 4.3 Challenges in Assessment of NFM Effectiveness

### 4.3.1 Data Availability and Uncertainty

One of the key limits to using modelling to assess effectiveness is data availability and uncertainty: even for key datasets such as rainfall and terrain. The availability and quality of these data are fundamental to the accuracy of modelling of flood risk. Many smaller streams that are relevant to NFM are not gauged and, even in well monitored countries, the distribution of rainfall stations is much broader than the scale of spatial variability of rainfall.

#### 4.3.1.1 The Complex Drivers of Flooding

A further challenge is that flooding can result from poorly known conditions antecedent to the rainfall event or poorly known characteristics of the event (e.g. the detailed spatio-temporal rainfall intensities, blockage of channels and culverts, subsurface flow, snowmelt). Even with recent advances in meteorological monitoring and prediction, the location and timing, size and duration of large rainfall events is still poorly known, and therefore the prediction of the hyper-local outcomes that characterise flooding and generate flood damage profiles is thus challenging. Soil and subsurface hydrology is also poorly known at policy relevant scales, and there are many green (natural) and grey (engineered) flood water stores in the landscape to take account of, some of which are also poorly known.

#### 4.3.1.2 Model Uncertainty

Model uncertainty is also a challenge In the case of Eco:Actuary the applied models are physically based, *i.e.* representations of physical processes that generate discharge, rather than empirical relationships between rainfall and discharge. Complex, physically-based models are more useful in scenario studies (since the physical principles can be applied to scenarios as well as baseline conditions, whilst empirical relationships usually cannot). Nevertheless, physically-based models make many assumptions which are difficult to test in real landscapes. The advantage of low cost, DIY (self build) approaches to monitoring is that more dense data collection networks can be developed that can provide data to help reduce model uncertainties. A hydrological model is usually calibrated, e.g. against discharge data of rivers and streams in order to achieve greater correlation between modelled and observed flows but calibration is a 'sticking plaster' that does not help understand the reasons for model disagreement. A key to effective use of models in NFM is to make them easy to apply, such that a number can be compared, alongside field measurements and a 'weights of evidence' approach to understanding outcomes can be developed.

### 4.3.1.3  Where the Assets at Risk Are

It is important to locate NFM where they have greatest influence on the most downstream assets at risk. Good asset maps for different classes of asset type are necessary but are not usually available; Eco:Actuary uses Open Street Map data for a variety of asset types combined with EO products representing nighttime lights and building height in order to map asset types and assess the scale of asset value within an asset class (with brighter, taller assets being of higher value, than shorter, darker ones). These are coupled with locally specified parameters representing the absolute max and min monetary value per asset type. For example, the asset type of greatest value is shown for each sq km. the Danube catchment in Fig. 4.5.

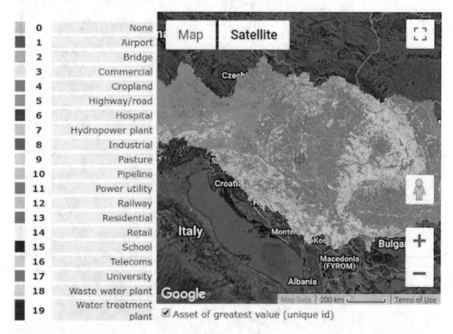

**Fig. 4.5** Eco:Actuary output: Asset type of greatest value, River Danube catchment. (Base Map: Map data ©2019 Google)

#### 4.3.1.4 Asset Valuation Uncertainty

To calculate avoided damage, accurate asset values and damage functions are crucial. Eco:Actuary replacement value and loss of function value have to be estimated on the basis of proxies (nighttime light, building height and asset type) since no other data are usually available in the public domain at the scales that NFM analyses are applied. This creates uncertainty in the absolute values produced by the valuation component, but the results will nevertheless be of the correct order of magnitude and scaled appropriately relative to each other. More readily available data are necessary to reduce uncertainty.

#### 4.3.1.5 The Scale of NBS

The impact of all interventions decays with distance downstream of the intervention. Thus, small scale NFM interventions in large catchments, such as leaky log dams will only have local impact and very significant interventions are required to have significant impacts downstream, particularly given data and model uncertainties. Figure 4.6 shows output from the Eco:Actuary PSS indicating the rapid decline

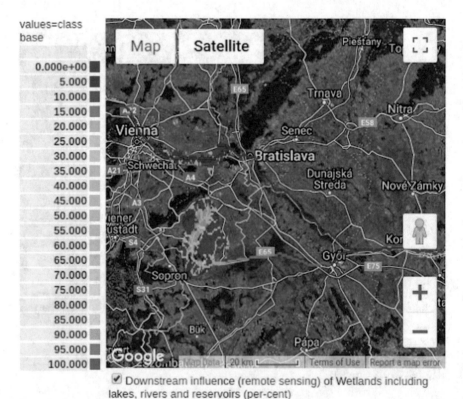

**Fig. 4.6** Rapid decay of downstream influence (%) of wetlands using Eco:Actuary Policy Support System. (Base Map: Map data ©2019 Google)

in influence of a key wetlands in the Danube with distance downstream: within only 40 km downstream, the wetland storage represents only 0.4% of the discharge. The Danube is 2860 km long so for most of it, the wetland has virtually no flood-relevant influence or at the least any influence is of a smaller magnitude than the errors in the data that we have available to assess it.

It is also important to consider the effectiveness of the sum of all interventions upstream, so interventions cannot be considered in isolation from each other. The location of NFM and networks of NFM are also important: analysis using the Eco:Actuary PSS shows that to be effective interventions have to match the scale of rivers. Small interventions will only have impact on small streams and on large rivers interventions must be substantial to make a positive hydrological impact. Substantial mean few large or many small interventions.

## 4.3.2   Ways Forward

Assessing the effectiveness of NFM is challenging, especially for the civil society organisation that lead on many of these investments. By providing easily used monitoring and modelling tools based on open-science and open data we have removed some of the barriers to more effective deployment of NFM as part of the flood mitigation toolkit. For large scale strategic analyses we provide the Eco:Actuary PSS and the Eco:Actuary Investment Planner to better understand the scales and magnitudes of intervention required to achieve specific levels of benefit in particular contexts.

For post-deployment assessment of the effectiveness of NFM, we provide the //Smart: system which allows users to cheaply and effectively measure water stores relative to volumes of flow downstream near assets at risk, for a range of intervention types, intervention settings and storm conditions. This has the potential to measure the real-time contribution of NFM to flow and flood mitigation. For smaller interventions this is a much more direct measure of their efficacy than modelling could ever be. The //Smart: system facilitates this through Internet-of-Things connected electronic monitoring of storage and its influence downstream using self-build FreeStation technology. However this is not possible for interventions that have not been installed yet so some combination of measuring and modelling may be necessary.

An alternative to monitoring and modelling is space-for-time substitution. A large number of NFM interventions have been installed around the world in similar settings and thus, in many cases, finding an analogous NBS in a similar setting to that proposed can provide a good indication of the likely efficacy of the proposed NBS. However this requires similar (and well known) conditions between the analogue and the proposed intervention as well as a number of years of data on the effectiveness of the existing intervention at times of flood. Some 'scaling' may be needed to account for differences in scale or context between the analogues. Databases of NBS such as the Oppla platform can facilitate this process.

## 4.4　Conclusion

This chapter has introduced newly developed methods for assessing the effectiveness of nature based solutions for flood mitigation, with a focus on low cost and open-access solutions. It highlights the opportunities and challenges of different monitoring and modelling methods, from physically based scenario models such as the Eco:Actuary PSS for assessing impacts of climate change and green infrastructure on flood risk to assets around the world, to using networks of low cost environmental monitoring stations for direct measurement of effectiveness of specific interventions. The latter has been used by a number of citizen-led Flood Action Groups in the UK to assess the effectiveness of the NFM interventions. More detailed applications of these methods are described in Chap. 12.

Empirical and physically based modelling is best suited to large scale strategic assessment of the optimal type, magnitude and locations of interventions whereas assessment of the effectiveness of specific interventions is best achieved through a measurement approach or – where that is not possible – space-for-time substitution with comparable sites for which the intervention has already been applied and its effectiveness tested.

NFM will be an important component of the global toolkit for future flood management in order to address persistent small-scale flooding on smaller rivers that are beyond the reach of governmental grey infrastructure flood protection schemes and also contribute to 'green-grey' solutions for larger rivers, recognising the many co-benefits of NFM for nature and society. It is, however, vital that NFM interventions are carefully designed and fully tested for their effectiveness and cost-effectiveness so as not to waste valuable flood mitigation resources and to ensure that these solutions do not themselves generate further problems. Only through accessible, low cost approaches can this be achieved since each NFM project is usually rather small scale, low budget and unique. Generic or expensive engineering assessments such as those carried out for some grey (engineered) infrastructure investments may not be suitable.

**Acknowledgements**　This project has received funding from the European Union's Horizon 2020 research and innovation programme under grant agreement No 730497.

Many thanks to all our supporters during the NAIAD project. Our //Smart:River supporters included Matt Butcher (Environment Agency), Archie Ruggles-Brise (Spains Hall Estate), Mike McCarthy, Phil Wragg and Geoff Smith (Shipston Area Flood Action Group), Rob Dejean (Hallingbury Marina), Mark Baker (Unum Ltd), Nigel Brunning (Johnstons Sweepers Ltd), Joanna Ludlow (Essex County Council), Dean Morrison and Ed Byers (South East Rivers Trust), and Stephen Haywood and Lucy Shuker (Thames21). Our //Smart:Soil supporters were John and Paul Cherry, Ian Waller, Tony Reynolds and Andrew Maddever, and their neighbouring farmers.

# References

Boardman J, Vandaele K (2020) Managing muddy floods: balancing engineered and alternative approaches. J Flood Risk Manag 13(1):e12578

Dadson SJ, Hall JW, Murgatroyd A, Acreman M, Bates P, Beven K, Heathwaite L, Holden J, Holman IP, Lane SN, O'Connell E (2017) A restatement of the natural science evidence concerning catchment-based 'natural' flood management in the UK. Proc R Soc A Math Phys Eng Sci 473(2199):20160706

Emori S, Taylor K, Hewitson B, Zermoglio F, Juckes M, Lautenschlager M, Stockhause M (2016) CMIP5 data provided at the IPCC Data Distribution Centre. Fact Sheet of the Task Group on Data and Scenario Support for Impact and Climate Analysis (TGICA) of the Intergovernmental Panel on Climate Change (IPCC), 8 pp

IPCC (2007) Climate change 2007: synthesis report. Contribution of Working Groups I, II and III to the fourth assessment report of the Intergovernmental Panel on Climate Change [Core Writing Team, Pachauri RK, Reisinger A (eds)]. IPCC, Geneva, 104 pp

Mulligan M (2015) Tropical agriculturalisation: scenarios, their environmental impacts and the role of climate change in determining water-for-food, locally and along supply chains. Food Secur 7:1133–1152

Pickett STA (1989) Space-for-time substitution as an alternative to long-term studies. In: Likens GE (ed) Long-term studies in ecology. Springer, New York

Thomas H, Nisbet T (2012) Modelling the hydraulic impact of reintroducing large woody debris into watercourses. J Flood Risk Manag 5(2):164–174

Wilkinson ME, Addy S, Quinn PF, Stutter M (2019) Natural flood management: small-scale progress and larger-scale challenges. Scott Geogr J 135(1–2):23–32

# Chapter 5
# Participatory Modelling for NBS Co-design and Implementation

**Raffaele Giordano and Alessandro Pagano**

**Highlights**

- Participatory and integrated tools provided a representations of NBS multi-dimensionality
- Co-benefits were considered in many cases even more important than the reduction of water-related risks, and were used as the main elements for the co-design of the NBS.
- In addition to socio-economic, technical and institutional barriers, NBS implementation claims to detect and overcome those related to the interaction between the various decision-actors.

## 5.1 Introduction and Conceptual Frame

Nature-based solutions (NBS) have become a valid alternative to grey infrastructures – i.e. hard, human-engineered structures (Palmer et al. 2015) – for coping with climate-related risks in urban and rural areas alike (Raymond et al. 2017; Calliari et al. 2019; Frantzeskaki 2019). The increasing success of NBS is due to their capacity to foster the functioning of ecosystems and to generate additional environmental, economic and social benefits that are considered as essential backbones of actions for climate-change mitigation and adaptation (Bain et al. 2016; Kabisch et al. 2016; Josephs and Humphries 2018). Nevertheless, the transition of the risk management system from the grey solutions toward NBS is still slow (Wihlborg et al. 2019). This is mainly due to the existence of several barriers to NBS implementation. Most of the works in the scientific literature demonstrate that physical

R. Giordano (✉) · A. Pagano
Istituto di Ricerca Sulle Acque del Consiglio Nazionale delle Ricerche (IRSA-CNR),
Bari, Italy
e-mail: raffaele.giordano@cnr.it

© The Author(s) 2023
E. López Gunn et al. (eds.), *Greening Water Risks*, Water Security in a New World, https://doi.org/10.1007/978-3-031-25308-9_5

barriers (i.e. the technical effectiveness of NBS in water-related risk reduction) are less important than those related to governance, socio-institutional and economic dimensions(O'Donnell et al. 2017; Calliari et al. 2019; Pagano et al. 2019; Giordano et al. 2020). Among the different barriers, this work focuses on two issues that need to be addressed in order to enable the NBS implementation, the low level of social acceptance and the collaboration barriers. The main scope of this work is to demonstrate the effectiveness of participatory modelling exercises in facilitating stakeholders' engagement in NBS design and implementation as viable approach for overcoming the above-mentioned barriers.

Stakeholders' needs and concerns represented the backbone of the adopted approach in the different NAIAD case studies. Therefore, efforts have been carried out to put stakeholders' problem understanding, risk perceptions and preferences at the core of the adopted approach. Before describing the methodological approaches, a few definitions are needed. Problem understanding refers to the mental construction of a certain issue to be addressed, in terms of main causes, impacts, objectives to be achieved and actions to be applied. Risk perceptions refers to the fact that people construct their own reality and evaluate risks according to their subjective perceptions. This type of intuitive risk perception is based on how information on the source of a risk is communicated, the psychological mechanisms for processing uncertainty, and earlier experience of danger (Renn 1998).

Neglecting the differences among values and perceptions held by different stakeholders, which in many cases are not well represented in the decision-making process, may lead to conflict, hampering the effective implementation of NBS. Moreover, stakeholders' engagement in participatory processes may be turned into an often controversial and futile process (Brugnach and Ingram 2012; Giordano et al. 2017a). Therefore, NBS design and implementation should be based on inclusive and equitable participatory processes, capable to ensure the active involvement of all different categories of stakeholders, and to reflect the diversity of meanings and interpretations that the inclusion of multiple actors brings (Brugnach and Ingram 2012; Cohen-Shacham et al. 2016).

Ambiguity analysis plays a key role in facilitating the stakeholders' engagement. Ambiguity refers to the degree of confusion that exists among actors in a group for attributing different meaning to a problem that is of concern to all (Weick 1995). Ambiguity, which can be considered as a form of uncertainty and indeterminacy (Brugnach et al. 2011; van den Hoek et al. 2014), is ineradicable in complex decision-making processes (Jasanoff 2007). Figure 5.1 shows how a different stakeholder involvement affects the decision-making process, and helps highlighting how neglecting the role of stakeholders' engagement in NBS design may lead to barriers to their implementation. On the one hand, ambiguity in problem understanding could cause discussions and conflicts in the initial stage of the participatory process, increasing the time required for making the decision, compared to the unilateral decision-making process. Nevertheless, addressing the ambiguity issues in the early phase of the process has a positive impact on the implementation phase, which is faster.

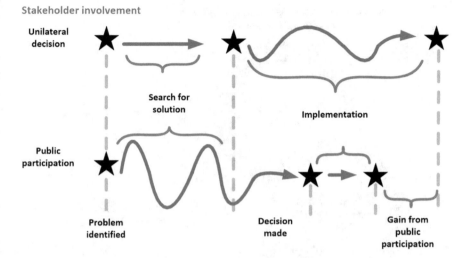

**Fig. 5.1** Impacts of the stakeholders' involvement on the decision-making process

Contrarily, neglecting the existence of different and equally valid problem framings (unilateral decision-making process) facilitates the identification of the most suitable solution according to one kind of knowledge – i.e. the technical knowledge – but conflicts will immediately arise, hampering the implementation phase and/or reducing the measure's effectiveness (Giordano et al. 2007; Giordano et al. 2017a).

Starting from these premises, we aimed at enabling the stakeholders' engagement, facilitating the dialogue, aligning divergences and reducing conflicts among different decision-makers due to the ambiguity in problem understanding and risk perception.

To this aim, a multi-steps process was applied in different NAIAD case studies. As shown in Fig. 5.2, the whole process was based on a continuous interaction with local stakeholders, and combined individual interactions and group discussion.

## 5.2 Applied Tools and Methods

Figure 5.3 below shows the different phases of the applied approach and the methods used. Three main phases can be defined for handling ambiguity in risk perceptions through the stakeholders' engagement in NBS design, i.e. (i) individual risk perception elicitation and analysis; (ii) detection of the main barriers to NBS co-design and implementation; (iii) trade-offs analysis and conflicts detection. Specifically, the analysis carried out in phases (i) and (ii) allowed to bring stakeholders and decision-makers in a participatory process whose main scope was to co-design effective interventions for reducing the water-related risks and producing

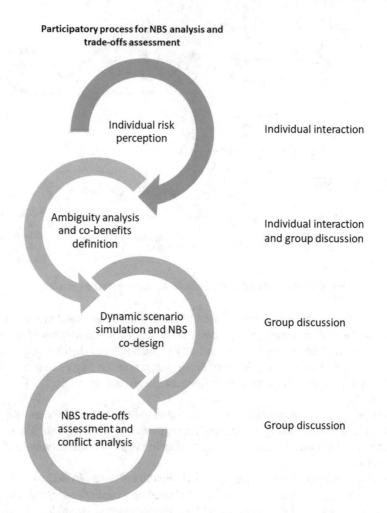

**Fig. 5.2** Combination of individual and group interactions in NAIAD implementation

the expected co-benefits. The methods applied in phase (iii) were meant to enhance the equity of the NBS implementation process.

Prior to describing the different methods, it is worth mentioning that a key preliminary activity has to be carried out in order to guarantee the success of the whole process that is the *selection of the stakeholders to be involved*. This is due to different reasons. Firstly, because the knowledge elicited by interacting with them is at the basis of the whole process (Jetter and Kok 2014).

Therefore, their representativeness needs to be taken into account. Secondly, the stakeholders-driven process is quite long and requires the stakeholders to go through different phases of individual inputs and group discussion. Therefore, the stakeholders' selection should also account for their willingness to commit themselves to the

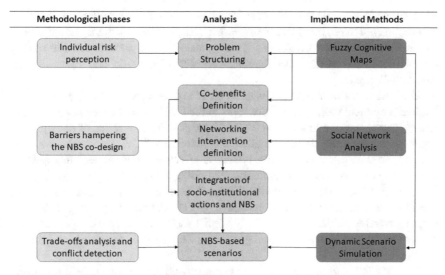

**Fig. 5.3** Different phases of the applied methodology and methods used

whole process. Efforts are required from the analyst in order to keep the stakeholders interested and motivated for the whole process duration. The "snow-ball" sampling approach demonstrated its usefulness in selecting the stakeholders. Basically, we started interacting with key stakeholders, characterized by a pretty high risk awareness and willing to cooperate. Then, other stakeholders were indicated by them during the interviews. In this way, we were capable to define gradually the set of representative stakeholders.

### 5.2.1  Individual Risk Perception and Co-benefits Definition

The first phase of the applied methodology aimed at *collecting and structuring stakeholders' risk perception and problem understanding*, in order to support the co-design of the most suitable NBS. To this aim, Fuzzy Cognitive Mapping (FCM) approach was applied. FCM is part of the Problem Structuring Methods, based on the assumption according to which the most demanding and troublesome task in problem solving often consists in defining the nature of the problem, rather than its solutions (Rosenhead and Mingers 2001).

FCM are defined as a "mirror" of the causes and effects that are inside the mind of decision makers (Montibeller et al. 2008; Kok 2009). FCMs can simulate the cause – effect relationships between the main variables in the model. Semi-structured interviews involving local stakeholders were carried out in order to collect the diverse risk perceptions (Olazabal et al. 2018). The interviews aimed at gathering stakeholders' understandings about: (i) the main elements affecting the water-related risks at local level; (ii) the direct and indirect expected impacts of the water-related risks; and (iii) the most important issues (social challenges) that need to be addressed in order to increase the effectiveness of the risk management actions and enhance the system conditions. Finally, stakeholders were required to specify the expected roles of the NBS in reducing water-related risks and addressing the social challenges.

The interviews were then analyzed in order to detect the keywords in the stakeholders' argumentation – i.e. the concepts in the FCM – and the causal connections among them – i.e. the links in the FCM. Table 5.1 shows a series of examples. In order to facilitate the development of the individual FCM, the interviews were designed in such a way as to make the cause-effect relations immediately identifiable in the stakeholders' argumentation. The collected knowledge was, hence, processed in order to obtain the individual FCM. The sentences were broken down into specific categories, i.e. (i) cause variables; (ii) effect variables; and (iii) relationships type (Kim and Andersen 2012). Table 5.1 shows an example of the stakeholders' argumentation analysis, allowing to detect the structural relationships for FCM development.

A FCM is composed by interrelated variables and directional edges, i.e. connections – representing the causal relationships between variables (Kok 2009). The connections are defined by the strength of the causal relationship between two variables. The connection strength indicates the stakeholder's perceived mutual influence of two variables (Ozesmi and Ozesmi 2003). The weight can be either positive or negative. A positive weight indicates an excitatory relation between two connected variables – i.e. the increase of one variable leads to the increase of the connected one – while a negative weight indicates an inhibitory connection – i.e. the increase of one variable leads to the decrease of the other. For more details about the process for building FCM from interviews, a reader could refer to (Santoro et al. 2019; Giordano et al. 2020; Gómez Martín et al. 2020). Figure 5.4 shows two examples of FCM developed for the Lower Danube case study.

Once developed, the individual FCM were analyzed in order to identify the most central elements in the stakeholders' risk perceptions by assessing the centrality degree measure, the so called "nub of the issue" (Eden 2004).

**Table 5.1** Examples of the analysis of the interviews for developing the structural relationships in the FCM

| Quotes from the interviews | | Cause variables | Effect variables | Relationship type |
|---|---|---|---|---|
| "Poor maintenance increases the flood risk due to the canal's effectiveness" | | Canal maintenance | Canal effectiveness Flood risk | Positive Negative |
| "The urban elements affecting the intensity of flood risk are mainly the lack of urban planning and citizens' behaviour, producing unregulated settlements" | Urban planning Citizens' behaviour | Unregulated settlements Flood risk | Positive Negative | |

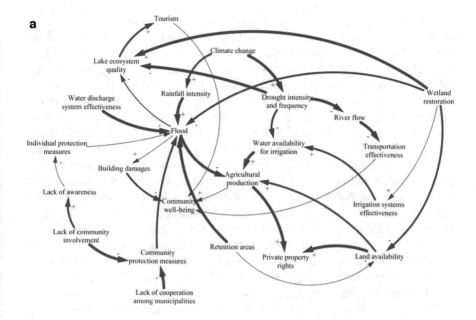

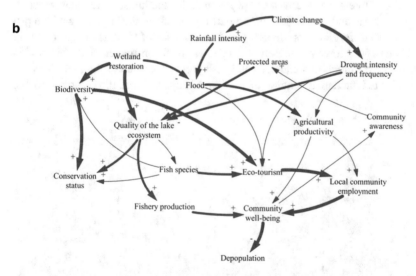

**Fig. 5.4** FCM developed using the stakeholders' interviews for the Lower Danube case study: (**a**) Bistret Municipality; (**b**) WWF Romania. The connections are characterized by different width according to the weight assigned by each stakeholder. The polarity of the connections is also represented (Giordano et al. 2020)

Then, FCM scenario analysis (Kok 2009) was applied in order to define the expected NBS impacts according to the stakeholders' problem understanding. The NBS impacts on the variables in the FCM were defined in a semi-quantitative way, considering a change in the variables within the interval [−1, 1] (for further details refer to (Gray et al. 2015)).

In this work, the comparison between the value of the variables in case of NBS implementation and without NBS allowed us to assess the stakeholders' expected impacts. Figure 5.5 shows the results of two FCM scenarios, i.e. with and without NBS. The larger is the difference between these two scenarios for the different variables, the stronger is the expected impact due to NBS implementation.

The proposed approach assumes that a stakeholder attributes a high importance to a certain variable if it is central in the FCM and if the NBS is expected to provoke a significant change in its state. The following Fig. 5.6 shows an example of the identification of the most important variables using the results of the FCM analysis.

Figure 5.6 shows the most important variables according to the stakeholders' problem understandings. These variables refer to: (i) barrier to risk management that need to be overcome (e.g. institutional cooperation and community risk awareness); (ii) socio-economic objectives (e.g. eco-tourism); (iii) ecosystem improvements (e.g. biodiversity). These variables represent the goals that, according to the stakeholders' perceptions, should be achieved while implementing measures for reducing water-related risks. That is, these variables were used in this work to describe the stakeholders' expected co-benefits. The results of this analysis were used for supporting the participatory modelling exercises for NBS co-design.

The experiences carried out in different case studies demonstrated that accounting for the different perspectives and problem understandings, rather than searching for the synthesis and consensus among participants since the beginning of the process, enhanced the richness, diversity and complexity of the knowledge collected for NBS design. Moreover, making the different decision-makers and stakeholders aware of the different, and equally valid, risk perceptions and problem understandings facilitated the debate among the participants and reduced the risk of conflicts.

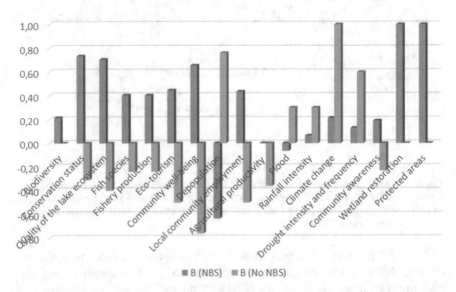

**Fig. 5.5** Simulated change of the FCM variables due to the NBS implementation (Giordano et al. 2020)

**Fig. 5.6** Infogram showing the differences among the stakeholders' problem understanding and risk perception. The centrality degree shows the most central variables in stakeholders' FCM; The impacts degree shows the expected impacts due to NBS implementation; The importance degree shows the most important variables in the stakeholders' FCM

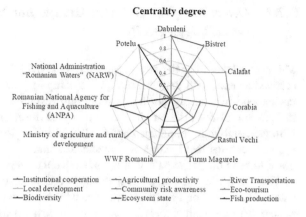

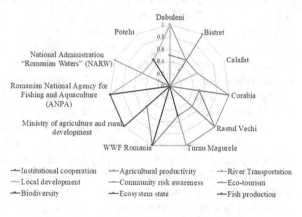

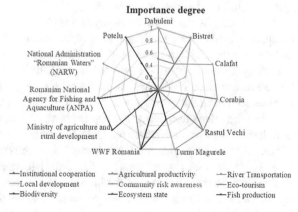

## 5.2.2   Detection of the Barriers Hampering NBS Co-design and Implementation

This section describes the efforts carried out in NAIAD for overcoming the collaborative barriers. NBS implementation is a complex issue, which effectiveness does not depend exclusively on the capacity and resources of the involved decision-makers, but also on the number and quality of the relationships with each other. However, ambiguity in risk perceptions (see previous section) may lead to collaboration structures that encourage stakeholders and decision-makers to avoid each other, turning the participatory process into a controversial and futile process.

There is a wide support in decision and conflict analysis for distinguishing two categories of conflict, i.e. (i) those provoked by existing differences among decision-makers over problem perceptions and preferences; (ii) conflicts due to disharmonious relationships among decision-actors due to lack of trust, also regardless to the problem at stake. Correlations have been detected between these two kinds of conflict in multi-actors decision-making environment. That is, conflicts may not occur between decision-makers with a rather different problem frames, but with good relationships. Conversely, highly similar opinions cannot ensure the absence of conflicts between decision-makers if they distrust each other (Liu et al. 2019).

Starting from these premises, the work carried out in NAIAD aimed at demonstrating that effective network of interaction in multi-actors decision-making environment could contribute at reducing the level of conflict due to differences in problem understandings and, consequently, could enable collective actions for NBS design and implementation. *Networking Interventions approach* has been applied in this work to enhance the existing network of interactions involving the different decision-actors and stakeholders (Valente 2012). Network Interventions are based on the diffusion of innovations theory, which explains how new ideas and practices spread within and between communities.

To this aim, Social Network Analysis (SNA) was applied in order to map the complex network of interactions taking place among the different decision-makers and stakeholders. SNA allows to represent the interactions as a linear graph, characterized by nodes - agents - connected through links of different strength. A link represents a connection between two actors. Its strength describes the importance of this connection in terms of frequency, level of trust, etc. SNA could detect weaknesses in the interaction network – e.g. problems of coordination, lack of information sharing and knowledge transfer, isolated agents, etc. – and support the definition of interventions for improving the cooperation among decision-makers.

In NAIAD, SNA implementation aimed at detecting the main elements negatively affecting the efficiency of the interaction network related to the NBS implementation, and at identifying the leverage elements, i.e. those elements nodes in the network that can be used for implementing interventions aiming at enhancing the cooperation among different decision-makers. To this aim, a stakeholders-based process was applied in NAIAD case studies. Specifically, a participatory network mapping exercise was organized involving institutional and non-institutional

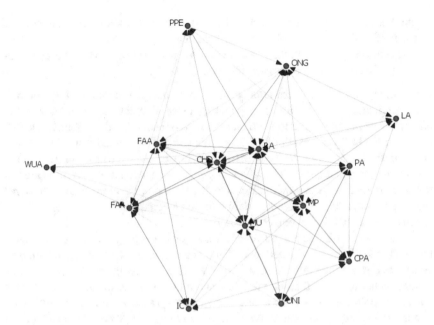

**Fig. 5.7** Map of interactions among different actors in the Medina del Campo case study

decision-actors. During the mapping exercise, participants were requested to mention the tasks that each actor in the list was required to carry out in risk management and NBS implementation. Links were drawn connecting actors and tasks. Then participants were requested to specify with whom the different actors were supposed to cooperate in order to carry out the defined tasks. Finally, the information was introduced in the map. Participants connected the different kinds of information with the tasks this information was supposed to support (Information x Task network), and the actors owning/using the information (Agent x Information network). Once the map describing the Agents-Information-Tasks connections was developed, participants were requested to assign an importance degree to each link according to their own understanding. Three different values were used in this phase, i.e. "High importance" (+++ in the map), "Medium importance" (++ in the map), "Low importance" (+ in the map). Figure 5.7 shows an example of interaction map.

A reader interested in learning more about the SNA methodological approach could refer to (Giordano et al. 2017b). This work aims at describing the potentialities of the method in overcoming the collaboration barriers hampering the NBS implementation.

Four different maps were developed during the participatory mapping exercise:

- the Agent X Agent map, describing the interactions among the different decision-makers and stakeholders;
- the Agent X Knowledge map, connecting the different pieces of knowledge (e.g. groundwater state) with the agents owning/using such knowledge;

- the Agent X Task map, connecting the different agents with the tasks they are required to carry out in water-risk management;
- The Knowledge X Task map, connecting the different tasks with the pieces of knowledge used/needed for carrying out such tasks.

Graph theory measures[1] were applied in order to analyze the network of interactions, to detect key vulnerabilities – that is, those elements that can hamper the effectiveness of the interactions among decision-makers – and to identify the key nodes for the interventions (Table 5.2). In this work, we assumed that a key vulnerability in the organization can be due to agents, information and tasks, or a combination of the three categories. The following graph theory measures were applied in the network analysis. For a more extensive description of the graph theory measures for the network analysis, a reader could refer to (Freeman 1978; Carley et al. 2007; Giordano et al. 2017b; Pagano et al. 2018).

As shown in the Table 5.2, different measures were aggregated in order to detect the key vulnerabilities in the network of interactions. Specifically, an agent could be considered as a vulnerable element if she/he has a low centrality degree – that is, weak connections with the other actors – and a high number of tasks to be carried out in the collective process for risk management. In these conditions, the agent would not be able to cooperate with the others and, thus, there is the risk of not fulfilling the tasks. Similarly, an agent with high "most knowledge" degree has access to a high number of pieces of important knowledge. Nevertheless, a low centrality degree in the Agent X Agent network means that this agent is poorly connected in the network, reducing the effectiveness of the knowledge flow within the network.

**Table 5.2** Graph theory measures for detecting key vulnerabilities in the network of interaction

| Network | Measure | Meaning |
|---|---|---|
| Agent X Agent Agent X Knowledge Agent X Task | Centrality degree Most knowledge Most task | An agent with a few and weak connections with the others would be capable to carry out important tasks and share key pieces of knowledge |
| Agent X Knowledge Knowledge X Knowledge Knowledge X Task | Most knowledge Centrality degree Most task | A piece of knowledge with a high centrality degree and high most task degree is key in the process because it enable the access to other pieces of knowledge and allow the fulfillment of several tasks. Nevertheless, a low Most knowledge measure means that it has a low level of access for many agents. |
| Agent X Task Task X Task | Most task Centrality degree | A task with a high centrality degree is needed in order to enable the fulfillment of the other tasks. A low Most task measure means that this key task is not cooperatively performed. The risk of failure is high, leading to the impairment of the other tasks. |

[1] They measure the level of connectivity in a graph and is expressed by the relationship between the number of links (e) over the number of nodes (v).

A piece of knowledge could represent a vulnerability if it is central in the process – i.e. enable the access to other kinds of knowledge and/or allow the fulfillment of important tasks – and it is not effectively shared within the network. Finally, a task could represent a vulnerability if it is carried out by a single agent and if it plays a key role in activating other important tasks. In these conditions, if the exclusive agent would fail in carrying out this task, the whole process will be affected.

The results of this analysis allowed to detect the key vulnerabilities in the interaction network that need to be tackled through the design and implementation of the network interventions. Table 5.3 shows the barriers detected in the Medina del Campo case study.

These vulnerabilities of the interaction network could lead to barriers to the NBS implementation. As a way of example, we could refer to the Medina del Campo case study. The selected NBS was the Managed Aquifer Recharge, which main goal was to enhance the quality of the groundwater, protecting the resources from the effects of the over-exploitation for irrigation purposes. The above described SNA-based methodology was applied in order to detect the vulnerabilities in the network of interactions and to assess their impacts on NBS implementation. The first vulnerability detected through the SNA was related to the agent Water Users Association (WUA). As learned during the first phases of the process in Medina, the formation of WUA could have a positive impact on the control of the territory and, thus, on the over-exploitation of the groundwater. The second key vulnerability in the SNA is the task "Water rights management", influenced by the farmers' risk awareness, and affecting the capability of the River Basin Authority to reduce the volume of groundwater used for irrigation purposes. The third key vulnerability in the network of

**Table 5.3** Key barriers to NBS implementation due to the network of interactions in the Medina Case Study

| Key vulnerabilities | Type | Meaning in the NBS process |
| --- | --- | --- |
| WUA | Agent | This agent is characterized by a quite low centrality degree in the Agent X Agent network. That is, it has few and weak connections with the other actors. It is supposed to carry out important tasks. |
| Water rights management | Task | This task has a high centrality degree in the Task X Task network. Therefore, it enables the fulfillment of other important tasks. Nevertheless, it is connected exclusively with the CHD. |
| Technical support for crop selection | Knowledge | This piece of knowledge plays a key role in carrying out the most important tasks (Knowledge X Task network). Nevertheless, it has a low Most knowledge degree in the Agent X Knowledge network and, thus, it is not effectively shared in the network. |
| GW state | Knowledge | This piece of knowledge has a high centrality degree in the Knowledge X Knowledge network, which means that is enables the access to other important pieces of information. Nevertheless, only few agents have access to it (Agent X Knowledge network). |

**Table 5.4** Networking interventions for NBS co-design and implementation

| Vulnerability | Actors | Knowledge | Tasks |
|---|---|---|---|
| WUA formation and water rights management | Municipalities Regional Authority Ministry | Information on the water right assignment process | Detect illegal groundwater exploitation Voting system in the River Basin Authority Enhance transparency of the process Create a register of the rights |
| GW state information | Ministry of environmental CHD Famers | Climate information Water footprint label for the products GW extraction costs Virtual water | Sustain rural eco-tourism Consumers awareness raising Implementing drought management plan GW metering |
| Technical support to farmers for crop changes | Technicians Regional Authority Universities and research centres Farmers organization | Crop water requirements Information drought resistant crops Market evaluation | Enhance CAP distribution Training on novel crops Crop-based water allocation Awareness raising on sustainable agriculture |

interactions is the technical support to farmers for enabling crop change, which depends on the reputation of the basin authority (which should be capable to provide effective information and technical support to select less water demanding crops), and affects the farmers' capabilities to reduce groundwater exploitation. Finally, the fourth key vulnerability in the network of interactions was the "GW state information". The availability of this information could enhance farmers' risk awareness and, thus, reduce the exploitation of groundwater.

The results of the SNA analysis were used to inform the debate among the different decision-makers and stakeholders aiming at co-designing the interventions for overcoming the detected collaborative barriers, as shown in Table 5.4.

### 5.2.3 NBS Scenario Simulation and Trade-Offs Analysis

As stated previously, the work carried out in different NAIAD case studies aimed at demonstrating the suitability of Participatory Modelling approaches in enabling the stakeholders' engagement for NBS co-design. Specifically, two modelling methods were applied in NAIAD case studies, i.e. *System Dynamic Model (SDM) and Fuzzy Cognitive Map (FCM)*. According to (Sterman 2000) System Dynamics Modelling is a set of conceptual tools that enables an improved understanding of the structure and the dynamics of complex systems, as well as of rigorous modeling methods that enable building formal simulations of complex systems to design more effective policies. SDM is widely used to analyze complex ('wicked') problems over time, taking into account their multi-dimensionality through the integration of qualitative and quantitative, 'hard' (e.g. technical) and 'soft' (e.g. social) variables. Both SDM and FCM were selected for three main characteristics that made them suitable for addressing the complex issues related to NBS co-design and assessment. Firstly, both SDM and FCM are based on System Thinking approach – i.e. the evolution of the modelled system is affected by the structure of the interconnections among the different elements. Therefore, these two modelling approaches were considered suitable for mapping and analyzing the complex web of interactions involving physical, ecological and socio-economic factors affecting the NBS effectiveness. Secondly, these methods were selected due to their capabilities to simulate the dynamic evolution of the system, accounting for the time dimension, whereas many other modelling approaches provide simply "snapshot" of the system state. Thirdly, both SDM and FCM allow the integration of stakeholders' and scientific knowledge and, in doing so, enhance the legitimacy of the developed model.

Several experiences were carried out in different NAIAD case studies using integrated modelling tools, in order to define a *bottom-up procedure for co-designing NBS and for supporting their assessment*. The basic idea behind the proposed approach is to focus on the identification, analysis and modelling of the co-benefits production, which is the key value added of NBS with respect to traditional grey infrastructures. The rationale of the proposed approach is the development of a sequence of individual and collective activities that should support assessing the effectiveness of strategies (i.e. a measure of a combination of multiple measures) in the production of the high-ranked benefits and co-benefits according to local stakeholders. The key advantage and element of innovation lies in the effort of proposing a solid procedure for eliciting and structuring local knowledge, collectively building a vision of the system under investigation, and simulating the effects of the most relevant scenarios. The collected and structured stakeholders' knowledge was, then, integrated with scientific and expert knowledge for simulating NBS scenarios. Figure 5.8a shows an example of integrated model (SDM) developed for the Glinscica case study. Figure 5.8b shows the dynamic FCM developed for the Lower Danube case study.

Both models were built, using a participatory approach, starting from the aggregation of individual mental models (discussed in Sect. 5.2.1). This process is not straightforward, given the ambiguity and the differences in problem framing, although several methods exist for the purpose. In this cases, stakeholders were directly invited to construct an aggregated map. The process started, with the support of the analysts, merging similar variables (e.g. the same concept expressed using different words). Then, a discussion between the stakeholders helped drawing the weighted connections among the variables and identifying potential additional variables and connections. The global structure was then further discussed to check for potential inconsistencies. Additional details are available in (Pagano et al. 2019; Gómez Martín et al. 2020; Coletta et al. 2021).

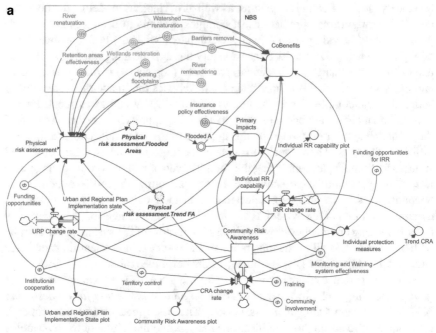

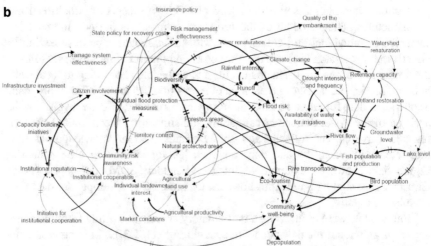

**Fig. 5.8** (**a**) Overview of the stock and flow model built in the Glinscica Case study (Pagano et al. 2019); (**b**) Aggregated dynamic FCM developed for the Lower Danube case study (Giordano et al. 2020)

The stock and flow model (one of the most common SDM tools) represented in Fig. 5.8a can be interpreted as an evolution of FCM. It includes a multiplicity of mathematical expressions governing the system, incorporated via flow diagrams and finally transformed using a simulation environment. Basically, FCM's variables and causal relationships are identified and translated into the common SDM sets, and hypotheses are formulated on the mathematical aspects, integrating multiple sources of information, such as expert consultation and scientific/grey literature. In the Glinscica river case it was adopted for a twofold reason: (i) the need to quantitatively integrate both 'hard' information (e.g. hydraulic models and data) and expert opinion on 'soft' variables (e.g. socio-economic aspects); (ii) the need to explicitly model the multi-dimensional implications of several NBS, in order to comparatively analyze their benefits and co-benefits with time.

The aggregated dynamic FCM (Fig. 5.8b) was used in the Lower Danube case study to develop a more simplified scenario analysis related to different sets of measures identified by the stakeholders. This analysis was highly relevant for a twofold reason: (i) to explicitly highlight the differences in stakeholders' perception of benefits and co-benefits production with time (using delays and multiple time steps for analysis), identifying the potential onset of trade-offs; (ii) to help decision-makers in the timely identification of such trade-offs, thus helping in the timely identification of conflicts and facilitating NBS implementation.

As already mentioned, the integrated models were, then, used for simulating NBS scenarios and assessing NBS effectiveness in reducing the water-related risks and producing the expected co-benefits, as shown in Fig. 5.9a. Specifically, the focus is on how stakeholders might differently evaluate the co-benefits, which depends on the individuals' benefits perception. Neglecting these differences and ignoring the consequences of trade-offs between values held by different stakeholders, which in many cases are not well represented in the decision-making process, may lead to conflict, and thus to policy resistance mechanisms.

The integrated models were also used for detecting and analyzing potential trade-offs among different beneficiaries. Starting from the 'Importance degree' that was attributed to specific co-benefits according to individual problem understanding, we assumed that there was a trade-off between two stakeholders if there was an unequal distribution of such co-benefits (i.e. an increase for a stakeholder, and a decrease for another). A stakeholder would therefore not fully capture NBS co-benefits if the value of her/his objective function associated to the NBS implementation would be lower than she/his expects. Figure 5.10 shows the comparison between the simulated and desirable (given individual risk perception and co-benefits analysis) NBS benefits for some of the stakeholders involved in the Lower Danube case study, based on the results of Fig. 5.9b. The axes in Fig. 5.10 plot the results for the 'short', 'medium' and 'long' term of analysis, which is useful to show how the difference between the simulated and perceived benefits and co-benefits may significantly change over time, in different stages of measures implementation.

Figure 5.10 shows that the simulated value is lower than the desirable one on the short-term axis for a subset of the involved stakeholders, which means that they all perceive a dis-benefit in the short term. This is mainly because the effectiveness of the selected strategy on variables such as community well-being and risk awareness is limited in the short term. In some cases, such as e.g. Corabia, the objective function is lower than the expected one in all time steps. This is mainly because these stakeholders gave a high importance degree to the co-benefits "agricultural productivity" and "river transportation". The model simulation showed that: (i) the former is expected to decrease in the medium and long terms due to the increase of the

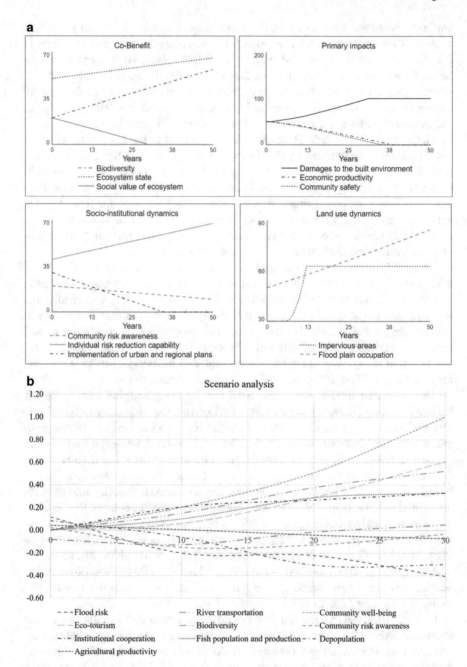

**Fig. 5.9** Dynamic evolution of NBS benefits and co-benefits using the SDM in the Glinscica Case study (Pagano et al. 2019) (**a**) and the FCM in the Lower Danube Case study (**b**) (Giordano et al. 2020)

**Fig. 5.10** Comparison between desirable and simulated benefits due to NBS implementation for single stakeholders, based on the results of the FCM in Fig. 5.9b. This Figure refers to a subset of the results obtained in the Lower Danube case study. The three axes represent the stakeholders' objective functions in the short-, medium and long-term

natural protected areas; and (ii) the implemented NBS was supposed to have a limited impact on the river flow and, consequently, river transportation.

In many cases, the stakeholders perceived a high benefit from the strategy implementation in the long term because of its positive impact on the eco-tourism and community well-being. The analysis showed that most of the potential conflicts can occur in the long term, and could involve mainly the stakeholders that assigned a high value to the agricultural productivity.

The results of the trade-offs analysis can be used by decision-makers to prevent potential conflicts and to facilitate the NBS implementation. The results demonstrate also that all stakeholders need to be informed in the early stage of the project implementation, in order to make them aware of the time lag needed for producing the expected co-benefits.

## 5.3 Concluding Remarks

This section is meant to discuss to what extent the activities carried out in the NAIAD case studies allow to demonstrate the suitability of the applied approaches for overcoming two of the key barriers that could hamper the effective implementation of NBS, i.e. lack of stakeholders' engagement and collaborative barriers among different decision-makers and stakeholders.

Participatory and integrated tools, such as SDM and FCM, helped overcoming one of the key limits of the existing frameworks for NBS assessment that is the lack of structured representations of their multi-dimensionality. Co-benefits were considered in many cases even more important than the reduction of water-related risks, and were used as the main elements for the co-design of the NBS. The NAIAD Case studies demonstrated the *relevance of using participatory tools to increase social acceptance towards NBS*, since the whole process helped breaking down some of the existing socio-institutional barriers, mainly related to the limited knowledge and to the partial involvement of the stakeholders in the discussion. The active participation of stakeholders throughout the process of NBS design is crucial in order to move beyond individual perception and problem understanding, and to support building a shared view of the problem under consideration. Additionally, defining a shared problem frame and group model facilitates interdisciplinary and cross-sectoral communication and collaboration.

The applied approaches for eliciting and structuring individual risk perceptions allowed to account for the diversities, and enabling an inclusive and equitable participatory process. Moreover, the experiences carried out in NAIAD demonstrated the importance of *making explicit to the stakeholders how the knowledge they provided during the different phases was actually used for developing the model*. This contributes to create a sense of ownership toward the developed model, facilitating the interaction with the stakeholders. Finally, by contributing to the model development, stakeholders became aware of this complexity and realized that NBS effectiveness is influenced by several elements, ranging from the physical to the socio-institutional ones. As result of this learning process, stakeholders and decision-makers selected several socio-institutional actions to be implemented as "supportive" measures for NBS effectiveness.

The approach based on SNA demonstrates that overcoming the barriers to collaboration and enhancing the effectiveness of the network of interactions, through the implementation of the networking interventions, could have positive impacts on the NBS implementation and effectiveness. Therefore, this work demonstrates that, in addition to socio-economic, technical and institutional barriers, NBS implementation claims to detect and overcome those related to the interaction between the various decision-actors. Methods for unravelling the complex network of interactions taking place among different decision-makers are needed.

Two key messages can be identified from the work carried out in the NAIAD case studies concerning the trade-off analysis. Firstly, differences among stakeholders concerning the definition of co-benefits to be produced through the NBS implementation and their importance could lead to trade-offs among different stakeholders. Therefore, the *trade-offs analysis requires methods and tools capable to handle the diversity in problem frames among the different stakeholders*, and suitable to simulate the dynamic evolution of complex systems. Secondly, trade-offs analysis claims for a clear understanding and modelling of the complex cause-effects chains affecting the NBS impacts on the system.

The results of the trade-offs analysis can be used for supporting decision-makers in the definition of actions/measures for reducing trade-offs and potential conflicts, supporting NBS acceptance. The work described in this chapter is mainly based on risk perceptions and problem understandings. Therefore, the results of the analysis could be useful for *enhancing the communication with the stakeholders, affecting their perceptions and enabling learning processes*. The described example shows the importance of raising stakeholders' awareness about the great potentialities of the NBS implementation in creating new community development opportunities due to the eco-tourism. Therefore, the communication should emphasize the production of the co-benefits, rather than being simply focused on the reduction of the climate-related risks.

The analysis of the activities carried out in the case studies demonstrated also some drawbacks of the applied methodology. Capturing and processing stakeholders' knowledge starting from individual inputs is time consuming and requires substantial efforts by skilled analysts for post-processing the information collected during the individual interviews. Moreover, efforts are required in order to avoid stakeholders' fatigue in taking part to the different phases of the methodological approach, and keep them engaged. The qualitative nature of the modelling approaches also represented a limit of the applied methodology. Although the structure of the developed models, based on causal connections, was easily understandable by the stakeholders, and used for supporting the debate, many perplexities were mentioned by the participants concerning the results of the scenario simulations. The participants seemed inclined to prefer quantitative evaluation, rather than qualitative results, specifically when they were required to comment the NBS capability to reduce climate-related risks. Therefore, efforts are still required to integrate qualitative modelling approaches with more quantitative, physically based models.

---

**Box 5.1 Key Lessons Learnt**

- Involving stakeholders in the development of integrated models could raise their awareness concerning risks and NBS effectiveness, affecting social acceptance;
- The applied approaches should aim at enhancing the potential richness, diversity and complexity of the collected knowledge, rather than searching consensus among participants;
- Dynamic approaches are required to assess NBS effectiveness, co-benefits production and trade-offs among different beneficiaries;
- Ineffective interactions among different decision-actors – i.e. limited exchange of information – could create barriers to NBS co-design and implementation.
- Enhancing communication on co-benefits production and potential trade-offs improves NBS acceptance.

---

**Acknowledgements**  The Authors would like to thank the local stakeholders of all NAIAD Demos for their support in the activities performed: without their invaluable contributions this work would have been impossible.

# References

Bain PG et al (2016) Co-benefits of addressing climate change can motivate action around the world. Nat Clim Chang 6(2):154–157. https://doi.org/10.1038/nclimate2814

Brugnach M et al (2011) More is not always better: coping with ambiguity in natural resources management. J Environ Manag 92(1):78–84. https://doi.org/10.1016/j.jenvman.2010.08.029

Brugnach M, Ingram H (2012) Ambiguity: the challenge of knowing and deciding together. Environ Sci Pol 15(1):60–71. https://doi.org/10.1016/j.envsci.2011.10.005

Calliari E, Staccione A, Mysiak J (2019) An assessment framework for climate-proof nature-based solutions. Sci Total Environ. Elsevier BV 656:691–700. https://doi.org/10.1016/j.scitotenv.2018.11.341

Carley KM et al (2007) Toward an interoperable dynamic network analysis toolkit. Decis Support Syst 43(4):1324–1347. https://doi.org/10.1016/j.dss.2006.04.003

Cohen-Shacham E et al (2016) Nature-based solutions to address global societal challenges. IUCN. https://doi.org/10.2305/IUCN.CH.2016.13.en

Coletta VR et al (2021) Causal loop diagrams for supporting nature based solutions participatory design and performance assessment. J Environ Manag. Elsevier Ltd 280(xxxx):111668. https://doi.org/10.1016/j.jenvman.2020.111668

Eden C (2004) Analyzing cognitive maps to help structure issues or problems. Eur J Oper Res 159(3):673–686. https://doi.org/10.1016/S0377-2217(03)00431-4

Frantzeskaki N (2019) Seven lessons for planning nature-based solutions in cities. Environ Sci Pol 93:101–111. https://doi.org/10.1016/j.envsci.2018.12.033

Freeman LC (1978) Centrality in social networks conceptual clarification. Soc Networks 1(3):215–239. https://doi.org/10.1016/0378-8733(78)90021-7

Giordano R et al (2007) Integrating conflict analysis and consensus reaching in a decision support system for water resource management. J Environ Manag 84(2):213–228. https://doi.org/10.1016/j.jenvman.2006.05.006

Giordano R et al (2017b) Modelling the complexity of the network of interactions in flood emergency management: the Lorca flash flood case. Environ Model Softw. Elsevier Ltd 95:180–195. https://doi.org/10.1016/j.envsoft.2017.06.026

Giordano R et al (2020) Enhancing nature-based solutions acceptance through stakeholders' engagement in co-benefits identification and trade-offs analysis. Sci Total Environ 713:136552. https://doi.org/10.1016/j.scitotenv.2020.136552

Giordano R, Brugnach M, Pluchinotta I (2017a) Ambiguity in problem framing as a barrier to collective actions: some hints from groundwater protection policy in the Apulia region. Group Decis Negot 26(5):911–932. https://doi.org/10.1007/s10726-016-9519-1

Gómez Martín E et al (2020) Using a system thinking approach to assess the contribution of nature based solutions to sustainable development goals. Sci Total Environ 738:139693. https://doi.org/10.1016/j.scitotenv.2020.139693

Gray SA et al (2015) Using fuzzy cognitive mapping as a participatory approach to analyze change, preferred states, and perceived resilience of social-ecological systems. Ecol Soc 20(2):11. https://doi.org/10.5751/ES-07396-200211

van den Hoek RE et al (2014) Analysing the cascades of uncertainty in flood defence projects: how "not knowing enough" is related to "knowing differently". Glob Environ Chang 24:373–388. https://doi.org/10.1016/j.gloenvcha.2013.11.008

Jasanoff S (2007) Technologies of humility. Nature 450(7166):33–33. https://doi.org/10.1038/450033a

Jetter AJ, Kok K (2014) Fuzzy cognitive maps for futures studies—a methodological assessment of concepts and methods. Futures 61:45–57. https://doi.org/10.1016/j.futures.2014.05.002

Josephs LI, Humphries AT (2018) Identifying social factors that undermine support for nature-based coastal management. J Environ Manag. Elsevier Ltd 212:32–38. https://doi.org/10.1016/j.jenvman.2018.01.085

Kabisch N et al (2016) Nature-based solutions to climate change mitigation and adaptation in urban areas. Ecol Soc 21(2):39. https://doi.org/10.5751/ES-08373-210239

Kim H, Andersen DF (2012) Building confidence in causal maps generated from purposive text data: mapping transcripts of the Federal Reserve. Syst Dyn Rev 28(4):311–328. https://doi.org/10.1002/sdr.1480

Kok K (2009) The potential of fuzzy cognitive maps for semi-quantitative scenario development, with an example from Brazil. Glob Environ Chang 19(1):122–133. https://doi.org/10.1016/j.gloenvcha.2008.08.003

Liu B et al (2019) Large-scale group decision making model based on social network analysis: trust relationship-based conflict detection and elimination. Eur J Oper Res 275(2):737–754. https://doi.org/10.1016/j.ejor.2018.11.075

Montibeller G et al (2008) Reasoning maps for decision aid: an integrated approach for problem-structuring and multi-criteria evaluation. J Oper Res Soc 59(5):575–589. https://doi.org/10.1057/palgrave.jors.2602347

O'Donnell EC, Lamond JE, Thorne CR (2017) Recognising barriers to implementation of blue-green infrastructure: a Newcastle case study. Urban Water J. Taylor & Francis 14(9):964–971. https://doi.org/10.1080/1573062X.2017.1279190

Olazabal M et al (2018) Transparency and reproducibility in participatory systems modelling: the case of fuzzy cognitive mapping. Syst Res Behav Sci 35(6):791–810. https://doi.org/10.1002/sres.2519

Ozesmi U, Ozesmi S (2003) A participatory approach to ecosystem conservation: fuzzy cognitive maps and stakeholder group analysis in Uluabat Lake, Turkey. Environ Manag 31(4):518–531. https://doi.org/10.1007/s00267-002-2841-1

Pagano A et al (2018) Integrating "hard" and "soft" infrastructural resilience assessment for water distribution systems. Complexity 2018:1–16. https://doi.org/10.1155/2018/3074791

Pagano A et al (2019) Engaging stakeholders in the assessment of NBS effectiveness in flood risk reduction: a participatory system dynamics model for benefits and co-benefits evaluation. Sci Total Environ 690:543–555. https://doi.org/10.1016/j.scitotenv.2019.07.059

Palmer MA et al (2015) Manage water in a green way water security: Gray or green? Manage water in a green way. Science 349(6248):584–585. https://doi.org/10.1126/science.aac7778

Raymond CM et al (2017) A framework for assessing and implementing the co-benefits of nature-based solutions in urban areas. Environ Sci Pol. Elsevier 77(June):15–24. https://doi.org/10.1016/j.envsci.2017.07.008

Renn O (1998) The role of risk perception for risk management. Reliab Eng Syst Saf 59(1):49–62. https://doi.org/10.1016/S0951-8320(97)00119-1

Rosenhead J, Mingers J (2001) Rational analysis for a problematic world revisited: problem structuring methods for complexity, uncertainty and conflict, 2nd edn. Wiley

Santoro S et al (2019) Assessing stakeholders' risk perception to promote nature based solutions as flood protection strategies: the case of the Glinščica river (Slovenia). Sci Total Environ 655:188–201. https://doi.org/10.1016/j.scitotenv.2018.11.116

Sterman J (2000) Business dynamics: systems thinking and modelling for a complex world. McGraw-Hill Higher Education

Valente TW (2012) Network interventions. Science 337(6090):49–53. https://doi.org/10.1126/science.1217330

Weick KE (1995) Sensemaking in organizations. SAGE Publishing

Wihlborg M, Sörensen J, Alkan Olsson J (2019) Assessment of barriers and drivers for implementation of blue-green solutions in Swedish municipalities. J Environ Manag. Elsevier 233(November 2018):706–718. https://doi.org/10.1016/j.jenvman.2018.12.018

# Chapter 6
# Economic Assessment of Nature-Based Solutions for Water-Related Risks

**Philippe Le Coent, Cécile Hérivaux, Javier Calatrava, Roxane Marchal, David Moncoulon, Camilo Benítez Ávila, Mónica Altamirano, Amandine Gnonlonfin, Ali Douai, Guillaume Piton, Kieran Dartée, Thomas Biffin, Nabila Arfaoui, and Nina Graveline**

**Highlights**

- This study combines the integrated cost-benefit analysis of NBS strategies aiming at reducing water risks in four case studies
- The cost of implementation and maintenance of NBS strategies is lower than the cost of grey solutions for the same level of water risk management, confirming their cost-effectiveness advantage
- Benefits in terms of avoided damages are however generally not sufficient to cover investment and maintenance costs
- Co-benefits represent the largest share of the value generated by NBS strategies
- The cost-benefit analysis of NBS strategies implemented in four case-studies, is positive in three case studies and negative in one

P. Le Coent (✉) · C. Hérivaux
G-Eau, BRGM, Université de Montpellier, Montpellier, France
e-mail: p.lecoent@brgm.fr

J. Calatrava
Department of Business Economics, Universidad Politecnica de Cartagena, Cartagena, Spain

R. Marchal · D. Moncoulon
Caisse Centrale de Réassurance, Paris, France

C. Benítez Ávila
Deltares, Delft, The Netherlands

M. Altamirano
Deltares, Delft, The Netherlands

Faculty of Technology, Policy and Management, Delft University of Technology, Delft, The Netherlands

A. Gnonlonfin
University Côte d'Azur, IMREDD, Nice, France

A. Douai
University Côte d'Azur, GREDEG, CNRS, Valbonne, France

## 6.1 Introduction

The economic assessment of NBS is a key step in the evaluation of NBS. Indeed, assessing the value of costs and benefits of NBS and being able to compare them to alternative strategies such as business as usual grey solutions is fundamental for decision makers to develop these solutions and eventually turn them into implementable Natural Assurance Schemes, with solid business models and business cases (see Chaps. 8 and 9). However, real case studies are relatively scarce, evidence is therefore needed to understand under which conditions it seems relevant for decisions makers to invest in NBS.

This chapter therefore presents the methodological framework developed for the economic assessment of NBS for water related risks, drought and flood, and its application to NAIAD case studies. Since the reduction of water related risks is the main aim of these NBS, we particularly elaborate on the methodologies that can be used to estimate the reduction of damage costs. Another specificity of NBS, as compared to grey solutions, is their capacity to produce additional environmental and social benefits: the co-benefits. We therefore also present the various methods that can be used for the monetary valuation of co-benefits. We also provide details on the elements that need to be considered for the evaluation of costs. Finally, cost-benefit analyses are implemented to help determining whether projects, such as NBS, improve social welfare from an economic standpoint and should therefore be considered for implementation by decision makers.

This methodological framework has been fully or partially applied to seven case studies of the NAIAD project. We conclude by some lessons learned from the implementation of the methodology as well as some of the key results of the economic assessment.

## 6.2 Methods

### 6.2.1 Overall Methodology of the Economic Assessment

The overall aim of the economic assessment methodology is to assess the economic value of alternative actions aiming at managing water risks, through a Cost benefit Analysis (CBA) (European Commission 2014), which provides an evaluation of the economic efficiency of a programme. Depending on the case study, one or several

G. Piton
Université Grenoble Alpes, INRAE, CNRS, IRD, Grenoble INP, IGE, Grenoble, France

K. Dartée · T. Biffin
Field Factors, Delft, The Netherlands

N. Arfaoui
Catholic University of Lyon, Lyon, France

N. Graveline
University of Montpellier, UMR Innovation, INRAE, Montpellier, France
e-mail: n.graveline@inrae.fr

alternatives are compared which incorporate different levels of NBS strategies and traditional grey infrastructure. The methodology is based and detailed in Graveline et al. (2017).

In this study, we mention as **NBS strategies**, alternative projects that generally include a combination of NBS and grey infrastructures. The principle of CBA is to compare an alternative with-the-project with a counterfactual baseline alternative without-the-project, generally referred to as the Business As Usual alternative (BAU). The CBA performed compares strategies without NBS (considered as the BAU strategy) with one or several strategies including NBS measures.

The CBA requires the estimation of all direct and indirect costs and benefits for the different NBS strategies under study. The following typology of monetary values associated with NBS strategies is considered:

– **Costs of implementation** are those that are necessary for the implementation and maintenance of the NBS included in the NBS strategies.
– **Opportunity costs** are those that are foregone with the NBS strategies, for instance areas that are taken out of production or land that is used for NBS and that cannot be used for other purposes such as the construction of building. They are the indirect costs of the NBS strategies.
– **Avoided damages** are the damages avoided due to the reduction of water risks generated by NBS strategies. Avoided costs are the primary benefit generated by NBS strategies aiming at reducing water risks.
– **Co-benefits** are the additional environmental, economic, and social benefits generated by NBS. In the CBA, we will focus on the ones that can be evaluated monetarily although they cover only part of the co-benefits generated by NBS strategies or only a portion of their overall value. The level of co-benefits varies between the different strategies.

In Fig. 6.1, we present a schematic representation of the CBA method applied to the evaluation of NBS aiming at reducing water risks. Table 6.1 presents the main information of the economic assessment implemented in the different case studies. The cost-benefit analysis was performed fully for the Lez, Brague and Rotterdam case studies and partially for the Medina case.

Several indicators can be calculated to carry out a CBA. In this study, we mainly report on the Benefit-Cost Ratio (BCR) that is estimated with the following formula, where $CB_t$ is the Co-Benefits in year t, $AD_t$ is the Avoided Damage in year t, $r$ is the discount factor, $C_t$ and $OC_t$ are implementations Costs and Opportunity Costs in year t and T is the time horizon of the assessment.

$$BCR = \frac{\sum_{t=0}^{T}\left[\dfrac{AD_t + CB_t}{\left(1+r\right)^t}\right]}{\sum_{t=0}^{T}\left[\dfrac{C_t + OC_t}{\left(1+r\right)^t}\right]}$$

$$(6.1)$$

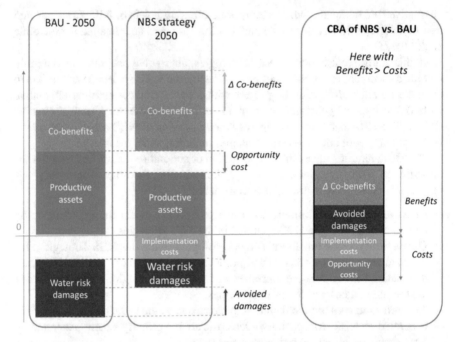

**Fig. 6.1** Description of the cost-benefit analysis approach for the economic valuation of NBS within NAIAD. (Modified from Graveline et al. 2017)

In this formula, benefits and costs are discounted with the discount factor r, in order to estimate their present value. This reflects the social view on how future benefits and costs should be valued against present ones: the highest the discount factor the more preference for the present. The European Commission recommends a discount factor ranging from 3 to 5% (European Commission 2014) whereas the Quinet report recommend a value of 2.5% (CGSP 2013). In practice, the discount factor varies in the different case studies, considering the discount factor prevailing in the evaluation of investment projects at the country level.[1]

A BCR superior to 1 means that a project is economically efficient, i.e. that it improves the economic welfare and that it should be eligible for investment by public funds. Decision makers may also compare different alternatives and invest in the alternative that present the highest BCR.

Other partial CBA indicators can be used which we focus on the primary benefit and consider only the direct cost of implementation. Although partial in economic terms, this indicator can be useful for decision makers.

---

[1] We did not include in the evaluation a risk adjusted cash-flow that would account for the risk of not ultimately producing the expected benefits. This step would be necessary for the building of a financeable investment project.

**Table 6.1** Description of the economic valuation implemented in the different case studies.

| Case study | Medina | Thames | Lower Danube | Lez | Rotterdam | Copenhagen | Brague |
|---|---|---|---|---|---|---|---|
| Country | Spain | United Kingdom | Romania | France | Netherlands | Denmark | France |
| Main water risk addressed | Drought and flooding | Flooding | Flooding | Flooding | Flooding | Groundwater Flooding | Flooding |
| Scale | Aquifer and city | Large Watershed | Large Watershed | Small watershed | Neighboorhood | City catchment | Small Watershed |
| NBS strategy evaluated | (1) Agriculture practice change (2) Aquifer recharge (3) River renaturation and floodplain restoration | (1) Leaky dams (2) Retention ponds (3) Conservation agriculture | (1) Wetland restoration (2) River renaturation (3) Retention areas (4) Reforestation | (1) NBS 1: Bioswales, city deproofing and permeable pavement, (2) NBS 2 = NBS1 + green roofs, open retention basins, vegetated bioswales, | (1) Grey: separated sewer system and permeable pavement (2) Hybrid: Separate sewer system with natural retention and infiltration at public squares, including aquifer storage (3) Green: Only green infrastructure for retention and infiltration | Urban river renaturation | (1) Grey: large retention dams, (2) NBS high: Reopening river corridors (3) NBS very high: restoring an integrated flood plain to the river. |
| Avoided damage | X | X (Non-monetary) | X (Partial) | X | X (Non-monetary) | X | X |
| Life Cycle Cost and opportunity costs | X | X | | X | X | | X |
| Co-benefits | X (Non-monetary) | X (Non-monetary) | X (Non-monetary) | X | X | | X |

$$Avoided\ damage\ /\ Cost(rate) = \frac{\sum_{t=0}^{T}\left[\dfrac{AD_t}{(1+r)^t}\right]}{\sum_{t=0}^{T}\left[\dfrac{C_t}{(1+r)^t}\right]}$$

(6.2)

Finally, a cost-effectiveness indicator, which indicates the cost incurred to achieve a given output could also be useful. This indicator in expressed in euro by a physical unit measuring the effectiveness of the measure such as m³ of water retention for flood management. This indicator is compiled only for individual NBS measures and not for NBS strategies (but see Bokhove et al. 2019 for a way to do so at the masterplan scale).

$$Cost\ effectiveness = \frac{\sum_{t=0}^{T}\left[\dfrac{C_t}{(1+r)^t}\right]}{Indicator\ of\ effectiveness}$$

(6.3)

In order to implement the CBA in the case studies, the following stepwise approach has been implemented. This chapter subsequently develops step 4 of this stepwise approach: the economic assessment methods. The details for the implementation of the other steps, especially the engagement with stakeholders, is described in Chap. 19 of this publication.

1. **Set scale and time horizon**: The spatial scale varies greatly depending on the case study: from a neighbourhood (Rotterdam), to a city catchment (Copenhagen), an aquifer (Medina del Campo) and to a river basin scale (Brague, Lower Danube, Lez and Thames). The time horizon at which the strategies are evaluated defines the number of years for which the benefits and costs are taken into account in the economic analysis. This time horizon varies depending on the type of investment and is usually set at the expected lifetime of the considered investment.

2. **Define and describe scenarios and NBS strategies**. This step is crucial for the analysis. The identification of scenarios and NBS strategies for water-related risks is undertaken using a participatory process involving the main stakeholders of the considered territory (See Chaps. 5, 7 and 19 for possible methods). Scenarios should be elaborated to determine the prevailing conditions along the time horizon (climate change, land use change) that may affect NBS impact. They are built based on a historical analysis of past trends and the identification of driving forces that may affect the territory under study. NBS strategies are the alternative combinations of NBS measures, developed to address water-related risks, which are compared in the economic analysis. NBS strategies were co-designed with stakeholders based on an assessment of water-related risks and the information available on the impact of NBS on risks and co-benefits. More

sophisticated approaches based on System Dynamic Modelling (See Chap. 5) were used to identify strategies responding to territorial challenges.

3. **Impact assessment**. The impact of NBS strategies needs to be established to subsequently assess the economic effects of these impacts. Given the focus of this study on water risks, a large effort of hydrologic and hydraulic modelling is undertaken to estimate the impact of NBS on water risks (see Chap. 4). Other more simple models are used in order to estimate the physical impact on co-benefits. The impact assessment also requires NBS strategies to be translated into usable inputs for physical modelling. This requires either a simple quantification of some physical variables associated with strategies such as total volume of water retention brought by NBS (for flood control) or in some case GIS modelling for the spatial setting of scenarios and strategies.

4. **Assessment of costs and benefits**: The details of the methods for the estimation of implementation costs, opportunity costs, avoided damages and co-benefits are presented in the following sections.

5. **CBA and sensitivity analysis**: Finalization of the CBA by compiling the BCR, according to the formula above, and carry out a sensitivity analysis.

## 6.2.2   Implementation Costs and Opportunity Costs Assessment

The evaluation of implementation costs was based on the development of guidelines based on the estimation of Life Cycle Costs (LCC) methodology. LCC, also named Total Cost of Ownership (TCO), consider the total cost of acquisition, use/administration, maintenance and disposal of a given item/service (Ellram 1995). The accurate identification of LCC provides the information needed to assess the magnitude of investments for keeping socio-technical system functionality over time. In our case, the expected functionality of NBS is framed in relation to avoiding damages from water-related risks (Denjean et al. 2017).

Therefore, the LCC methodology focused on identifying the generating activities and cost determining factors to maintain the main functionality of NBS, avoiding water-related damages. Cost generating activities can be grouped into five LCC components namely: capital expenditures (CAPEX), operating and minor maintenance expenditures (OPEX), capital maintenance, expenditure on direct support, expenditure on indirect support and cost of capital. Table 6.2 presents the general framework for assessing LCC.

The LCC methodology can be used as a framework to evaluate costs to be integrated in the CBA. In Rotterdam, the three strategies (grey, hybrid and green) were set to meet the same level of flood risk reduction, as requested in the LCC framework. On the other hand, Medina del Campo, Brague, Thames and Lez assessed the LCC components of NBS strategies that emerged from different iterations between technical analysis for meeting policy goals and stakeholder consultations rather than the definition of specific levels of service.

**Table 6.2** General framework for assessing LCC with examples of costs. (as presented in Altamirano and de Rijke 2018)

| LCC component | Cost elements | Cost drivers |
|---|---|---|
| 1. Capital expenditure | **Planning, design and construction:** Hydrological assessment, bio-engineering, earth removal and recharge with machinery, concrete channelization, bed widening | **Function & level of service Design:** Sheer stress determines the level of service, bio-engineering method **Location-specific conditions:** Hydrology and climate conditions **Socio-economic conditions:** Property prices, salaries |
| 2. Operating and minor maintenance expenditure (OPEX) | **Maintenance, monitoring, operations:** Vegetation maintenance, water quality monitoring, environmental quality monitoring | **Function & level of service Design:** Sheer stress factor that determines the level of service **Location-specific conditions:** Hydrology and climate conditions **Socio-economic conditions:** Salaries |
| 3. Capital maintenance | **Asset renewal, replacement and rehabilitation:** Post-disaster riparian vegetation reconstruction, River bed cleaning | **Function & Level of Service Design:** Sheer stress factor that determines the level of service **Location-specific:** Probability of hazard occurring. Measures to reduce the vulnerability of NBS to hazards **Socio-economic conditions:** Salaries |
| 4. Expenditure on the direct support | **Activities directed to local-level stakeholders, users or user groups:** Increase hazard knowledge and risk awareness | **Existing technical and institutional capacity** |
| 5. Expenditure on indirect support | **Activities not directly linked to an asset:** Risk awareness in urban planning | **Institutional environment- existing legal/economic barriers for implementation** |
| 6. Cost of Capital | **Financing costs:** Interests, dividends | **Risk profile of project:** Capability of implementing actor to mitigate risks (past experience) |

Throughout the case studies, the cost assessment focused on the estimation of LCC components 1 to 3. Rotterdam case study was able to mobilize some directly estimated cost figures as an essential solution of the NBS strategy had been actually implemented in a pilot project. Other costs estimates in Rotterdam and other studies, relied on the transfer of cost parameters from literature, national databases of market prices and expert opinions that allowed the estimates of cost per units of surface (or volume) of individual NBS measures, composing NBS strategies. These costs were then extrapolated to the size of each measure within NBS strategies, to estimate the overall cost associated with NBS strategies. Therefore, the estimation of costs present rather large range of uncertainty depending on the origin of the costs.

NBS implementation usually requires large-scale land use change. Not accounting for opportunity costs arising from land use change would artificially advantage NBS strategies as compared to grey strategies. When NBS are implemented on private land, the cost can be integrated in the capital expenditure related to land purchase, but what about when NBS are mainly implemented on public owned land? The European Commission guidelines on CBA for investment projects (European Commission 2014) mentions that *"Many public investment projects use land as a capital asset, which may be state-owned or purchased from the general government budget. Whenever there are alternative options for its use, land should be valued at its opportunity cost [...]. This must be done even if land is already owned by the public sector. If it is reasonable to assume that market price captures considerations about land's utility, desirability and scarcity, then it can generally be considered reflective of the economic value of land."* The question of whether a land may have an alternative use remains largely subject to interpretation when public roads, parking lots or sidewalks are concerned. In the Lez and Rotterdam case, we applied a conservative approach which is to consider land market prices as a proxy of opportunity costs, although alternative possible use of this land is not always clear. In Brague, NBS strategies involve privately owned land use change. In this case, land acquisition costs were included in investments costs while additional opportunity costs were estimated based on profits private land-owners could have obtained from the use of this land (estimated by revenues they could have perceived over this land).

Although the estimation of opportunity costs is fundamental in the CBA framework, the LCC framework normally focuses on making explicit the actual expenses to be assumed by project sponsors for implementing the NBS project. The inclusion of opportunity cost therefore does not appear necessary in this framework. In line with this argument, CBA excluding opportunity costs from overall costs were therefore computed to complete the evaluation.

## 6.2.3  Assessment of Avoided Damages

### 6.2.3.1  Overall Approach to the Assessment of Avoided Damage

Brémond et al. (2013) define "damage" as a negative impact of a natural hazard on a socioeconomic system and "cost" as the monetary valuation of such damage. The damages from natural hazards, and their costs, can be classified in tangible -easy to quantify in monetary terms- and intangible – difficult or even impossible to measure, as they comprise non-market values – (Merz et al. 2010; Brémond et al. 2013; Meyer et al. 2013).

The overall approach to the estimation of the avoided damage associated with NBS comprises two main steps: (1) estimation of the relation between water related hazards and damages (catastrophe risk models (CAT) model); and (2) estimation of the impact of NBS strategies on the modification of hazard (droughts or floods) through physical models. The combination of these two steps leads to the estimation of damages under different NBS strategies and without these (Business As Usual, BAU). The difference between damages in the BAU and NBS strategies provides an estimation of the avoided damage (Fig. 6.2), which is expressed in Mean Annual

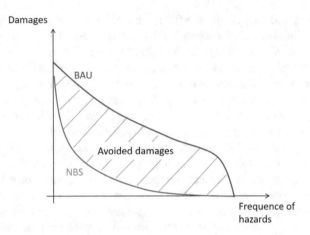

**Fig. 6.2** Avoided damages brought by NBS

(Avoided) Damage (MA(A)D) and can be integrated in the overall economic evalu-ation. In the different case studies, this approach was implemented according to the methodological framework as described in Calatrava et al. (2018).

### 6.2.3.2 Estimating the Relation Between Hazard and Damage Costs: The CAT Model Framework

The CAT model aims to establish the costs of a hazard based on its magnitude (event intensity) and the vulnerability of the elements at risk (Naulin et al. 2016). This involves the following steps (Merz et al. 2010; Foudi et al. 2015).

1. Characterisation of the hazard event;
2. Assessment of the exposure of the assets/elements at risk;
3. Vulnerability analysis to define the damage functions/models.
4. Calculation of the value of the damage cost.

Consequently, the structure of CAT models relies on three units: hazard, vulnerabil-ity and damage. This CAT model framework has been applied in all the case studies with methodological differences due to their particularities and to each area's data availability.

The **characterisation of the flood hazard event** (Fig. 6.3 hazard unit) in the different case studies has been done by using or adapting hazard models previously developed by project partners. For example, the Lower Danube case study used a hydrological model, to recreate a past event at a large scale. After this, the same parameters were used to simulate the maximum hazard intensity at a smaller scale instead of using a different hydrological model (see Chap. 10).

The **assessment of assets exposure** (Fig. 6.3 vulnerability unit) consists of the identification, localization and classification of those elements at risk that would be affected by a hazard and the estimation of their value (Merz et al. 2010). It was done

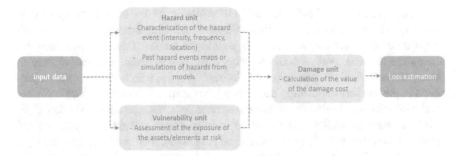

**Fig. 6.3** Structure of any CAT models. (Adapted from Merz et al. 2010; Foudi et al. 2015)

by combining the GIS layers of flood maps with the layers of assets/elements at risk to identify those affected by the hazard (Foudi et al. 2015). Elements at risk were then pooled into homogeneous classes of assets (residential, commercial, industrial, agricultural, infrastructures, etc.) for which the assessment is done (Moncoulon et al. 2014). Most case studies used micro-scale approach (i.e. identifying individual elements) using public databases, except for the Rotterdam case study that performed a meso-scale assessment at the neighbourhood level. The typology of elements used depended on the available sources of data (see Chap. 15).

The next step in a CAT model is the **vulnerability analysis** (Fig. 6.3 damage and avoided damage unit), which consists of establishing a relation between the hazard intensity and the damage caused to each type of element at risk. This implies developing and/or using water stage-damage functions (DF) for each type of element at risk (flood damage models). Damage functions/models for each type of asset can be developed either (i) hypothetically from "what-if analysis" using expert assessment or conceptual/abstract functions (synthetic approach); (ii) empirically, using data on real damage losses from past events; or (iii) a combination of both (Merz et al. 2010; Cammerer et al. 2013).

The empirical approach for developing damage functions (DF) used in both the Lez and Brague case studies relies on the use of insurance data (policies and claims), as in Moncoulon et al. (2014) and Naulin et al. (2016) (see Chaps. 13 and 15). The function is established from direct tangible insured losses for residential homeowners. The damage curves are obtained from historical geolocalised flood claims data for runoff and overflow hazards (Moncoulon et al. 2014). The observed damages are defined by the destruction rate ($DR$), obtained by dividing the amount of insurance claims by the insured value. The damage curves are established by estimating a relation between DR and the hazard intensity, expressed in cubic meter per second ($m^3/s$) for runoff and in water depth (m) for river flooding. The calibrated flood damage functions are applied to all exposed assets in the property exposure portfolio to estimate the total insured losses. Thus, the simulated costs for each individual element at risk are compared to the real costs of the event. It allows the validation of the DF and the reduction of uncertainties (Cammerer et al. 2013).

If there are no previously estimated DF available for the area of study, or data available to build damage curves, the option is transferring damage functions developed for other areas. However, the transferability of damage functions is limited. An alternative is using synthetic DF developed for larger spatial extents, as long as these are available for similar types of elements at risks, which is the approach proposed in the Medina del Campo case study (see Chap. 11). The French case studies used ad-hoc DF calibrated by CCR on insurance data, but also country-wide DF for public assets (see CGDD 2018), while Medina Demo considered synthetic DF defined at country level. In the Copenhagen case study, they used aggregated insurance data for city of Copenhagen from which a simple unit DF (damage/m$^2$) was derived for the period 2006–2012, including the 2011 cloudburst event. The aggregated value for damage caused by both surface and groundwater flooding was then applied as an estimate for damage as a result of groundwater flooding (see Chap. 17).

Last, the **calculation of damage costs** for each element at risk is done by combining the outputs of the hazard models (hazard intensity parameters, such as water depth) and the vulnerability assessment (damage functions/models). The total damage for each individual element at risk is obtained by multiplying its asset value by the relative damage (not necessary if absolute damage functions are used). The total damage from a given flood hazard is obtained aggregating the individual damage across all elements at risk of all types. This canonical approach is the one used in the Lez, Brague, Copenhagen and Medina del Campo case studies. In Rotterdam, damage cost estimates were taken from ex-post assessment of previous hazard events.

An alternative approach to CAT models based on DF is using vulnerability indicators to assess at coarser scale (mesh of 250 m/500 m/1 km) the most at-risk areas (Papathoma-Köhle 2016). The method combines detailed geographically-based input layers ranging from physical (e.g. flood depth layer or any other hazard maps) to socio-economic ones (e.g. land use cover) into the raster calculator. The outputs of the GIS-method are validated with information gathered from field surveys and literature review. It allows estimating damage by averaging the water depth in the mesh. The developed damage curves for other case studies are then applied to the water depth. This methodology has been applied in the Lower Danube case study.

### 6.2.3.3 Estimating the Impact of NBS on Hazards

The assessment of the damage costs avoided through the implementation of NBS strategies requires modelling the change in the physical damages caused by the hazard both with and without the NBS. This involves the different units in the CAT model, as it involves modelling how NBS strategies would change the intensity and location of the hazard (hazard unit) and the level of exposure and vulnerability of elements at risk (vulnerability unit), thus resulting in a change of the estimated damage cost (damage unit). The effect of NBS on physical damages was done using the CAT model in the Thames, Copenhagen and Brague case studies. The Rotterdam case study assumed that the three strategies would result in equivalent hazard reduction, so modelling was not useful for differentiating between strategies, while the

Lez and Medina Del Campo encountered computational difficulties for the physical modelling of NBS effects. The Lez case study took justified assumptions about the reduction in physical damages resulting from the proposed NBS instead. In addition to the different NBS strategies considered, several case studies also considered future climate scenarios, and the Brague and Lez scenarios also considered future urbanisation prospects in the area of study.

#### 6.2.3.4  Assessing Avoided Damages

The assessment of the avoided damages for NBS strategies is done by comparing the damage cost estimated under the BAU and the NBS strategies, i.e. by modelling the impact of the hazard under both the NBS and BAU strategies and then estimating the difference in the corresponding damage costs. This can be done either for a specific hazard event or for different events with different return periods. The latter consists in combining the CAT model with the probability of occurrence of different hazard events to obtain damage-probability curves, which relate the damage caused by each potential event with its probability of occurrence, as in Fig. 6.1 (Foudi et al. 2015). The Mean Annual Avoided Damage (MAAD) is calculated from damage-probability curves. The difference between the MAAD for both the BAU and NBS strategies yields the avoided damages resulting from the implementation of the latter. The MAAD was calculated in the Medina, Lez and Brague case studies, while Copenhagen and Rotterdam calculated the avoided damages for specific past hazard events.

### 6.2.4  Co-benefits Assessment

The IUCN's definition of NBS stresses on their multiple benefits including addressing societal challenges, providing human well-being and biodiversity benefits (Cohen-Shacham et al. 2016). We group here the multiple benefits of NBS under the concept of co-benefits to stress on additional benefits to the primarily benefit of water-related risk reduction. According to the emerging field of value pluralism approaches (Gómez-Baggethun and Barton 2013; Elmqvist et al. 2015; Jacobs et al. 2016; Costanza et al. 2017), co-benefits can be assessed with several types of indicators: biophysical, socio-cultural and monetary. For the purpose of this Chapter, which focuses on the economic valuation of NBS strategies through a CBA, we present here the methods used by three case studies that conducted a full economic valuation of the co-benefits expected from NBS strategies: the Brague, Lez and Rotterdam case studies. These three evaluations were carried out in two stages, with (1) the identification of co-benefits and (2) the monetary valuation of those co-benefits. As recommended by Nesshöver et al. (2017), a strong involvement of local stakeholders was organised throughout the process in order to integrate their perceptions and knowledge.

### Identification of Co-benefits

In the three case studies, the identification of co-benefits strongly relied on the organisation of focus groups or workshops, in which potential benefits of NBS strategies were discussed with local stakeholders. Existing co-benefits classifications and frameworks were used as a basis for discussion:

- the Millennium Ecosystem Assessment (MA 2005), the Common International Classification of Ecosystem Services (CICES) (Haines-Young and Potschin 2018), and the Intergovernmental Science-Policy Platform on Biodiversity and Ecosystem Services (IPBES) framework (Díaz et al. 2015) for the identification ecosystem services;
- the EKLIPSE framework (Raymond et al. 2017) for the identification of challenges areas in urban contexts.

Those frameworks were finally combined in order to embrace a wide range of potential co-benefits and specific local issues. In the Lez case study for instance – a watershed with 50% of natural areas and 30% of agricultural areas- the use ecosystem services classifications was chosen (MEA and IPBES), while in the Rotterdam case study – an urban neighbourhood – the EKLIPSE framework was used as a core framework for co-benefits identification due to its comprehensive coverage of urban issues (see Chaps. 14 and 16).

### Co-benefits Valuation

The three case studies used different types of monetary valuation methods described in details by Herivaux et al. (2019):

*Direct valuation approaches (market price and cost-based methods)* were used in the Rotterdam and Lez case studies, to valuate seven co-benefits, namely climate mitigation through carbon storage, air quality regulation, water cycle regulation, urban regeneration, human health and wellbeing, and aesthetic amenities. These approaches rely on two main steps (Fig. 6.4).

- Step 1: the quantification of the level of ecosystem services provided by the NBS strategy in non-monetary terms (e.g., annual carbon sequestration expressed in t-eqCO$_2$/year; water availability expressed in m$^3$/year) derived from models, functions or reference values obtained in similar contexts;
- Step 2: the monetary valuation of the change in the co-benefit level derived from market prices (when market exists), replacement costs (costs required to provide a similar ecosystem service with a human engineered solution) or avoidance costs (costs that would occur if the ecosystem service were lost).

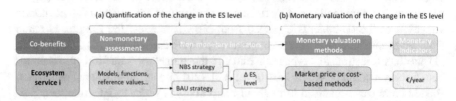

**Fig. 6.4** Stepwise approach for ES valuation when direct valuation approaches are used. (Adapted from Herivaux et al. 2019)

These approaches provide biophysical and monetary indicators for each ecosystem service. However, they do not reflect the total economic value of these ecosystem services, as they only capture direct use values. Results should thus be considered as lower bound estimates.

*Stated preference approaches,* namely the Contingent Valuation Method and the Discrete Choice Experiment (Johnston et al. 2017), were used in the Brague and the Lez case studies. These approaches rely on representative surveys of the population to estimate people's willingness to pay (how much they would contribute in terms of fee or tax increment) for a hypothetical modification of the environment (here the implementation of NBS strategies). In both cases, the survey gives the opportunity to evaluate the preferences of the population for different NBS strategies, their flood risk perception and the importance the population grants to ecosystem services. They provide socio-cultural and monetary indicators for different NBS strategies and associated bundles of ecosystem services, without seeking to evaluate ecosystem services one by one. Results obtained with such approaches reflect the total economic value (including non-use values) associated with NBS strategies: they cannot be easily added to those obtained with direct valuation methods, as this would lead to double counting.

*A benefit transfer approach* was also used in the Brague case study to value the co-benefits expected from the NBS strategies. This approach consists in the "application of values and other information from a 'study' site where data are collected to a 'policy' site with little or no data" (Rosenberger and Stanley 2006).This approach requires relatively sophisticated econometric models to determine variables that have an effect on the overall value of ecosystem services related to NBS. A Meta Regression Analysis (Arfaoui and Gnonlonfin 2019) was performed with 187 monetary estimates from 52 studies evaluating the Willingness to Pay for river restoration measures (restoration of the river stream, restoration of the floodplain, restoration of riparian vegetation, ecosystem-based management practices) and their ecosystem services (food and material provision, local environmental regulation, global climate regulation, habitat quality and species diversity protection, recreational services, aesthetic appreciation). The transfer value function obtained was applied to the Brague case study, considering the characteristics of NBS strategies and the co-benefits that local stakeholders considered as relevant (see Chap. 13).

## 6.3  Key Results of the Economic Assessment of NBS for Natural Assurance Schemes

We present in Table 6.3 the results of four out of the seven economic assessment, since only these four case studies estimated monetarily at least the costs and the avoided damages brought by NBS strategies.

Some key results can be drawn from the analysis these economic assessment.

**Table 6.3** Overview of results of the economic analysis in the Lez, Brague, Rotterdam and Medina case studies (See Chaps. 11, 13, 14 and 16 for details)

| Case study | Brague | | Rotterdam | | | Lez | | Medina |
|---|---|---|---|---|---|---|---|---|
| Strategy | Grey | NBS High | Grey | Hybrid | Green | NBS 1 | NBS 2 | Conservation Agriculture |
| Time Horizon | 2070 | 2070 | 2070 | 2070 | 2070 | 2040 | 2040 | 2070 |
| Cost of NBS scenarios (M€) | **169**[a] 87–270 | **61** 45–97 | **8.2** | **8.1** | **6.9** | **52** 39–65 | **120** 92–148 | **97** |
| Opportunity cost (M€) | **0.6** 0.5–0.7 | **19** 14–35 | **0** | **4.9** | **13.0** | **210** 135–285 | **318** 239–396 | |
| Avoided damages (M€) | **13** 5–19 | **14** 6–21 | **0.7**[b] 0.4–0.9 | **0.7**[b] 0.4–0.9 | **0.7**[b] 0.4–0.9 | **3.4** | **29** | **309** |
| Co-benefits (M€) | – | **68**[c] 25,103 | **0.004** | **6.0** | **10.7** | **287** | **363** | |
| Avoided damage/ Cost (rate) | **0.1** 0.1–0.1 | **0.2** 0.1–0.2 | **0.08** 0.05–0.11 | **0.08** 0.05–0.11 | **0.09** 0.06–0.13 | **0.07** 0.05–0.09 | **0.24** 0.2–0.3 | **3.2** |
| BCR | **0.1** 0.1–0.1 | **1.0** 0.2–2.1 | **0.08** 0.05–0.11 | **0.50** 0.46–0.53 | **0.57** 0.56–0.58 | **1.3** 0.8–1.7 | **1.0** 0.7–1.2 | **3.2** |
| BCR exc opportunity costs | **0.1** 0.1–0.1 | **1.3** 0.3–2.8 | **0.08** 0.05–0.11 | **0.82** 0.79–0.85 | **1.65** 1.61–1.68 | **6.0** 4.5–7.5 | **3.5** 2.7–4.3 | |

[a]Economic value are expressed in net present value in millions € discounted over the time horizon of the evaluation. Average values are indicated in bold and value ranges reflecting uncertainty (when evaluated) are indicated in normal font

[b]The avoided damages were not specifically assessed in during the assessment in Rotterdam. The approach was to size NBS strategy so that they reach an equivalent level of service. A study carried out in the same neighbourhood could nevertheless be used to estimate the magnitude of avoided damages in the Rotterdam case study although the methodology was different from the one used to dimension strategies

[c]The valuation of co-benefits with the contingent valuation methodology includes people preference for flood protection. Part of this value is therefore redundant with estimated avoided damages

In the cases where a grey strategy has been evaluated, the **cost of implementation and maintenance of grey solutions is higher than the cost of NBS strategies for the same level of risk management**. This confirms previous information mentioning that **NBS may be more cost effective solutions as compared to grey solutions** as they are less costly. This is particularly highlighted in the Rotterdam and the Brague case, since for the same level of avoided damage, the NBS solutions are 15% and 63% less costly than the grey solutions, respectively.

The cost-effectiveness of individual NBS measures has also been investigated in some case studies. The cost of different measures is compared for the same level of service or a proxy of this level of service: for example for floods the cost/m³ of water retention. This analysis shows a **very large heterogeneity of cost-effectiveness of individual NBS measures**. For example, in the Lez case study, the cost-effectiveness of green roofs is extremely low because of the large cost of green roofs and their limited water storage capacity (see Chap. 14 for more details). In the Thames case study, a similar assessment reveals a similar heterogeneity with £2.9/m³ of water retention for conservation agriculture, £16.8/m³ of water retention for retention ponds and £61.8/m³ of water retention for leaky dams (see Chap. 12).

In urban areas, taking into account the **opportunity costs of NBS can totally change the appreciation of their cost advantage**. In Lez and Rotterdam, two urban cases, land price is used as a proxy of opportunity costs, even though NBS are developed on public areas. Although this cost is not actually spent, it provides an estimation of the value associated with the fact that this space cannot be used for other profitable uses. For example, NBS may take space that may not be available for real estate development. Considering that NBS require a large spatial extent as compared to traditional grey strategies, the inclusion of opportunity costs has a strong weight in the overall cost estimation, especially in urban areas where land cost is high.

In cases where flood risk reduction is the main objective of the NBS (Lez, Rotterdam and Brague), **benefits in terms of avoided damages are not sufficient to cover capital expenses and operation and maintenance costs**. This result needs to be nuanced because our estimations of avoided damage take into account only a share of the damages avoided thanks to protection measures; with a focus on insured damages (public damages are included in the Lez estimation). Indirect damages, such as the macro-economic impact of floods, due to their effect are not considered, although these costs can be highly significant. The assessment also does not consider the potential of protection measures on other non-monetary but essential indicators such as the capacity to reduce the exposition (number of residents in flood prone areas), life protection, injuries or post-traumatic stress. In addition, authorities may have an obligation to deliver a certain level of flood protection, regardless of whether costs are superior to expected avoided damages. The cost-effectiveness advantage of NBS mentioned above therefore remains a key advantage in these contexts. It is to be mentioned that in the case of Medina, which addresses agriculture drought risk, the benefits associated with the change of agriculture crops to reduce drought exposure, in terms of avoided drought damages, overcomes the cost of this change.

**Co-benefits represent the largest share of the value generated by NBS strategies.** This is the case in all three studies in which co-benefits have been monetarily estimated. This result does not depend on the method used for the estimation of co-benefits, since revealed preference methods have been used in the Lez and Brague case studies, while direct valuation has been used in Rotterdam.

There are **no clear-cut conclusions on the overall economic efficiency of NBS** in our assessments. Indeed, NBS strategies have a BCR higher than 1 in Lez, Brague and Medina, which means these NBS strategies would be worth the investment, whereas it is <1 in Rotterdam, whatever the strategy. The picture is more positive if we exclude opportunity costs from the economic analysis. Interestingly however, for Brague and Rotterdam, the economic efficiency of NBS strategies is nevertheless much higher than the one for grey strategies. The Benefit Cost ratio should however not be the only criteria considered. For example, the NBS- strategy in the Lez, has the highest Cost-Benefit ratio however the rate of avoided damages on implementation cost is extremely low, since this strategy has very limited effect on flood protection.

## 6.4 Discussion-Conclusion

This chapter presents a methodological framework for the economic assessment of NBS and its application to seven case studies. Results reveal that NBS aiming at solely reducing water risks cannot be automatically assumed to be economically efficient. It is therefore fundamental to carry out thorough case specific economic valuations of a diversity of strategies, involving NBS, grey and hybrid solutions, in order to identify the most adequate strategy for water risk management and to address territorial challenges. In a context of limited public resources, economic valuation can help identifying the adequate solution to address water risks, the one that maximizes the net benefit for society.

The economic valuation of NBS strategies requires a large effort for the design of strategies. This step requires the participation of stakeholders and preliminary modelling approaches. It is of fundamental importance because the quantification of the physical characteristics (e.g. retention capacity, number of trees) is the basis for the estimation of their costs and benefits.

In our applications, cost estimates mainly rely on the transfer of existing values evaluated in other projects. This reliance on a diversity of sources gives rise to a high level of uncertainty. Costs can indeed vary greatly depending on the exact feature of the NBS and on local contexts. The development of local references for the estimation of costs in all European countries would improve the precision of cost estimation and facilitate greatly cost estimations. The estimation of opportunity costs based on land price, which has been used here, is an upper bound. Some of these areas may indeed not have other profitable use (e.g. sidewalks). The estimation of opportunity costs may indeed need further investigation in the future.

In order to assess the avoided damages as a result of NBS strategies, both simple, straightforward methods and advanced models are necessary to fully estimate the effect of NBS on the intensity and spatial extent of hazards, especially when assessments are carried out at the catchment scale. The evaluation of avoided damages also depends on the availability of data to be able to link the reduction of hazard to a reduction of damages. The detailed estimation of the economic benefits of NBS related to water risk reduction in several case studies is therefore a key contribution of this book. Collaboration with the insurance sector to provide expertise on damage evaluation and data on damages has been instrumental and should be pursued in future studies.

Our methodology also provides a framework for monetary valuation of co-benefits. A diversity of approaches was used to evaluate co-benefits (direct valuation, value transfer, stated preferences approaches) that all require advanced skills in environmental economics. The implementation of this step is key considering the magnitude of co-benefits in NBS benefits and should not be overlooked in the evaluation of NBS. This step can be challenging as it may be difficult for some stakeholders to accept the principle of the monetary valuation of co-benefits. This requires careful explanation to stakeholders that emphasize the limits and the advantages of monetary valuation techniques and their complementarity with other environmental valuation methods. Another challenge is that some methods used in the monetary valuation of co-benefits, such as contingent valuation or choice experiments, require the implementation of surveys with samples of residents. These surveys include the presentation to citizens of alternative water risk management measures and to collect their preference on this matter. This "public consultation" may be considered a delicate issue for certain stakeholders, such as decision makers, that may want to control the way this type of information is revealed to the general public. Using these methods may therefore require lengthy negotiations with stakeholders.

Our economic assessment methodology also has several limits. A large share of these limitations are inherent to every economic analysis with ecological or environmental variables. The multiplicity of models required for the estimation of the different cost and benefits increases their overall uncertainty. On the one hand, it is the relative magnitude of costs and benefits that should be compared rather than the precise values that we have presented. On the other hand, only indicators that could be evaluated monetarily were included in this study. Other indicators such as non-monetary impacts on water risks, co-benefits that could not be or partially be valued monetarily such as social and environmental indicators are important in the decision making process for the development of NBS. The implementation of the economic assessment of NBS should therefore be complemented with the implementation of a Multi-Criteria Decision Analysis (MCDA) such as the one described in Chap. 7. Finally, the cost-benefit analysis only aims at evaluating whether aggregated benefits are higher than aggregated costs. This does not preclude from distributional issues, i.e. the existence of population that benefit from the project and others that lose, for example due to the expropriation of citizens. A project that yields positive economic returns may therefore face the opposition from some stakeholders. These approaches are therefore complementary with approaches focusing on social acceptance and the design of soft measures to facilitate the implementation of NAS such as the one presented in Chap. 5.

# References

Altamirano MA, de Rijke H (2018) Costs of infrastructures: elements of method for their estimation. DELIVERABLE 4.2: EU Horizon 2020 NAIAD Project, Grant Agreement N°730497 Dissemination

Arfaoui N, Gnonlonfin A (2019) The economic value of NBS restoration measures and their benefits in a river basin context: a meta-analysis regression. FAERE Working Paper 09

Bokhove O, Kelmanson MA, Kent T, Piton G, Tacnet JM (2019) Communicating (nature-based) flood-mitigation schemes using flood-excess volume. River Res Appl 35:1402–1414. https://doi.org/10.1002/rra.3507

Brémond P, Grelot F, Agenais A (2013) Review Article: economic evaluation of flood damage to agriculture – Review and analysis of existing methods. Nat Hazards Earth Syst Sci 13:2493–2512. https://doi.org/10.5194/nhess-13-2493-2013

Calatrava J, Graveline N, Moncoulon D, Marchal R (2018) DELIVERABLE 4.3: economic water-related risk damage estimation. EU Horizon 2020 NAIAD project, Grant Agreement N° 730497

Cammerer H, Thieken AH, Lammel J (2013) Adaptability and transferability of flood loss functions in residential areas. Nat Hazards Earth Syst Sci 13:3063–3081. https://doi.org/10.5194/nhess-13-3063-2013

CGDD (2018) Analyse multicritère des projets de prévention des inondations – Guide méthodologique 2018. Commissariat général au développement durable [online]. Available from: https://www.ecologique-solidaire.gouv.fr/sites/default/files/Th%C3%A9ma%20-%20Analyse%20multicrit%C3%A8re%20des%20projets%20de%20pr%C3%A9vention%20des%20inondations%20-%20Guide.pdf

CGSP (2013) Rapport quinet: évaluation socioéconomique des investissements publics

Cohen-Shacham E, Walters G, Janzen C, Maginni S (2016) Nature-based solutions to address global societal challenges. IUCN

Costanza R, de Groot R, Braat L, Kubiszewski I, Fioramonti L, Sutton P, Farber S, Grasso M (2017) Twenty years of ecosystem services: how far have we come and how far do we still need to go? Ecosyst Serv 28:1–16

Denjean B, Denjean B, Altamirano MA, Graveline N, Giordano R, Van der Keur P, Moncoulon D, Weinberg J, Máñez Costa M, Kozinc Z, Mulligan M, Pengal P, Matthews J, van Cauwenbergh N, López Gunn E, Bresch DN, Denjean B (2017) Natural Assurance Scheme: a level playing field framework for Green-Grey infrastructure development. Environ Res 159:24–38. https://doi.org/10.1016/j.envres.2017.07.006

Díaz S, Demissew S, Carabias J, Joly C, Lonsdale M, Ash N, Larigauderie A, Adhikari JR, Arico S, Báldi A, Bartuska A, Baste IA, Bilgin A, Brondizio E, Chan KMA, Figueroa VE, Duraiappah A, Fischer M, Hill R, Koetz T, Leadley P, Lyver P, Mace GM, Martin-Lopez B, Okumura M, Pacheco D, Pascual U, Pérez ES, Reyers B, Roth E, Saito O, Scholes RJ, Sharma N, Tallis H, Thaman R, Watson R, Yahara T, Hamid ZA, Akosim C, Al-Hafedh Y, Allahverdiyev R, Amankwah E, Asah TS, Asfaw Z, Bartus G, Brooks AL, Caillaux J, Dalle G, Darnaedi D, Driver A, Erpul G, Escobar-Eyzaguirre P, Failler P, Fouda AMM, Fu B, Gundimeda H, Hashimoto S, Homer F, Lavorel S, Lichtenstein G, Mala WA, Mandivenyi W, Matczak P, Mbizvo C, Mehrdadi M, Metzger JP, Mikissa JB, Moller H, Mooney HA, Mumby P, Nagendra H, Nesshover C, Oteng-Yeboah AA, Pataki G, Roué M, Rubis J, Schultz M, Smith P, Sumaila R, Takeuchi K, Thomas S, Verma M, Yeo-Chang Y, Zlatanova D (2015) The IPBES Conceptual Framework – connecting nature and people. Curr Opin Environ Sustain 14:1–16

Ellram L (1995) Total cost of ownership: an analysis approach for purchasing. Int J Phys Distrib Logist Manag 25:4–23. https://doi.org/10.1108/09600039510099928

Elmqvist T, Setälä H, Handel SN, van der Ploeg S, Aronson J, Blignaut JN, Gómez-Baggethun E, Nowak DJ, Kronenberg J, de Groot R (2015) Benefits of restoring ecosystem services in urban areas. Curr Opin Environ Sustain 14:101–108

European Commission (2014) Guide to cost-benefit analysis of investment projects, EUROPEAN C

Foudi S, Osés-Eraso N, Tamayo I (2015) Integrated spatial flood risk assessment: The case of Zaragoza. Land Use Policy 42:278–292. https://doi.org/10.1016/j.landusepol.2014.08.002

Gómez-Baggethun E, Barton DN (2013) Classifying and valuing ecosystem services for urban planning. Ecol Econ 86:235–245. https://doi.org/10.1016/j.ecolecon.2012.08.019

Graveline N, Joyce J, Calatrava J, Douai A, Arfaoui N, Moncoulon D, Manez M, Ryke M de, Kozinj Z (2017) General Framework for the Economic assessment of NBS and their insurance value. DELIVERABLE 4.1: EU Horizon 2020 NAIAD Project, Grant Agreement N°730497 Dissemination

Haines-Young R, Potschin MB (2018) Common International Classification of Ecosystem Services (CICES) V5.1 and guidance on the application of the revised structure. Available from www.cices.eu

Herivaux C, Le Coent P, Gnonlonfin A (2019) Natural capital and ecosystem services to valuate co-benefits of NBS in water related risk management. DELIVERABLE 4.4: EU Horizon 2020 NAIAD Project, Grant Agreement N°730497 Dissemination

Jacobs S, Dendoncker N, Martín-lópez B, Nicholas D, Gomez-baggethun E, Boeraeve F, Mcgrath FL, Vierikko K, Geneletti D, Sevecke KJ, Pipart N, Primmer E, Mederly P, Schmidt S, Aragão A, Baral H, Bark RH, Briceno T, Brogna D, Cabral P, De Vreese R, Liquete C, Mueller H, Peh S, Phelan A, Rincón AR, Rogers SH, Turkelboom F, Van Reeth W, Van Zanten BT, Karine H, Washbourne C (2016) A new valuation school: integrating diverse values of nature in resource and land use decisions. Ecosyst Serv 22:213–220. https://doi.org/10.1016/j.ecoser.2016.11.007

Johnston RJ, Boyle KJ, Adamowicz W (Vic), Bennett J, Brouwer R, Cameron TA, Hanemann WM, Hanley N, Ryan M, Scarpa R, Tourangeau R, Vossler CA (2017) Contemporary guidance for stated preference studies. J Assoc Environ Resour Econ 4:319–405. https://doi.org/10.1086/691697

MA (2005) Millenium Ecological Assessment. Millennium ecosystem and human well-being: a framework for assessment

Merz B, Kreibich H, Schwarze R, Thieken A (2010) Review article assessment of economic flood damage. Nat Hazards Earth Syst Sci 10:1697–1724. https://doi.org/10.5194/nhess-10-1697-2010

Meyer V, Becker N, Markantonis V, Schwarze R, Van Den Bergh JCJM, Bouwer LM, Bubeck P, Ciavola P, Genovese E, Green C, Hallegatte S, Kreibich H, Lequeux Q, Logar I, Papyrakis E, Pfurtscheller C, Poussin J, Przyluski V, Thieken AH, Viavattene C (2013) Review article: Assessing the costs of natural hazards-state of the art and knowledge gaps. Nat Hazards Earth Syst Sci 13:1351–1373. https://doi.org/10.5194/nhess-13-1351-2013

Moncoulon D, Labat D, Ardon J, Leblois E, Onfroy T, Poulard C, Aji S, Rémy A, Quantin A (2014) Analysis of the French insurance market exposure to floods: a stochastic model combining river overflow and surface runoff. Nat Hazards Earth Syst Sci 14:2469–2485. https://doi.org/10.5194/nhess-14-2469-2014

Naulin JP, Moncoulon D, Le Roy S, Pedreros R, Idier D, Oliveros C (2016) Estimation of insurance-related losses resulting from coastal flooding in France. Nat Hazards Earth Syst Sci 16:14

Nesshöver C, Assmuth T, Irvine KN, Rusch GM, Waylen KA, Delbaere B, Haase D, Jones-Walters L, Keune H, Kovacs E, Krauze K, Külvik M, Rey F, van Dijk J, Vistad OI, Wilkinson ME, Wittmer H (2017) The science, policy and practice of nature-based solutions: an interdisciplinary perspective. Sci Total Environ 579:1215–1227. https://doi.org/10.1016/J.SCITOTENV.2016.11.106

Papathoma-Köhle M (2016) Vulnerability curves vs. Vulnerability indicators: Application of an indicator-based methodology for debris-flow hazards. Nat Hazards Earth Syst Sci 16:1771–1790. https://doi.org/10.5194/nhess-16-1771-2016

Raymond CM, Frantzeskaki N, Kabisch N, Berry P, Breil M, Razvan M, Geneletti D, Calfapietra C (2017) A framework for assessing and implementing the co-benefits of nature-based solutions in urban areas. Environ Sci Pol 77:15–24. https://doi.org/10.1016/j.envsci.2017.07.008

Rosenberger RS, Stanley TD (2006) Measurement, generalization, and publication: sources of error in benefit transfers and their management. Ecol Econ 60:372–378. https://doi.org/10.1016/J.ECOLECON.2006.03.018

# Chapter 7
# Designing Natural Assurance Schemes with Integrated Decision Support and Adaptive Planning

**Laura Basco-Carrera, Nora Van Cauwenbergh, Eskedar T. Gebremedhin, Guillaume Piton, Jean-Marc Tacnet, Mónica A. Altamirano, and Camilo A. Benítez Ávila**

**Highlights**

This Chapter provides an illustration on how to support the decision-making process of selecting and applying NBS. In this chapter you will learn:

- Which decisions need to be taken in the analysis, selection, design and implementation of NBS, considering the common steps in strategic planning;
- Which tools, methods and models can be used to support the decision-making process;
- How to integrate all the information, data and results to reach a robust strategy that fulfils the requirements for decision-making;
- How NBS solutions can bring benefits in terms of DRR, water resources management (WRM) and CCA;
- The substantial co-benefits that NBS can bring in terms of economic growth, service provision and social equity, while protecting the environment.

L. Basco-Carrera (✉) · E. T. Gebremedhin · C. A. Benítez Ávila
Deltares, Delft, The Netherlands
e-mail: laura.bascocarrera@deltares.nl

N. Van Cauwenbergh
IHE Delft Institute for Water Education, Delft, The Netherlands

G. Piton · J.-M. Tacnet
Université Grenoble Alpes, INRAE, CNRS, IRD, Grenoble INP, IGE, Grenoble, France

M. A. Altamirano
Deltares, Delft, The Netherlands

Faculty of Technology, Policy and Management, Delft University of Technology, Delft, The Netherlands

© The Author(s) 2023
E. López Gunn et al. (eds.), *Greening Water Risks*, Water Security in a New World, https://doi.org/10.1007/978-3-031-25308-9_7

## 7.1　Introduction: Integration of DRR, WRM and Climate Change Adaptation Planning

Natural Assurance schemes mainly deal with issues coming from three arenas that address hydrological risk: Disaster Risk Reduction (DRR), Water resources management (WRM) and Climate Change Adaptation (CCA). Whereas all approaches look at nature-based solutions to reduce hydrological risk, the larger framework in which NBS are used differs. Main differences between DRR, WRM and CCA are in relation to objectives and scope (see Table 7.1). The distinction is important since it has important implications on how problems are assessed, and which kinds of solutions are proposed. To understand how the different approaches can be integrated in NAS, this section presents each of them in a nutshell to then discuss how they can be integrated, contributing to the design of NAS.

### 7.1.1　DRR in a Nutshell

Disaster risk management (DRM) is a framework used to respond to disasters at local, municipal, and national level. The goals of DRM are (Warfield 2020):

1. To reduce or avoid losses from hazards;
2. To assure prompt assistance to victims and;
3. To achieve rapid and effective recovery.

**Table 7.1** Similarities and differences between DRR and WRP contexts

| Disaster Risk Reduction (DRR) | Water Resources Planning (WRP) | Climate change adaptation (CCA) |
|---|---|---|
| **Differences** | | |
| Anticipate and prevent disaster consequences (ex-ante), combined with ex-post activities such as response and recovery | Purely ex-ante, forward looking approach, with combination of development and adaptation actions | Combined responsive and preventive action with both short- and long-term effects of climate change |
| Objective is reduction of disaster risk | Objectives are multiple (and possibly competing) | |
| Minimizing risk at its core | Maximizing benefits of the water resources system at its core | |
| **Similarities** | | |
| Involving stakeholders | | |
| Use of models and tools to understand the water system | | |
| Cross-sectoral activities requiring understanding of institutional and stakeholder environment | | |
| Cyclical exercise: involving multiple scenarios | | |
| Look at combination of adaptation and mitigation Focus on adaptation | | |

In order to achieve these goals, DRM follows a process of four steps from mitigation, preparedness, response to recovery (see Fig. 7.1). Specifically, one can distinguish pre-impact and post-impact assessment phases. Specifically, the existing links between Drivers-Pressures-States that occur beforehand can be evaluated in terms of Preparedness and Mitigation, especially when concerning the application of NBS as protective and mitigating measures for risk reduction, as well as other measures.

DRM utilizes DRR and combines the principles of mitigation and preparedness with a management perspective through the added principle of response. DRM includes the management of risk and disaster and is a framework to establish policy and administrative mechanisms related to emergency response (Baas et al. 2008).

Whereas the majority of DRR activities focus on response, and therefore present a mainly post disaster approach, the typical strategic planning in water management is a forward looking or pre-disaster approach that aims to create a strategic position in the future. This is based on understanding of the current challenges and identification of pathways and action plans to overcome all possible identified problems.

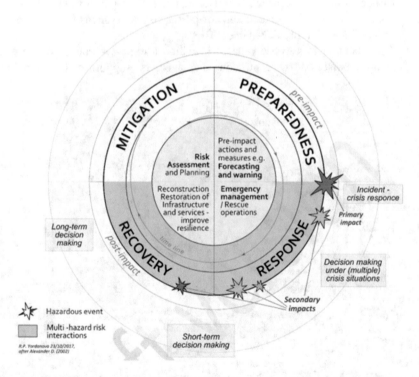

**Fig. 7.1** The disaster management cycle principles and the spiral principal redrawn after (Alexander 2002). The importance of the existence of multi-hazard impact occurring prior or after an event is depicted on diagram

### 7.1.2    Integrated Water Resources Management (IWRM) in a Nutshell

In water systems, strategic planning usually aims at reaching several objectives linked to the social, economic and environmental dimensions of the water system:

- The main purpose is to ensure the sustainable exploitation of water resources in support to the production of goods and services required to meet national and regional demand objectives;
- Systematic procedures to generate a synthesis of information in such a manner as to gain insight into the nature and consequences of possible management strategies;
- In a risk context, planning will target the present and future risks and develop strategies for both mitigation and adaptation.

In this regard, IWRM provides the guiding principles to achieve water security for all by means of strategic planning, or also called master planning (Fig. 7.2). Water security is the capacity of a population to safeguard sustainable access to adequate quantities of acceptable quality water for sustaining livelihoods, human well-being, and socio-economic development, for ensuring protection against water-borne pollution and water-related disasters, and for preserving ecosystems in a climate of peace and political stability (UN-Water 2013).

IWRM planning is a cyclic process in which a logical sequence of steps is implemented bolstered by continuous management support and stakeholder

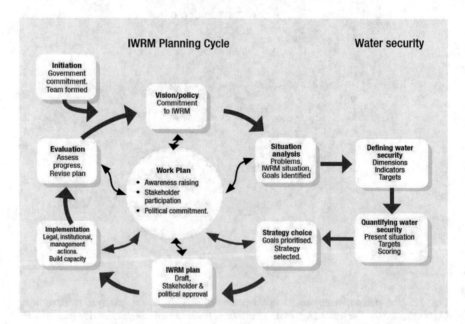

**Fig. 7.2** IWRM planning cycle to achieve water security. (Source: Van Beek and Arriens 2014)

involvement (see Fig. 7.2). The expected outcome of the IWRM process is a concrete plan, approved and implemented by decision-makers (for example the government) and stakeholders. Following the process, decision makers and stakeholders will get a good understanding of an area's water system, its performance, and the importance and benefits of managing the resources in a sustainable manner. It will serve as a roadmap for longer term initiatives needed to achieve the overarching objectives of (i) sustainable environment, (ii) social equity and (iii) economic growth.

### 7.1.3   Climate Change Adaptation in a Nutshell

Climate change adaptation is focused on adjusting or adapting to the actual or expected future climate. The main objective is to reduce vulnerability to harmful effects of the changing climate (such as sea-level rise and increased frequency and intensity of weather events). In doing so, CCA follows a number of steps similar to the stages in strategic water resources planning (WRP) (Fig. 7.2) and mirroring more general problem structuring planning methods. The steps followed by the climate-adapt tool proposed by the EU Climate adaptation community (Prutsch et al. 2014) are shown in Fig. 7.3.

### 7.1.4   Integration: Merging Approaches and Different Policies

The approaches introduced above, are guided by a series of EU and global policies such as the Sendai framework for DRR, the EU water policies (e.g. Water Framework Directive, Floods Directive and river basin and drought management plans) and

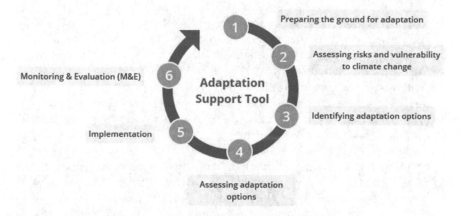

**Fig. 7.3** Climate-ADAPT tool. (Adopted from https://climate-adapt.eea.europa.eu)

policies focusing on climate adaptation specifically (e.g. Paris agreement, EU Adaptation strategy). In addition, NBS related policies as the EU strategy Green Deal and the Sustainable Development Goals play an important role when shaping responses under before mentioned policies.

The main similarities and differences between the three approaches are summarized in Table 7.1. The differences are mainly related to the scope, objectives and anticipative character (or not) of the approach. Because of these differences in the objectives and approach between DRR and IWRM, their coordination is challenging. However, both approaches have in common that they deal with complex decisions and involve multiple methods and actors. The latter is managed using stakeholders' engagement processes (*see Chap. 5 Giordano et al., this volume*), use of multiple value capturing and method integration. Whereas these are common approaches, large part of the integration approach in NBS will be coloured by the context of a given case study. In terms of contexts, we distinguish:

- Time (e.g. rapid response vs strategic planning) and spatial scales;
- Decision-making contexts;
- Thematic focus;
- Institutional and business or investment readiness levels.

**Role of Stakeholders in Integrated Decision Support and Adaptive Planning of NAS**

The participatory nature of the proposed planning approach addresses broad societal and scientific calls for democratizing decision making in DRR (e.g. Okada et al. 2018; Samaddar et al. 2017), CCA (Cvitanovic et al. 2019) and natural resource management (e.g.Grimble and Chan 1995; Van Cauwenbergh et al. 2018) in general.

Throughout the entire planning process discussed in this chapter, involvement of stakeholders is key to a number of issues. First of all, it helps to assure a good understanding of the often complex issues and to handle trade-offs in a societal acceptable way. However, stakeholder involvement is also necessary to anticipate and adapt to a number of implementation issues to avoid producing results that those potentially impacted will not support. Stakeholder involvement brings both knowledge and preferences to the planning process—a process that typically will need to find suitable compromises among all decision-makers and stakeholders if a consensus is to be reached.

Choices about managing water and other natural resources trade-offs involve more than hydrology and economics. They involve people's values, ethics, and priorities that have evolved and been embedded in societies over thousands of years (Priscoli 2004). International policies, e.g. the Dublin principles and Aarhus convention, drive governments to engage stakeholders as an explicit operationalization of involving people's values, ethics and priorities, in line with principles of democracy and transparency. These principles have been adopted by the main policies and institutional frameworks in the fields of DRR, CCA and WRM mentioned earlier.

## 7.2 Strategic Planning Framework

### 7.2.1 Definition and Main Steps

The strategic planning framework approach is a forward looking or ex-ante approach that aims at creating a strategic position for the future (Deltares 2020; Loucks and Van Beek 2017). This is based on the understanding of current challenges in the identification of measures and action plans to overcome them. In water systems, strategic planning aims at achieving numerous objectives, by looking at the socio-economic and environmental dimensions of the water system. It provides a systematic procedure to generate a synthesis of information, so we can develop an effective and efficient water management plan. The framework is illustrated in Fig. 7.4. It consists of five phases namely: (i) inception phase, (ii) situation analysis, (iii) strategy building, (iv) action planning and (v) implementation. The engagement of stakeholders and decision-makers is key to ensure the sustainability and ownership of the planning and decision-making process and the outcomes from the process. In each of the phases, relevant stakeholders and the extent of their engagement must be identified. Also, there are various methods and tools that can be used to carry out each of the steps. The tools and methods used to design and monitor natural assurance schemes will be explained in detail in Sect. 7.3 of this chapter.

A brief description of the five phases of the master planning framework is given below and represented graphically in Fig. 7.4:

- **Inception/Scoping:**
- Inception is the first step in adaptive planning. It defines the boundary conditions, establishes the objectives and specifies the limitations. This requires the involvement of all decision makers and setting up the circumstances or enabling conditions under which a solution or plan is created for the decision makers to discuss. The analysis includes a thorough investigation of the existing policy mechanisms, institutional frameworks, problems, measures of success and the available data.
- **Situation Analysis:**
- It focuses on data collection and modelling. Using the conditions and frameworks from the previous step, the natural resource, socio-economic and administrative system are described. These systems components are usually captured in models, in close collaboration with stakeholders to ensure the same understanding of the system.
- A structured analysis is needed to identify present and future problems, which provide the necessary tools to identify measures to address these problems. E.g. a scenario analysis is made, often linked to socioeconomic development pathways and climate change, to prepare for problems that may arise in the future.
- **Strategy Building:**
- The most promising measures are combined into strategies, which are assessed in detail. The results are a set of selected strategies that are presented to

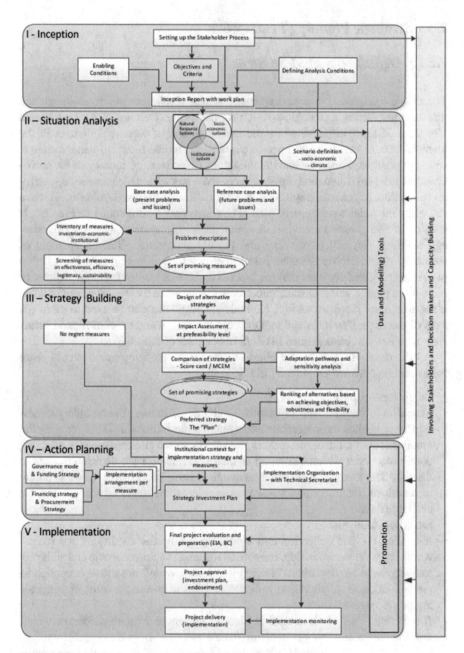

**Fig. 7.4** Master planning framework. (Adopted from Deltares 2020)

decision-makers to select a preferred strategy (*in the section below, the adaptive management analysis for the selection and evaluation of alternative strategies will be further discussed*).

- **Action Planning:**
- After the selection of the preferred strategy, this phase focuses on its translation to concrete actions. The involvement of various stakeholders needs careful planning and coordination. Action planning is not intended to be static or prescriptive, it leaves room for decision-makers to further discuss and taking into account their own responsibilities. This last point is key, as this stage should assign concrete actions. This phase includes the funding and budgetary requirements for implementation.
- **Implementation:**
- This phase focuses on the implementation of the strategies selected according to the action plan devised. It includes the actual creation of construction measures and its subsequent monitoring and evaluation.

## 7.2.2 Towards the Strategic Planning of NBS for Adaptive Management

Anticipating future uncertainties in the system, the strategic planning process incorporates a number of elements that allow decisions to be adapted to a new situation in the future. In the situation analysis step, scenario definition presents an important exercise in exploring possible futures that will affect the system to varying degrees. When building strategies, these are checked for their flexibility and robustness in view of these future scenarios and adaptive pathways can be defined. The latter set out pre-defined routes of response when changes in the system are manifested. More general for the purpose of NAS, the implementation of NBS can bring a high degree of uncertainty related to their functioning regarding the future socio-ecological systems under climate and other changes as well as their possible impacts. There is therefore the need for adaptive capacity of the planning and management itself (see box below).

> When there are many plausible scenarios for the future, it may well be impossible to construct any single static policy that will perform well in all of them. It is likely, however, that the uncertainties that confront planners will be resolved over the course of time by new information. Thus, policies should be adaptive - devised not to be optimal for a best estimate future, but robust across a range of plausible futures. Such policies should combine actions that are time urgent with those that make important commitments to shape the future and those that preserve needed flexibility for the future. (Daniels and Walker 2001)

**Management of Uncertainty in Planning and Implementation of NBS**

A review of NBS literature by Dourojeanni (2019) identified a wide range of drivers and barriers that enable or impede the implementation, uptake and mainstreaming of NBS and natural assurance schemes. Roughly, the barriers can be categorized into 4 groups: (1) institutional and regulatory barriers, (2) absence of clear evaluation of NBS performance, (3) funding and financing barriers and (4) knowledge and acceptance barriers. The set of barriers are intimately related to uncertainties, which can be classified in barriers related to uncertainties in the natural and technical system and those related to political-legal, economic/financial and institutional issues, i.e. the social system.

To capitalize on the drivers of NBS implementation and overcome barriers hindering their integration in climate adaptation plans, management of uncertainty is key (Dourojeanni 2019; Van Cauwenbergh et al. 2020). Considering the risk context, planning will target the present and future risks and develop strategies for both mitigation and adaptation. Strategies are developed to achieve the goals and are a combination of management interventions that can be either infrastructural (e.g. flood protection through building of dikes or hydro-forestry measures), economic (e.g. water pricing, pollution taxes) or institutional (e.g. water allocation schemes, pollution control, land use planning). In order to operationalize these goals, we propose the framework of planning as a systematic procedure to generate a synthesis of information in such a manner to gain insight into the nature and consequences of possible management strategies. For a detailed discussion on management of uncertainty within the NAIAD project refer to Van Cauwenbergh et al. (2020).

### 7.2.3   Towards Implementation: Financing Framework for Water Security

The implementation of NBS requires bringing together the diversity of value expectations between authorities, proponents and investors to translate NBS strategies into implementable projects. In this regard, the 'D7.3 Handbook for the Implementation of Nature-based Solutions for Water Security" (Altamirano et al. 2020), the Financing Framework for Water Security (FFWS) (Altamirano 2017) for structuring NBS implementation arrangements (*Chap. 9 Altamirano et al., this volume*). The FFWS guides the design of an implementation arrangement – choosing from a wide range of project delivery and finance options that vary from purely public governance options up to the creation of markets for private initiatives. In a nutshell, the FFWS adapted to the implementation of Ecosystem-based DRR defines a process for defining funding and governance structure of a NBS strategy for a sustainable financing and implementing strategy. In this regard, the 'D7.3 Handbook for the Implementation of Nature-based Solutions for Water Security differentiates between funding and financing. Funding refers to the question of who ultimately will pay for the investments made. Funding could come from three generic sources: Taxes, Tariffs and Transfers (3 T) (OECD 2009). Financing, on the other hand, refers to mustering the up-front resources needed to be repaid over time by the funding. This simple but fundamental clarification avoids the mistaken idea that private/commercial (i.e. repayable) finance could be a substitute for a shortage of internally generated project revenues. Upon this clarification, the FFWS for implementing Ecosystem-based DRR develops five business cases to evaluate public investment, following the Five Cases Model of the UK HM Treasury (Government of the United Kingdom 2018).

- **Strategic** – is there a compelling case for change?
- **Economic** – does the recommended measure optimise public funding?
- **Commercial** – is the proposed measure achievable and attractive in the marketplace?
- **Financial** – is the spending proposal affordable?
- **Managerial** – how will the proposal be successfully delivered?

The five business cases for the context of Ecosystem-based DRR elaborates on the expected levels of risk reduction to be sustainably delivered, the type of transaction to govern the service delivery, and the enabling institutional setting. Hence, the Handbook details the process where NBS proponents make explicit:

(i) how the implementation of NBS measures enables a paradigm shift towards resilient and sustainable economic growth (Theory of Change[1]) in a given institutional setting;

(ii) a hierarchy of services and their levels at which specific target groups are willing to pay using 3 T;

(iii) the characterization of these services as economic goods susceptible to be transacted as public procurement, markets or other hybrid organizational forms;

(iv) the funding of revenue inflows and the costs outflows of the project, for identifying the need and opportunities for front-end financing; and,

(v) the project owner in-house and procurement capabilities, to make predictable cashflows by delivering on-time and maintaining levels of service over the project life cycle.

The implementation of the FFWS requires a deep understanding of the institutional enablers and constraints. An institutional understanding of the contextual embeddedness proves the professional criteria to assess the feasibility of implementing 3Ts, the availability of financing instruments, and the contracting and procurement possibilities.

---

[1] Theory of Change is a comprehensive description and illustration of how and why a desired change is expected to happen in a particular context. It is focused in particular on mapping out or "filling in" what has been described as the "missing middle" between what a program or change initiative does (its activities or interventions) and how these lead to desired goals being achieved.

## 7.3 Tools and Methods Used in the Planning Phases

### 7.3.1 Why Do We Need Tools for Planning?

Each step of the planning framework results in specific decisions (see Fig. 7.5). To support these decisions, various tools and methods are available for specific steps of the planning framework including e.g. existing decision-support methods such as multicriteria decision analysis. This section gives an overview of some of the tools and methods that can be used in the planning steps. We also explain how, and which tools and methods were applied for different Case studies (*see Chaps. 10, 11, 12, 13, 14, 15, 16, and 17, this volume*).

It is important to point out that in a multi-stakeholder context, each stakeholder is likely to have his or her objectives, value system and preferences (*e.g. Chap. 5 Giordano et al., this volume*). As the planning process progresses, and strategies and measures become more detailed, new questions will be raised reflecting the different perspectives from each of the partners. Whatever be the planning step, some stakeholders may base their decision on technical information and will thus rely on advanced technical tools and expert knowledge to interpret indicators. Meanwhile, understanding the perspective of other stakeholders is required to push forward the planning and overcome the divergent phase typical in strategy building. Simplified decision support tools, developed by experts that are both educational yet and as rigorous as possible, are often needed to share technical knowledge, explore possible strategies and enter the convergent phase where misunderstanding and misalignment are hopefully overcome to find possible trade-offs. Multi-criteria and decision aid methods may help this convergent phase in the decisions of each stakeholder. (Tacnet et al. 2019) provide descriptions and examples of the different tools

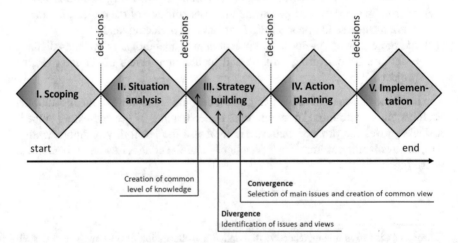

**Fig. 7.5** Steps of planning framework ending with decisions by stakeholders, each step experiencing phases of knowledge creation, divergence and convergence toward new decision and next step. (Adopted from Deltares 2020)

available ranging from simple and educational, to advanced or for aiding decisions and communication.

## 7.3.2   Finding the Right Tool at the Right Phase

At each step of the planning framework, various questions must be addressed to end up with concrete decisions. This requires a great variety of methods and tools which have to be carefully selected to address the needs decision makers and stakeholders have, while considering the capacity available (e.g. data, technical expertise, communication support) and potential bias and trade-offs with methods/tools. Below we illustrate with some examples how methods and tools with varying levels of participation can be used in the different planning steps of NAS.

### 7.3.2.1   Inception/Scoping

The definition of objectives (e.g. risk reduction, ecosystem services), along with evaluation criteria to monitor and evaluate those objectives, is of utmost importance for the final configuration of NBS and NAS in the plan. Objectives are often derived from a mix of policy prescriptions (e.g. Sendai, EU WFD, Climate Adaptation Strategy etc.) and issues on the ground. The way in which objectives are articulated should ideally be supported by a participatory process.

Stakeholder interviews and workshops can be conducted to agree on objectives and prepare a comprehensive workplan specifying all the activities that need to be carried out further to achieve the defined objectives. Some of the participatory tools and methods that can be used for stakeholder workshops are serious gaming, fuzzy cognitive mapping, system dynamics by means of Group Model Building or Mediated Modelling to mention just a few. For a complete overview of the participatory modelling methods and tools used in NAIAD, refer to Tacnet et al. (2019). For a general view of the most used methods and tools used in the context of participatory modelling, refer to Voinov and Basco-Carrera (2018).

### 7.3.2.2   Situation Analysis

This step should provide the decision makers with a complete understanding of the natural resources, socio-economic and institutional systems; existing and potential future problems and possible measures and interventions for further analysis. The tools and methods that can be used to achieve this can be more top-down or expert-based (e.g. hydrological modelling, hydraulic modelling, damage modelling, forest fire models, social network analysis, and institutional analysis) or explicitly involve stakeholders in the generation of system understanding (e.g. system dynamics). For a more comprehensive overview of possible methods and tools for situation analysis refer to Deltares (2020).

### 7.3.2.3　Strategy Building

A strategy is built when decision makers have an optimal combination of potential measures that contribute to achieving the defined objectives. From several alternatives, a preferred strategy is chosen based on an overview of the expected effectiveness of its constituting measures considering previously agreed objectives, and assessed using criteria and indicators. To assess a strategy (e.g. a combination of grey/green infrastructure with regulatory incentives and community-based operation), a combination of biophysical and socio-economic methods and tools are used to provide the indicator scores of alternatives (*Chap. 8, Altamirano el al., this volume*). Given the important role of NBS in NAS, this integrated assessment will consider ecosystem services. To support the interpretation of integrated assessment feeding from different models and tools, a meta-model can be used. To select the preferred strategy, cost-effectiveness, cost-benefit analysis and multi-criteria analysis are relevant tools (see Tacnet et al. 2019) and can involve stakeholders (e.g. Van Cauwenbergh et al. 2018). In addition different methods and tools can be used to support the negotiation and management of potential conflicts when making these sometimes controversial choices.

For adaptive planning, decision makers can use adaptive pathways methods such as Decision Trees and Dynamic Adaptation Policy Pathways (DAPP) to enable them to consider future uncertainties for making decisions, and as a result ensure that the preferred strategy is robust or flexible to address possible futures (Deltares 2020; Tacnet et al. 2019).

### 7.3.2.4　Action Planning

To operationalize the preferred strategy and assure the decisions on paper can be implemented on the ground, action plans need to be defined. These plans list a number of concrete actions, services and their instrumentalization by means of governance, funding, and financing strategies, and procurement strategies. Given the multiple and varied stakeholder involved in the implementation and (more so) operation of NBS in NAS, their involvement in defining the rules and responsibilities is crucial. In that sense, the definition of business cases for public and/or private investments can be supported by system dynamic models using Group Model Building or Mediated Modelling. In general, this step can be supported by tools for partnership development and consolidation.

### 7.3.2.5　Implementation

For a smooth implementation and monitoring and evaluation of NAS, some level of collective action is needed. Participatory monitoring and citizen science can support collective action and increase overall awareness and ownership to support long lasting implementation. Given the intrinsic uncertainty in NBS and NAS, their planning and implementation should be seen as an adaptive process in which decision makers and stakeholders can continuously assess the situation and determine the best way to proceed, either moving forward or moving backward if necessary. This cyclic and iterative process can be supported using participatory monitoring and evaluation.

### 7.3.3   Linking Case Studies to the Planning Framework

Varied disciplines should always be involved to work in parallel, i.e., in a multi-disciplinary approach, and perform their assessment and work in tight collaboration, i.e. an interdisciplinary approach. When integrating stakeholder and lay knowledge into an assessment, the latter becomes a trans-disciplinary approach. To our experience, richer results and assessments emerge from those trans-disciplinary approaches. However, an extra effort is needed in terms of communication, clarification of concepts and capacity building between stakeholders and experts involved. In our vision, integration relies on trans-disciplinary approaches and thus a clear understanding of all the methods used, and their potential bias toward certain value systems is necessary.

In the next section we describe a representative example of the different disciplines and methods used in the different Case studies. At the start, a great variety of methods and tools were identified to support the design, operation and monitoring of natural assurance schemes and specific activities associated to DRR and the planning phases. These tasks were performed by a NAS case study team, composed of different experts, decision makers and in some instances the stakeholders themselves. Depending on the case study context, specific NBS purposes and local conditions, each case study team co-defined the models, methods and tools to be used for each activity. It is therefore important to highlight that the participatory or collaborative modelling approaches and methodologies used, conform a Natural assurance toolbox. It was then up to each case study to decide which combination of tools, methods and approaches to use in each activity.

The disciplines included in the natural assurance Case studies are broadly categorized into three assessment pillars (biophysical, economic and social) and the integration consists of:

| | |
|---|---|
| Geography | Economy |
| Ecohydrology | Decision sciences |
| Hydrology and hydraulics | Sociology |
| Civil engineering | Political sciences |
| Safety and reliability analysis | Climate science |

Most Case studies made use of hydrological and hydro-dynamic modelling to assess the biophysical system. In Lez, Lodz, Lower Danube and Copenhagen Case studies the modelling was combined with the collection and use of spatial data (*see Chap. 14, Le Coent et al., Chap. 10 Scrieciu, and Chap. 17, Jørgensen et al., this volume*). The Eco-Actuary tool was used for monitoring and modelling ecosystem services in Lower Danube and Copenhagen Case studies (*see Chap. 4 and Chap. 12 Mulligan et al, and Chap. 10 Scrieciu, and Jørgensen et al., this volume*). Other Case studies like Brague, made use of a wide variety of biophysical modelling tools: numerical modelling, hydraulics and wildfire modelling, in combination with hydrological and hydro-dynamic modelling (*see Chap. 13 Piton et al., this volume*). Decision-making methods and safety reliability approaches were used to design a framework for NBS' effectiveness assessment. A whole chain ranging from NBS' physical to economic features has been proposed providing results through a pluri-disciplinary cost-effect-consequence analysis.

In terms of economics and decision sciences, cost-benefit analysis and to a lesser extent multi-criteria decision analysis are the predominant methods used to develop Natural Assurance Schemes. NAS Case studies with a higher advanced Technology Readiness Levels (TRL) such as Medina del Campo and Rotterdam were able to advance further the five business cases; going beyond the economic business case towards the financial and commercial business cases. In specifics, both Case studies applied the NAS canvas to develop the business model (i.e. commercial business case). Rotterdam also applied the participatory value evaluation (i.e. participatory budgeting) to refine Life Cycle Costs (LCC) calculations. Finally, both applied the FFWS to develop suitable implementation arrangements (i.e. funding, financing and procurement) (*see Chap. 8 Mayor et al., Chap. 9 Altamirano el al. and Chap. 16 Dartee, this volume*). Other Case studies with a lower TRL like Lez and Thames also applied LCC (*see Chap. 14 Le Coent et al. and Chap. 12 Mulligan, this volume*).

Finally, all Case studies used methods and tools for involving decision makers and/or stakeholder in the different modelling and planning activities. Participatory modelling was used in all Case studies except for Lez, which used a participatory scenario planning method. In the Thames Case study, participatory monitoring was also used to obtain data for the development of Eco-Actuary (*see Chap. 4 Mulligan et al., this volume*).

In sum, participatory modelling including some monitoring was widely spread throughout all Case studies and in all planning phases. The study shows that those Case studies that focused primarily in the first planning phases of designing a natural assurance scheme, spend considerable time and efforts modelling the bio-physical system. For strategy building, most Case studies made use of multi-criteria decision analysis and cost-benefit analysis, which both helped decision makers and stakeholders to define possible measures and build strategies. Action planning, however, requires additional methods and tools that relate to business models and implementation arrangements that define funding and financing strategies. It can be observed that these implementation arrangements require a high level of TRL, such as the Medina del Campo and Rotterdam Case studies.

Figure 7.6 provides an overview of the various disciplines and methods applied in natural assurance Case studies. An exemplary case of the application of the strategic planning framework and the use of models and tools for the case of Medina is presented in the next section.

## 7.4 Lessons Learned and Recommendations

Showcasing how to manage water-related risk with strategies mostly relying on NBS is a key objective of natural assurance schemes in the nine Case studies located across all Europe (*detailed in Chap. 2, this volume*). NBS are potentially powerful measures to use because these solutions provide multiple benefits, risk mitigation being only one of them. Each Case study site being peculiar and given the variety in scales from cities to entire catchments, there is not a one-NBS-strategy-fit-all. To identify which NBS strategy fits a given site, is a complex and often iterative process, involving multiple actors. To help guide this process, we propose the structure of strategic planning containing several standard steps well known by planning experts. These steps are not to be taken as a strict and sequential structure for the planning and design of NAS, but as a comprehensive recipe of elements needed to gain the necessary information and support to identify and implement NAS. As such, the proposed framework is applicable to different contexts across the globe. Importantly, the choice of methods and tools to support the decision process needs to be tailored to the capacity, needs and (political and other) preferences in a given context.

Whereas the multiple benefits related to NBS strategies provide an advantage over conventional or "grey" strategies, the drawbacks of NBS are their intrinsic uncertainty, difficulties in measuring the co-benefits and explicit role for a broader range of stakeholders. However, engaging all relevant partners is necessary and the use of multiple and transdisciplinary knowledge proves to be more efficient in the long run than top-down approaches tailored by experts who may miss key concerns of particular stakeholder groups. In addition, specific attention must be paid to checking that required functions for NBS are fulfilled: to reduce risk, physical effectiveness is the first mandatory objective to reach. In most cases, NBS will be used in combination with others more classical "grey" techniques within hybrid strategies. For the design of natural assurance schemes, participatory approaches from the water resources management field, need to be merged with approaches coming from the risk reduction community and the climate change adaptation community.

Among the key success factors for the natural assurance schemes in the Case studies discussed here, we found (Fig. 7.6):

- Alignment of key stakeholders and their objectives must be crystal clear;
- Project boundaries and responsibilities of partners must be stated;

Commitment of key stakeholders (champions and personal ambition).

Efforts to meet these success factors or facilitate their emergence in early stages of the decision process, will increase the projects' likelihood to go to full implementation. And whereas their absence might not impede a project to go ahead initially, our experience is that it will slow down full implementation or emerge at a later stage.

Our results also point to some important implications for NBS uptake. For one, we saw that decision support models and tools were only marginally used during the planning and implementation process in the case studies. Findings suggest that for

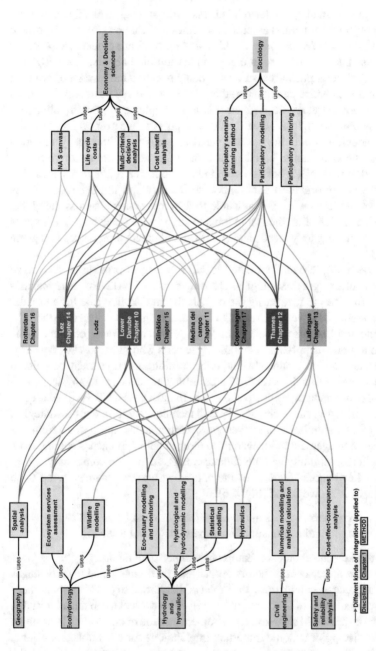

**Fig. 7.6** Overview of the integration of disciplines and methods used in the natural assurance Case studies Thirdly, when an important (dis-)benefit of a NBS strategy is identified, be it a co-benefit or the main benefit – which is a matter of perspective –a rigorous assessment should be performed regarding the related improvement or alteration (dis-benefit). In our experience, this might require advanced models or extensive surveys. The experts applying these methods should then tailor the communication of results to the audience: with higher levels of detail for other experts and targeted stakeholders and in a more educational way with simpler indicators for other stakeholders. In the Brague case study (Chap. 13, this volume) for instance, the engagement and support of the French Water Agency, which is key for funding, was conditional to the outlook on a significant improvement of the river's ecological status

NBS uptake it is far more important to have willingness and commitment from the key stakeholders. Nevertheless, the need for evidence might arise towards upscaling, calling for support by above mentioned methods and tools. In our experience, when used, these tools were considered useful by the stakeholders involved.

Secondly, we found that co-benefits can be a driver for success when the funding is available, a clear owner of the NBS project exists and there is a concretized level of service. In the case of Rotterdam, the case study which was the most advanced has been fully implemented, the NBS' ability to generate cheaper water supply for the sport arena nearby, leveraged the needed support for funding and ownership, with flood reduction and recreational value as co-benefits functioning as leverage for the willingness and acceptability of the project by other stakeholders. In cases where the added value of the NBS is not clearly linked to an existing operator (entity that directly receives the benefits and can take care of the Operation and Maintenance of NBS to deliver the agreed service), co-benefits might have to play a stronger role and it remains a question whether these co-benefits can do that. In all case studies where full cost-benefit analysis could be performed, co-benefits, i.e., all benefits other than risk reduction, outnumber in value the mere avoided damages. Thus, co-benefits might weigh more than risk reduction in the final decision balance. It is therefore worth paying attention to co-benefits, involving stakeholders willing to optimize the strategy to increase co-benefits while still meeting the risk reduction objectives. Natural assurance schemes as previously defined, based on our case study learning would therefore benefit from incorporating both the risk reduction element as well as the co-benefits identification through co-design, as key elements for success (*see Sect. 7.1 on the conceptual framing*).

Despite the low TRL level of this case study and the project being at its starting phase, an advanced eco-hydrological assessment was performed to assess how various strategies improve or alter the functionality, artificiality and adjustments of the river hydrology and morphology. The assessments were then aggregated in a unique indicator, the Morphological Quality Index (MQI) (Rinaldi et al. 2013), ranging between 0 (river totally altered) and 1 (fully natural river). While the French Water Agency was interested in the details, the mere improvement in MQI score was also helping other stakeholders understand where particular strategies were better than others regarding the environmental perspective. Indicators such as MQI can ultimately be used by decision makers in multi-criteria decision-making methods, after weighting of criteria, to trace and explain how decisions were taken. Transparency in these decision phases helps finding and maintaining the engagement of stakeholders (and potential future support).

Finally, we made several observations on the aspect of integration that underlies successful planning and implementation of NBS. Case study analysis shows a reality where objectives and related indicators are driven by sectoral interests. This makes that what is defined as a benefit or co-benefit depends on the viewpoint of the stakeholders involved. In the Rotterdam case, the decision-making on the NBS was defined by the leading organization (related to mandate and funding) and the clear risk/benefit cycle (involving the Evides water company and Sparta football club stadium) proved crucial to facilitate that decision-making (see point above). The case shows that institutional coordination is a key barrier to implementation (and that this is happening even within the municipality). Finally, we observed that in order to mainstream the NBS, evidence of performance across (co) benefits is needed. However, little to no monitoring incentives or interest exists. Learning across different NAS and mainstreaming of NBS in NAS requires considering financial feasibility, the soundness of economic incentives as well as monitoring and evaluation from the start of a project.

# References

Alexander DE (2002) Principles of emergency planning and management. Terra Publishing, Herdfordshire

Altamirano MA (2017) A financing framework for water security. Oral presentation, water futures workshop, during the XVI world water congress, Cancun, Mexico

Altamirano MA, Benítez Avila C, de Rijke H, Angulo M, Arellano B, Dartée K, Peña K, Nanu F, Mayor B, Lopez-Gunn E, Marchal R, Pengal P, Scrieciu A, Mori JJ (2020) Handbook for the implementation of nature-based solutions for water security: guidelines for designing an implementation and financing arrangement, DELIVERABLE 7.3: EU Horizon 2020 NAIAD project, Grant Agreement N°730497 dissemination

Baas S, Ramasamy S, Dey DePryck J, Battista F (2008) Disaster risk management systems analysis. A guide book, FAO, Rome, Italy

Cvitanovic C, Howden M, Colvin RM, Norström A, Meadow AM, Addison PFE (2019) Maximising the benefits of participatory climate adaptation research by understanding and managing the associated challenges and risks. Environ Sci Policy 94:20–31. https://doi.org/10.1016/j.envsci.2018.12.028

Daniels SE, Walker GB (2001) Working through environmental conflict: The collaborative learning approach. Westport, CT: Praeger

Deltares (2020) Analysis framework for water resources planning and implementation. Internal report

Dourojeanni P (2019) Understanding the role of uncertainty in the mainstreaming of nature-based solutions for adaptation to climate change (Master's thesis). In Water Management: Vol. MSc. IHE Delft

Government of the United Kingdom (2018) The green book. Central Government guidance on appraisal and evaluation, HM Tresury

Grimble R, Chan MK (1995) Stakeholder analysis for natural resource management in developing countries. Nat Resour Forum 19:113–124

Loucks DP, Van Beek E (2017) Water resource systems planning and management: an Introduction to methods, models, and applications. Springer

OECD (2009) Managing water for all. An OECD perspective on pricing and financing. Key Messages for Policy Makers

Okada N, Chabay I, Renn O (2018) Participatory risk governance for reducing disaster and societal risks: collaborative knowledge production and implementation. Int J Disaster Risk Sci 9:429–433. https://doi.org/10.1007/s13753-018-0201-x

Priscoli JD (2004) What is public participation in water resources management and why is it important? Water Int 29(2):221–227

Prutsch A, Felderer A, Balas M, König M, Clar C, Steurer R (2014) Methods and tools for adaptation to climate change. A handbook for provinces, regions and cities. Environment Agency Austria, Wien

Rinaldi M, Surian N, Comiti F, Bussettini M (2013) A method for the assessment and analysis of the hydromorphological condition of Italian streams: the Morphological Quality Index (MQI). Geomorphology 180-181:96–108. https://doi.org/10.1016/j.geomorph.2012.09.009

Samaddar S, Okada N, Choi J, Tatano H (2017) What constitutes successful participatory disaster risk management? Insights from post-earthquake reconstruction work in rural Gujarat, India. Nat Hazards 85:111–138. https://doi.org/10.1007/s11069-016-2564-x

Tacnet JM, Van Cauwenbergh N, Gomez E et al (2019) DELIVERABLE 5.4 integrative modelling framework and testing in the DEMOs: a global decision-aiding perspective (Main report: parts 1-5). EU Horizon 2020 NAIAD project, Grant Agreement N°730497

UN-Water (2013) What is water security? Infographic October 2013

Van Beek E, Arriens WL (2014) Water security: putting the concept into practice. GWP TEC Background Paper(20)

Van Cauwenbergh N, Ballester Ciuró A, Ahlers R (2018) Participatory processes and support tools for planning in complex dynamic environments: a case study on web-GIS based participatory water resources planning in Almeria, Spain. Ecol Soc 23(2). https://doi.org/10.5751/ES-09987-230202

Van Cauwenbergh N., Dourojeanni P., Basco-Carrera L. et al (2020) DELIVERABLE 5.7. Guidelines for the definition of implementation and investment plans for adaptation. EU Horizon 2020 NAIAD Project, Grant Agreement N°730497

Voinov A, Basco-Carrera L (2018) Tools and methods in participatory modeling: selecting the right tool for the job. Environ Model Softw 109:232–255

Warfield C (2020) The disaster management cycle. https://www.gdrc.org/uem/disasters/1-dm_cycle.html. Accessed Jan 2020

**Chapter 8**

# NAS Canvas: Identifying Business Models to Support Implementation of Natural Assurance Schemes

**Beatriz Mayor, Elena López Gunn, Pedro Zorrilla-Miras, Kieran Dartée, Thomas Biffin, and Karina Peña**

**Highlights**

- The NAS canvas enabled to elicit together with the stakeholders the value proposition of NAS and the components required to build a business model.
- The NAS canvas is flexible and replicable to any NAS or NBS strategy regardless of the stage or the context.
- One of the main difficulties in building business models is to engage indirect beneficiaries within the pool of payers and funders.
- Legislation can become either a critical enabler or a barrier for the development and implementation of business models for NAS.

## 8.1 Introduction

Extreme weather events and water challenges have ranked within the top three greatest risks to the global economy for the last 5 years, according to the World Economic Forum annual assessments (WEF 2019). Around 70–90% of the economic losses caused by floods across Europe between now and 2050 can be attributed to the increase in the value of assets in floodplain areas, with the rest attributed to climate change (EEA 2016). Conventional infrastructural measures are expensive – the investment needed in water infrastructure over the next 15 years has been estimated at 22 trillion dollars, which is more than half of the total expected infrastructure investment demand (USD 41 trillion) (WEF 2019). As discussed in

B. Mayor (✉) · E. López Gunn · P. Zorrilla-Miras
ICATALIST S.L., Las Rozas, Madrid, Spain
e-mail: beatriz.mayor.rodri@gmail.com

K. Dartée · T. Biffin · K. Peña
Field Factors, Rotterdam, The Netherlands

© The Author(s) 2023
E. López Gunn et al. (eds.), *Greening Water Risks*, Water Security in a New World, https://doi.org/10.1007/978-3-031-25308-9_8

previous chapters, there is a realisation on the relevance to move earlier into the disaster management cycle while helping to adapt to climate change, by mainstreaming and normalizing NBS as an alternative or complement to conventional grey solutions to prevent or reduce risks, thus increasing resilience and response capacity to water related hazards. However, NBS are facing several specific barriers for scaling up, including the difficulty to access funding and financing schemes from the lack of real examples providing evidence on their capacity and viability, and thus provide investor confidence and lower investment risks. Furthermore, making this type of projects attractive for private and impact investors requires a clear identification and quantification of the value proposition provided by these solutions, as well as a strong business case that ensures return of investment, particularly in the mid to long term. Most NBS projects fail to develop such a business case partly due to the limited data and evidence on the range of benefits provided by NBS, and their respective value. These projects also need to assess how the value generated – in our case by natural assurance services converted into viable schemes- through risk reduction and additional co-benefits can be captured and generate a series of revenue streams that makes them financially viable, similar to the business models developed for private projects providing goods and services. Identifying the "business model" for an NBS project – including a quantified value proposition, the elements required to deliver this value (resources and stakeholders), the costs of delivering this value, the range of beneficiaries and potential pool of clients and the associated possible revenue streams – will be an essential step to build a convincing business case that reduces the perceived risk by investors, also identifying the possible mix of funding sources to cover the whole range of lifecycle costs and also consider the opportunity costs.

In order to support the identification of possible business models for NAS projects, taking into account their particularities like providing public goods and services, the NAS canvas framework has been developed, as well as a template that allows a clear visual representation, entitled Natural Assurance Schemes canvas. The NAS canvas framework and template are built on the basis of a pluralistic approach to the value proposition in a relational manner, considering the whole range of different values (i.e. risk reduction and co-benefits) and spanning the public, collective and private domains. In other words, they display the components, actors and roles involved in the business model, as well as the relations between them following a market service provision logic (supply $\rightarrow$ service $\rightarrow$ demand). This chapter presents the NAS canvas framework and tool, as well as the co-design process followed for its application to the case studies. It also discusses the transversal findings derived across case studies, as well as the lessons learnt from the application co-design process, with views to replication and upscaling of the tool.

## 8.2    The NAS Canvas Conceptual Framework

The NAS canvas framework has been developed to guide the identification of the whole set of values generated by both NAS projects and NBS strategies and the set of elements and actors required to capture this value and turn it into a business or marketable service. The framework is aimed to sequentially identify and describe three aspects: (i) the co-design process and modules involved in the provision of climate adaptation (including natural risk reduction) services by an NBS or set of measures (NBS + soft/hybrid/grey measures), from both the supply and the demand side; (ii) the actors involved and their potential roles; and (iii) how the value of these services can be translated into revenues or funding resources required for the execution and maintenance of the measures. Hence, it can be used for the identification of potential business models and the required elements for NBS implementation, but also serves as a comprehensive framework to integrate the different steps from problem identification all the way to project design and implementation arrangements to accelerate NBS uptake for risk reduction and co-benefits (the assurance value). It also helps collect, organise and diagnose the type of information required and available in a way that is useful to engage and convince different stakeholder, particularly problem owners and potential investors to stimulate interest and potential buy in and collective momentum for this type of initiatives.

The NAS canvas framework is an adaptation of the traditional business model canvas, tailored to the specificities of DRR and climate adaptation services, and their contextual framework. It is composed of 8 clusters that go through the different steps required to identify the elements composing a business model for the commercialization of a product or, in this case, a service (see Fig. 8.1). The business model canvas is traditionally used to support companies and businesses to identify and structure their value proposition and the elements required to develop a strong and viable business model for the delivery of a product or service to the market. The most acknowledged business canvas is the one proposed by Osterwalder and Pigneur (2010).[1] The NAS canvas builds on this traditional business canvas model, and expands it, tailoring it to account for the specificities of climate adaptation services, DRR and the development of NAS schemes from potential NBS strategies. To do so, a review of the latest business model canvases for nature and NBS was carried out to identify the state-of-the-art advances in this field. Among the identified approaches (Topoxeus and Polzin 2017; Coles and Tyllianakis 2019; Mc. Quaid 2019; Somarakis et al. 2019), the 'PPP canvas' developed by the Inclusive Business Hub was considered the most applicable and aligned with our purpose, as it kept all the original canvas elements and adapted it for ecosystem services provided by nature, thus accounting for non-tangible and non-marketable values. It thus inspired the introduction of two new elements into the traditional canvas: a distinction between direct and extended beneficiaries (components 10A and 10B), impact (component 15), marked in purple font in Fig. 8.2. The canvas was expanded to

---

[1] Available at https://strategyzer.com/canvas

| CLUSTER C. SUPPLY | CLUSTER A. FLOW OF ES SERVICES | CLUSTER E. DEMAND | |
|---|---|---|---|
| **STEP 4. WHO IMPLEMENTS** *Identify the stakeholder who takes the responsibility or initiative of implementing the solution* | **STEP 1. PROBLEM TO BE ADDRESSED** *Describe the main problem derived from exposure to water-related risks to be addressed* | **STEP 9. WHO OWNS THE PROBLEM** *List the stakeholders affected by the problem.* | |

**STEP 5. KEY MEASURES** — *List of measures composing the strategy, and a breakdown of main measures to be undertaken.*

**STEP 2. VALUE PROPOSITION**

| 2A. Main service and value *Identify the main risk reduction service and the associated value as avoided damage costs (see chapter 6)* | 2B. Other services and values *List the co-benefits and associated values estimated using the method in chapter 6.* |
|---|---|

**STEP 10. CUSTOMER SEGMENTS**

| 10A. Direct Beneficiaries *Stakeholders who benefit from the main value (avoided damages)* | 10B. Clients *Stakeholders who would actually pay for the service* | 10C. Extended Beneficiaries *The range of stakeholders who benefit indirectly from the other services or co-benefits* |
|---|---|---|

**STEP 6. KEY RESOURCES** — *Type of resources needed to implement the measures, e.g. funding, knowledge, people and capacity, legal frame, political support, other, ....*

**CLUSTER B. REGULATORY CONTEXT**

**STEP 3. REGULATION** — *Main legislation enabling or hampering the implementation of the solution. It is also relevant to highlight regulatory gaps.*

**CLUSTER F. ABILITY/WILLINGNESS TO PAY**

**STEP 11. REVENUE STREAM** — *Identify the potential income streams that could emerge from the willingness of the different type of customers to pay for the services/value generated*

**STEP 7. KEY PARTNERS** — *key stakeholders you need to engage with to obtain the resources, or to undertake the implementation, e.g. subcontracts, knowledge partners, financing partners, governance partners*

**CLUSTER E. SUPPLY-DEMAND INTERACTIONS**

**STEP 13. CUSTOMER RELATIONSHIPS** — *Describe the type of communication between service provider and clients, e.g. direct assistance, transaction, automated (through a platform), community, etc.*

**STEP 12. FUNDING COMING FROM**

12A. Tariffs service use tariffs charged to the clients

12B. Taxes taxes charged to the users or indirectly to all citizens

12C. Transfers funds provided by the state or other international entities coming from public or international aid funds

12D. Private investments private money invested by external private investors or donors, philanthropy investments or investments by private stakeholders (i.e. farmers).

**CLUSTER D. COST STRUCTURE**

**STEP 8A. Life Cycle Costs** — *Costs of implementing the NBS measure Including capital, operation and maintenance (see chapter 6)*

**STEP 14. CHANNELS** — *Means of communication between service provider and clients, e.g. mail, email, workshops, online platform, etc....*

**STEP 8B. Opportunity costs** — *Avoided benefits from implementation of alternatives (see chapter 6 in this volume)*

**CLUSTER H. IMPACT**

**STEP 15. IMPACT THROUGH KPIS**

*List the different biophysical, economic or social impacts expected from the implementation of the measure, by using a list of key performance indicators. These can be identified through the method described in Blasco et al. (2022, chapter 7 in this volume). The impact values should be assessed applying the methods in chapters 4 and 5.*

**Fig. 8.1** NAS canvas. Colour legend: **Red**: Traditional business Canvas; **Purple:** PPP business canvas; **Green:** NAIAD project framework. (Source: own elaboration)

## ROTTERDAM NAS STRATEGY: URBAN WATER BUFFER, SPANGEN, ROTTERDAM

### CLUSTER C. SUPPLY

**CLUSTER C. SUPPLY**

**4. WHO IMPLEMENTS**
Municipality of Rotterdam

**5. KEY MEASURES**
Construction of retention tanks, biofilter and infiltration and recovery wells. Monitoring of water quality to maintain infiltration permit. Automated operation.

**6. KEY RESOURCES**
Stakeholder engagement; technical expertise; climate adaptation policy programmes.

**7. KEY PARTNERS**
**Operators:** Evides Water Utility;
**End-users:** inhabitants, Sparta Stadium;
**Regulator:** Water Authority Delfland
Implementing partners: TKI Urban Waterbuffer Consortium

### CLUSTER A. FLOW OF ES SERVICES

**1. PROBLEM TO BE ADDRESSED**
Water nuisance during heavy rainfall and lack of green in the neighbourhood

**2. VALUE PROPOSITION**
Capturing and storing rainwater to avoid water nuisance, while creating a new local source for high quality freshwater supply with one integral solution..

**2A. Main service and value**
1400 m³ of retention to reduce local flood risk = €700,000[3]
Alternative local water source, supplying 15,000 m³ of filtered rainwater per year to Sparta football stadium = €682,000[7]

**2B Other services and values**
Reduced pumping requirements: €75,000[4]
Energy savings from reduced water production: €44,000[2]
Emissions savings: €8,000[5].
Increased greenery and improved aesthetics resulting in higher property values: €561,000[6]
Increased water awareness

### CLUSTER E. DEMAND

**CLUSTER E. DEMAND**

**9. WHO OWNS THE PROBLEM**
Municipality and inhabitants

**10. CUSTOMER SEGMENTS**

**10A. Direct Beneficiaries**
Local residents,
Local Businesses

**10B. Clients**
Municipality of Rotterdam,

**10C. Extended Beneficiaries**
End user of the water (local football club)

### CLUSTER F. REVENUE STREAMS

**CLUSTER F. REVENUE STREAMS**

**11. REVENUE STREAMS**
End user of water (football club) - pay for water supply
Residents - increased tax revenue through higher property values
Water Utility - In kind contributions: operational responsibility in exchange for exploitation of water well

**12. FUNDING**
12A. Tariffs - Water tariff of €0.91/m³ to cover operational costs
12B. Taxes – Indirect funding through water taxes
12C. Transfers – Innovation fund (case specific)
12D. Private

### CLUSTER B. REGULATORY CONTEXT

**CLUSTER B. REGULATORY CONTEXT**

**3. REGULATION**
**Broader context -** Municipal water Plan (2011); Municipal Water Plan (revised, 2012), Rotterdamse Adaptatiestrategie (2013); Resilient Rotterdam and Water Sensitive Rotterdam Programs
**Specific relevant regulations -** Groundwater legislation (see STOWA rapport 35A, 2015 – *Ondergrondse Waterberging Infiltratiebesluit bodembescherming 2009*)

### CLUSTER G. SUPPLY-DEMAND INTERACTIONS

**CLUSTER G. SUPPLY-DEMAND INTERACTIONS**

**13. CUSTOMER RELATIONSHIPS**
Long term partnerships, Consortium Agreements, Operation & Maintenance contracts

**14. CHANNELS**
Direct communication, stakeholder meetings, documentation and reporting on system performance.

### CLUSTER D. COST STRUCTURE

**CLUSTER D. COST STRUCTURE**

**8A. Life Cycle Costs (50 year timeframe)**
Capital expenses: €1,245,876
Regular maintenance: €425,089
Capital maintenance: €16,326
Direct support: €10,000
Indirect support: €18,000
Total: €1,715,291

**8B. Opportunity costs**
Land where 90 m² biofilter is located cannot be used for other purposes (land value of €372/m²) = €33,480

### CLUSTER H. IMPACT

**CLUSTER H. IMPACT**

**15. KPIS**
Number of separated sewer overflows to surface waters / year; Target: < 2 per year
Amount of infiltrated/stored water; Target: 30,000 m³/year
Amount of water supplied; Target: 15,000 m³/year
Quality of water after treatment; Target: Ongoing compliance with legislation on water quality for infiltration and reuse.

**Fig. 8.2** NAS canvas applied to the Rotterdam case study described in Dartee et al. (2023, Chap. 16 in this volume). (1) Estimates from site data. (2) Value over 50 years, energy use of avoided water production = 0.35 kWh/m³ (Vewin 2017), electricity = €0.17/kWh. (3) Value of retention = €500/m³ (Gemeente Rotterdam 2019). (4) Value over 50 years, avoided pumping costs = €0.05/m³ (Arcadis 2019). (5) Value over 50 years, electricity = 569 g $CO_2$/kWh (Moro et al. 2018), carbon price of €54/ton (High Level Commission on Carbon Prices 2017). (6) 4% increase on 100 households nearest to UWB (Bervaes 2004), average property WOZ valuein Spangen = €140,233 (Gemeente Rotterdam 2019). (7) Value over 50 years, water tariff = €0.91/m³. (Source: own elaboration)

incorporate the essence and elements of the economic framework developed by Le Coent et al. (2023, Chap. 6 in this volume; Graveline et al. 2017). The new components coming from the NAS framework are distinguished with green font in Fig. 8.2. This resulted in the NAS canvas framework, which allows to capture the whole set of co-values, actors and contextual settings inherent to NBS strategies that will ultimately determine and condition the structure and feasibility of a NAS business model. The framework is composed of 8 clusters as shown in Fig. 8.1.

- **Cluster A. Flow of natural assurance services**, which describes the problem to be addressed, and the value proposition distinguishing between main value (risk reduction) and other values (co-benefits).
- **Cluster B. Regulatory context**, which lists the main regulatory context, supporting or conditioning the implementation.
- **Cluster C. Mapping the supply,** which identifies the main implementing actors, measures, resources (human, knowledge or economic), and partners required to provide the service.
- **Cluster D. Mapping the costs of the service**, which identifies the main financial costs, distinguishing between lifecycle costs (implementation and operation and maintenance costs) and opportunity costs, as defined in Le Coent et al. (2023, Chap. 6 in this volume).
- **Cluster E. The demand**, which identifies the main problems owners, i.e. people that suffer the problem, who turn into beneficiaries of the solution. These breakdown into direct beneficiaries, clients and indirect beneficiaries, as explained in Fig. 8.2.
- **Cluster F. Mapping ability/willingness to pay,** which makes the connection with how the willingness to pay by the different groups of beneficiaries, can turn into potential revenue streams or funding sources to support the implementation and maintenance of the solution. Funding sources can be of four types: (a) tariffs paid for the use of the service; (b) taxes for indirect payment for the service; (c) transfers from the government or international institutions with public funds; (d) private investment by donors, investors or private users.
- **Cluster G. Mapping the supply-demand interaction,** which identifies the type of relationship established between the service provider and the client, as well as the channels through which communication takes place.
- **Cluster H. Impact,** which displays the expected impact from the implementation of the measures through a series of quantified key performance indicators spanning environmental, social and economic aspects.

## 8.3 Applied Tools and Methods: How the NAS Canvas Is Used

The NAS canvas template was developed as a visual representation of the NAS business model components. This template has been applied in nine case studies for the different NAS strategies considered. Figure 8.1 shows the NAS canvas template indicating with the different colours the source of the components, i.e. the traditional business canvas, the PPP canvas or NAS's economic framework. In the figure, the components within the clusters described in the previous section are numbered as sequential steps to follow in a specific order to facilitate its use, and a description of the expected pre-filled content instructions provided in each box for all separate components. To apply the tool, the intended user should follow the steps and fill in the information requested. The user will immediately notice how each step builds on the previous steps, following a specific logic that allows the sequential identification of the required information.

The information needed as input to fill in the NAS canvas for the case studies comes from the methodologies and assessments described in the previous chapters in this volume. Figure 8.2 illustrates the actual application of the NAS canvas to the Rotterdam NAS and NBS strategy as described in Dartée et al. (2023, Chap. 16 in this volume). The Rotterdam case study has the most complete and detailed information to fill in the NAS canvas since it has already been fully implemented and it is in the co-design process being replicated to another country, allowing to contrast and complete the assessments with accurate estimations based on empirical evidence.

## 8.4 A Staged Approach in the NAS Canvas Implementation

The application of the NAS canvas framework to the NBS strategies in the case studies to develop NAS, was done in several phases. This included a co-design and collaborative approach as highlighted in the stakeholder protocol described in Lopez-Gunn (2023, Chap. 2 in this volume), and Van Cauwenbergh et al. (2023, Chap. 19 in this volume) – of qualitative and quantitative completion and collaborative validation, following the sequence described in Fig. 8.3.

**Phase 1** During Phase 1, each case study applied the NAS canvas framework in a linear table format to identify and qualitatively describe all the elements required to build a successful business model for the strategy. The description is completed

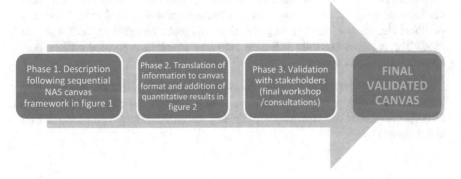

**Fig. 8.3** Sequence of NAS canvas framework application to the case studies. (Source: own elaboration)

with the quantitative results from the economic analysis (based on Chap. 6) looking first at the main service and value through avoided damage costs (step 2A), co-benefits and results from the valuation (2B), cost structure (8), and impact indicators with KPIs (step 15) from the biophysical and social analyses (based on Chaps. 4 and 5) (see Fig. 8.1).

*Reflection and lesson learnt from the implementation of phase 1*: some case studies applied the full economic analysis and some could only do it partially. In the second case, a qualitative estimation of the information was provided based on the case study team knowledge, which was validated with the stakeholders (e.g. river basin agency staff, etc.). In the particular case of "Other service and values" (step 2B in Fig. 8.2), different methods were selected by each case study to carry out the co-benefits valuation as reported by Le Coent (2023, Chap. 6 in this volume). Therefore, the values provided for the co-benefits were expressed through different indicators, units and approaches, some quantitative and some qualitative.

**Phase 2** During the second phase, the detailed description was revised by the canvas development team and transferred into the canvas format. Several cases had specificities that needed to be addressed through iterations and discussions with the case study leaders regarding the type and depth of the information required.

*Reflection and lessons learnt from the implementation of phase 2:* In most cases the "customer relationships" (step 13) was the most difficult to understand and apply. This has identified the importance in the future to develop a typology of potential customer relationships to help the usability of the canvas to other cases.

**Phase 3** The third phase consisted in validating the resulting NBS strategies into the NAS canvas with the case study stakeholders. This was done in a workshop planned within the stakeholder protocol (see Chaps. 2 and 18) or through alternative consultations with critical stakeholders. A standardized validation exercise was carried out, which entailed splitting the workshop participants into as many groups as strategies to be validated, ensuring the presence in each group of a varied representation of stakeholders that are most knowledgeable to the measures in a given strategy. An A1 printout of the strategy's canvas was used in each group leading the stakeholders step by step in a facilitated co-design process, to complete and validate the relevant information. In some cases, stakeholders were asked to rank the most probable element within the group (i.e. the most probable agent to pay for the service). The results were fully validated canvases which incorporated stakeholders' knowledge and perceptions.

*Reflection and lesson learnt from the implementation of phase 3*: some case studies could not validate the canvas in a workshop. Instead, the canvas was validated through one-to-one consultations with the key stakeholders (e.g. in the case of Lodz and Thames).

## 8.5 Common Factors and Lessons Learnt from NAS Canvas Application to Case Studies

### 8.5.1 Lessons Learnt from the Case Studies

The **main value of the case study strategies** is the disaster risk reduction capacity, valued through the avoided damages or insurance value (see glossary of terms). All case study strategies were mainly focused on natural hazards. The majority on flood risk reduction, with the exception of the Medina case study which focused on addressing drought risk. However, in some case studies there was an additional environmental objective for the selected measures that was prioritized by stakeholders, sometimes even higher than the risk reduction itself. This was the case of the aquifer stabilisation and wetland recovery in Medina case study, or biodiversity recovery in natural areas in Glinščica case study. Therefore, these objectives had to be included as main value and main selling points that naturally stirred the interest of potential implementers. This highlights the importance of the multi-value or multi-functionality nature of NBS, which constitutes one of the strongest comparative advantages as compared to grey solutions. Among the **other values (co-benefits)**, all the NAS strategies across case studies provided all three types of co-benefits (i.e. environmental, social and economic) regardless of the type of strategy. In addition to environmental benefits, the creation of jobs, the emergence of additional economic measures through new businesses, or the attraction of tourism are all important common features that need to be valued, valorised and turned into revenues to increase the viability of the scheme and its operation and long term maintenance, both in urban interventions where these benefits are more localized and at the territorial scale. However, the quantification of these values ex-ante is extremely complex, as is reflected in the canvas in Fig. 8.3.

The range of **measures** implemented included a mix of pure NBS (Lower Danube, Glinščica, Brague, Lez, Lodz), a mixture of grey and NBS measures (Thames, Copenhagen, Rotterdam) and a mix of NBS and soft or management measures (Medina). In the case of Medina NBS Strategy 2, which combined crop changes towards drought resilient species (NBS) with groundwater extractions control and creation of WUAS (management measures). This combination proved particularly effective (see Chaps. 6 and 11). Furthermore, it allowed aligning the environmental goals set by the EU Water Framework Directive (through management measures to reduce water abstractions) with the risk reduction and economic sustainability goals facilitated by the NBS.

The range of **resources** required for the implementation of the measures pivot around four main types: funding, knowledge and capacities, stakeholders' engagement, political will, and an enabling regulatory environment. Accordingly, the main **partners** to be involved include representatives from all the stakeholder groups in most cases, from citizens, farmers or service users (i.e. water users), through to governmental and management institutions.

Regarding common factors across critical supply-demand components of the business models – namely **who implements, who benefits and the funding sources** -, a comparative analysis between case study strategies by scale clusters (see Chap. 2 for the classification) was done in order to consider similar scales and somehow similar types of interventions.

Large scale case studies with spatially distributed interventions – namely Thames, Medina and Lower Danube – show a mix between public and private driven implementation and funding, with larger/common use infrastructure being promoted by public institutions (mainly water and land use management agencies), and smaller spatially-spread and individual use/application measures being implemented and funded by landowners or farmers. Funding relies partly on landowner investment capacities and partly on public funding. In this case, this would be facilitated through access to external support from e.g. EU funds or other international bodies as a complement. As a result, one of the perceived barriers is the lack of cooperation and coordination and the reluctance from individual private actors who do not see a clear flow of benefits from implementation (or incentives). Therefore, providing a more explicit list of benefits, as well as additional support or clear incentives for individual private actors through different mechanisms may help get closer to a viable implementable project, including e.g. the compensation or payment for the co-benefits generated. Across the world these incentives have included a range of options like for example, subsidies from cities or regional governments to support these investments, backing to the maintenance expenses, or to the abatement of surface water charges/fees, among others (Ossa-Moreno et al. 2017). In the case of farmers, Payments for Environmental Services have been used widely to support farmers to adopt pro-environmental practices. However, in the context of NAS schemes, payments to reduce flooding risk have not been widely developed so far based on the avoided damages and co-benefits as the NAS propose, with a source of revenue coming from the anticipated avoided damages and costs.

Medium scale case studies – namely Lez, Brague and Glinščica – focused on the river catchment or sub-catchment and surroundings within a smaller area of influence. These cases report groups of municipalities and water management institutions as the main problem owners and potential implementing agents, and therefore a stronger public role. Hence, funding is mainly focused on public sources through specific (and innovative) tax mechanisms like the GEMAPI tax[2] in France, government funds (including national funds), and external funding from international organizations (e.g. EU funding). An interesting and pioneering example in this sense is the Barnier fund in France (see Marchal et al., this volume Chap. 3), which shows the active role played by the insurance sector with a mandatory contribution to fund NBS to reduce risks, as investors that buy into prevention aware of the magnitude of potential future losses and the benefit from early action (to prevent is

---

[2]The GEMAPI tax is a recent tax levied at the municipal level to fund measures aiming at the prevention of floods and the management of aquatic ecosystems. This tax was created to support the transfer of this competence from the State to Municipalities, undertaken in the framework of the decentralization process.

better than to cure). Part of this investment could also be into the assurance value of ecosystems to deliver their resilience dividends. Meanwhile, in the case of Glinščica, external funding and perceived interest are considered as the critical drivers to determine the type of agent finally taking the initiative to implement the strategy (either as an NGO, a government or a private entity).

At the small city scale – namely Copenhagen, Rotterdam and Lodz-, most initiatives identify the municipalities as main promoters along with some private investments by neighbourhood communities, private sector or businesses in certain cases (i.e. Lodz). Funding strategies include indirect funding through citizen taxes, external funds from international organisations (e.g. EU grants), or community investments. It is interesting how in the case of Lodz, some public funds from the Municipality have been allocated as 'civilian budgets' to citizen organisations, such as 'Housing cooperatives', to undertake some of the interventions benefitting collectives within a certain part of the city. This kind of public-private partnerships have been important to engage the citizens and speed up the implementation of NBS in buildings in the city.

Finally, the role of **legislation** stands out as a critical element that can play either as a driver or a barrier depending on the context. The EU legislation (particularly the Water Framework Directive and Floods Directive) is found to be a strong driver and support for the implementation of NBS in most of the case studies. Another highly mentioned set of rules include the land use and rural/urban development agendas that push for innovation towards sustainable development. These pieces of legislation provide the enabling frame favouring the introduction of NBS within the investment and intervention programmes, as their comparative environmental benefits usually align with their overall strategic objectives. On the contrary, dialogues with case studies' stakeholders revealed that strict rules and protocols on public procurement at the national and municipal level play against public initiatives to invest in NBS. Such protocols and the associated eligibility standards are usually designed for well known traditional infrastructures with short term returns of investment, which often cannot be met with NBS even if the net final benefits are higher. This can hamper the initiative of both interested administrations willing to test solutions, and proactive ones aiming to upscale and mainstream successful pilots, that hold back due to cumbersome or even unsolvable bureaucratic burdens.

## 8.5.2   Lessons Learnt from the Modular Co-design Process: Transferability of the Method

The application of the NAS canvas to several case studies, regardless of the context and project stage, showed the flexibility and replicability of the tool, which can be applied to any NAS and NBS strategies in different contexts. Furthermore, the tool could be also applied to NBS strategies that are not primarily aimed at risk reduction, such as climate change adaptation, by changing the main problem to be addressed and its main value. The tool is easy to use and is focused towards scientists, technicians, project promoters and public bodies who are interested in exploring possible business model alternatives for an NBS strategy or a specific NBS in a particular project (including hybrid options mixing green and grey). The project stage should determine the level of detail of the information to be included in the canvas. In the earlier stage of the project, a qualitative description may be sufficient while a fully quantified characterization should be pursued for projects in the last stages of the co-design process (see Chap. 9). The context should determine the complexity required for the various components, such as the regulatory framework, the implementing partners, the governance and institutional arrangements, and the impact indicators to be estimated. For instance, in the case of developing countries, where the biophysical data or records on disaster damages may be scarce or non-existing (UNISDR 2014), the level of detail or accuracy of the value proposition and impact estimates may be lower. Tools like eco:actuary are particularly well suited for these contexts (see Chap. 4 this volume). This may also occur with projects in an early stage for which there are still some design uncertainties (see Chap. 19 on readiness levels). In these cases, the usefulness of the canvas as a tool is to provide a comprehensive and structured set of elements to guide promoters in designing an operational business model, by eliciting the value and impact of the NAS. At this stage, it can help in diagnosing the information gaps and missing elements required to build the business model, that will be also required further on as a basis to develop the business case for investors (see Chap. 9 this volume). The co-development and co-design process at the heart of the tool working hand by hand with the stakeholders can help raise awareness and buy in. It can help to elicit and document in a structured format the needs, interests and potential roles of each stakeholder. This in turn can be critical as shown in Chap. 5 (this volume) to identify trade-offs and strengthen synergies as well creating the conditions for collective action. This can be critical to engage stakeholders to invest resources (time, financial, knowledge, …) in the process of gathering the missing information, thus lowering collectively the transaction costs that often hamper smaller projects. Meanwhile, it may also help to structure a robust justification on the information needed with a view to apply for funding from e.g. an international body to undertake the preliminary assessments required for a feasibility study (like in the case of Europe a natural capital financing facility).

## 8.6 Conclusions

Overall, the application of the NAS canvas to nine case studies enabled us to elicit together with the stakeholders, in a co-design process, the value proposition of a wide range of NAS schemes based on a range of NBS strategies for different contexts. We also built a map of actors and actions required to pave the way towards their implementation. This single, visual compilation of the expected values, required resources, actors and roles, possible funding streams, regulatory framework and battery of indicators to measure performance, provides with a strong and comprehensive foundation to help showcase the feasibility and potential impacts from an NBS intervention, and advance towards developing the full business case and implementation arrangements (see Chap. 9). Meanwhile, from a co-design and process perspective, the application of the NAS canvas to the case studies, regardless of the context and project stages, showed the flexibility and adaptability of the tool. This could help with the replicability of NAS, enhancing the potential to develop NBS strategies and specific NBS and hybrid interventions to different contexts.

A few transversal highlights came out from the horizontal analysis of business models for NBS strategies across case studies.

First, it is an important lesson from the application of the tool that a key aspect is **to also engage indirect beneficiaries** within the pool of payers and funders, since often wider society benefits from these NAS schemes. This is in line with the role played by co-benefits in the value proposition of NBS strategies, and the fact that most of the value generated is related public goods and services, which often do not have a market. Most business models are oriented towards the generation of a good that has a market and a stated willingness to pay by clients, this in turn makes the revenue stream and capacity for reimbursement much clearer for potential investors. The fact that risk reduction and most co-benefits are public goods and/or are highly dispersed makes this valuation of willingness to pay more complex as well as its transformation into effective revenue streams.

This work also pinpoints the critical role that regulation can play in setting better rules of the game, acting as a lever for collective action aligning incentives, or making it possible to align incentives, rather than become a burden or a barrier. The legislative provisions provide the enabling frame that give investor confidence and stability, and with the new taxonomy of sustainable finance, as a strong message to tip the balance in favour of the introduction of NAS schemes as potential investments and intervention programmes. Therefore, its formulation and application at the national and local level and accompanying procedures (e.g. procurement and licensing) need to be adapted to include new types of interventions like NAS Schemes.

# References

Coles NA, Tyllianakis E (2019) Deliverable D7.3NBS Market Potential through Synergies at International Level: business plan case studies and scope for international mainstreaming. EU Horizon 2020 Think Nature project, Grant Agreement No 730338
EEA (2016) Flood risks and environmental vulnerability. Exploring the synergies between flood-plain restoration, water policies and thematic policies. Luxembourg: Publications Office of the European Union
Graveline N, Joyce J, Calatrava J, Douai A, Arfaoui N, Moncoulon D, Mañez M, De Ryke H, Zdravko K (2017) DELIVERABLE 4.1 : General Framework for the economic assessment of NBS and their insurance value. EU Horizon 2020 NAIAD Project, Grant Agreement N°730497

Mc. Quaid S (2019) Nature-based solutions business model Canvas Guide volume. EU Horizon 2020 Connecting Nature project, Grant Agreement No 730222

Ossa-Moreno J, Smith KM, Mijic A (2017) Economic analysis of wider benefits to facilitate SuDS uptake in London, UK. Sustain Cities Soc 28(2017):411–419

Osterwalder A, Pigneur Y (2010) Business model generation. Wiley & Sons, Canada

Somarakis G, Stagakis S, Chrysoulakis N (eds) (2019) ThinkNature nature-based solutions hand volume. ThinkNature project funded by the EU Horizon 2020 research and innovation programme

Topoxeus HS, Polzin F (2017) DELIVERABLE 1.3 part V: characterizing nature-based solutions from a business model and financing perspective. EU Horizon 2020 Naturvation Project

UNISDR (2014) Progress and challenges in disaster risk reduction: a contribution towards the development of policy indicators for the post-2015 framework on disaster risk reduction

WEF (2019) World economic forum: global risks report, 2019

# Chapter 9
# Closing the Implementation Gap of NBS for Water Security: Developing an Implementation Strategy for Natural Assurance Schemes

**Mónica A. Altamirano, Hugo de Rijke, Begoña Arellano, Florentina Nanu, Marice Angulo, Camilo Benítez Ávila, Kieran Dartée, Karina Peña, Beatriz Mayor, Polona Pengal, and Albert Scrieciu**

## 9.1 Introduction and Conceptual Frame

Evidence recorded over the last decade indicates that we are about to reach or have already reached a tipping point related to climate change. The Global Commission on Adaptation (GCA) (2019) report stated: "Climate change is one of the greatest threats facing humanity, with far-reaching and devastating impacts on people, the environment and the economy". The frequency of extreme events keeps increasing. In terms of overall losses, 2017 was the second-costliest year ever for natural disasters. Overall losses in 2017 (US$ 330 bn) were far greater even than those in the extreme years of 2005 and 2008. Only in 2011 higher loss figs. (US$ 350bn) have been recorded and they were related to the Tohoku earthquake and floods in Thailand. The share of insured losses (US$ 135 bn) is the highest figure in the period from 1980 to 2017.

M. A. Altamirano (✉)
Deltares, Delft, The Netherlands

Faculty of Technology, Policy and Management, Delft University of Technology,
Delft, The Netherlands
e-mail: monica.altamirano.dejong@gmail.com

H. de Rijke · B. Arellano · F. Nanu · M. Angulo · C. Benítez Ávila
Deltares, Delft, The Netherlands

K. Dartée · K. Peña
Field Factors, Rotterdam, The Netherlands

B. Mayor
Icatalist, Madrid, Spain

P. Pengal
Institute for Ichthyological and Ecological Research (REVIVO), Dob, Slovenia

A. Scrieciu
REVIVO, Institute for Ichthyological and Ecological Research, Ljubljana, Slovenia

© The Author(s) 2023                                                                              149
E. López Gunn et al. (eds.), *Greening Water Risks*, Water Security in a New
World, https://doi.org/10.1007/978-3-031-25308-9_9

Munich Re NatCatSERVICE recorded 710 relevant loss events, which is above the average of 605 events per year of the last decade and much higher than the average of 490 events over the last 30 years (Munich Re 2018). According to the GCA, rising seas and greater storm surges could force hundreds of millions of people in coastal cities from their homes and generate losses of more than USD 1 trillion yearly by 2050 in coastal urban areas. Meanwhile, a 2016 World Bank report indicates that the impacts of Climate Change will be channelled primarily through the water cycle and that water scarcity could cost some regions up to 6% of their GDP.

In the context of a climate and water crisis, and intensified by the Covid19 crisis awareness about the need to rethink our economic development paradigm has increased. Against this context the potential of Ecosystems and Nature-based Solutions (NbS) as important pieces of a new regenerative economic model and as important allies to mitigate water risks is being more and more recognised.

This approach is understood as the enriching of the traditional infrastructure planning process with green and hybrid (green and grey) solutions along with traditional grey infrastructure. Green infrastructure is defined by the World Bank (2019) as a subset of nature-based solutions (NbS) that intently and strategically preserves, enhances, or restores elements of a natural system to help produce higher-quality, more resilient and lower-cost infrastructure services. Green infrastructures are multi-functional and adaptive, making them a promising and robust long-term solution. Due to their characteristics, they can contribute to climate adaptation as well as to climate mitigation. They can provide a cost-effective approach to address deep uncertainty related to climate change by avoiding or delaying lock-in to capital-intensive infrastructure, allowing for flexibility to adapt to changing circumstances (OECD 2013).

The challenge is that, while the potential NBS and green infrastructure is increasingly acknowledged, and it is well positioned in the political agenda of the European Commission, Multilateral Development Banks and governments, the reality is that in many regions the implementation of these solutions at watershed scale remains latent. In most cases green infrastructure is still implemented solely as pilot projects removed from mainstream procurement strategies. Even in countries at the forefront, like Peru, where funds are being collected to invest in watershed protection and ecosystem conservation for water supply, the implementation of projects at scale is still an operational and procurement challenge.

In order to close this implementation gap this chapter presents guidelines to develop an implementation and financing strategy for natural assurance schemes, and for the implementation of Nature-based Solutions for Water Security in general (Altamirano et al. 2021). Following the Financing Framework for Water Security (Altamirano 2017, 2019) these principles have been further tailored and developed with additional elements to fit the innovative nature of NBS projects for which important evidence and information gaps remain, e.g. the expected and typical cash-flow and risk profiles of green and hybrid (green-grey) projects and the levels of service they can guarantee over time.

Summarising, our aim has been to develop a methodology that supports and enables the proponents of green infrastructure to structure and shape their project proposals as investable propositions, in a way and a language that appeal to either public or private investors. Our approach offers an interface between the project delivery and finance community and the water resources planning and watershed conservation communities.

In this chapter we present the basic methodological elements of our approach and the process it involves, as well as an illustration for one of the three demonstration cases we have supported to develop an implementation strategy. This is one of the three EU case studies where the framework was appplied to develop an implementation strategy. The three EU case studies are micro-wetlands in Rotterdam in the Netherlands, Medina del Campo Groundwater body (GWB) in Spain and Potelu wetland in the lower Danube in Romania which are presented in Chaps. 11 and 16 (this volume). To finalize conclusions and recommendations about what is needed to move ahead towards implementation at scale of NBS for water security in Europe are presented.

## 9.2   Financing Framework for Water Security

An important goal in relation to natural assurance schemes was also to enable the step from adaptive planning towards investment planning. For plans and projects of any type to be able to access funding and financing, it is essential to justify why the proposed investment optimises the use of scarce public and/or private funds. It is also very important to provide evidence that shows that the proposed investment(s) in NBS and the way these NBS will be procured will optimise Value for Money (VfM). In other words, the case for investment needs to be made.

The Financing Framework for Water Security supports the aforementioned objective by setting in motion a process that bridge the existing gap between adaptive planning and investment planning phases. In the adaptive planning phase both the strategic case – the need for change- and the economic case- on why the preferred strategy – NAS- will optimise the use of scarce public funds (see Chap. 6 Le Coent et al., this volume) are made. The framework then guide within the investment planning phase the further definition of the commercial case: how to organise the program so as to make its implementation achievable and attractive for market players (large companies as well as SME's); the financial case: is the program affordable for the local and national economy? And the management case: how could these concepts and the entire program be delivered successfully and by whom? (Public, Private, and civil society actors).

A crucial element towards the development of the five business cases: strategic, economic, as well as commercial, financial and management business case (see Fig. 9.1), is the development of a suitable implementation arrangement per measure. The FFWS guides the stakeholder involved in a planning process in designing an implementation arrangement for water security projects and natural assurance schemes including the development of a governance structure, a funding strategy, a financing and procurement strategy. This means considering a number of elements, namely: (a) the transaction (e.g. type of good and financial as well as physical project characteristics), (b) the level of service required over time and (c) the institutional setting (stakeholders, strengths of local government, private sector and community, the incentives created by formal and informal institutions and the

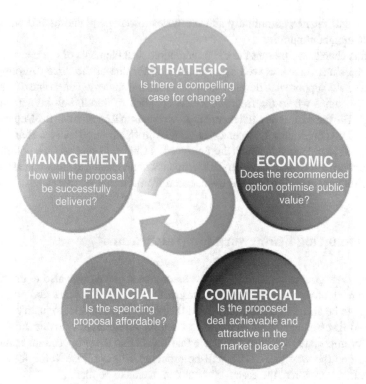

**Fig. 9.1** The five business cases for public investments. (Source: UK HM Treasury 2018. More information available at: https://assets.publishing.service.gov.uk/government/uploads/system/uploads/attachment_data/file/749086/Project_Business_Case_2018.pdf)

insurance and re-insurance schemes that apply) – and considering lessons learned from best practices worldwide, they can choose from a wide range of project delivery and finance options that vary from purely public governance options up to the creation public governance options up to the creation of regulated markets for private initiatives and innovative business models to emerge. The implementation arrangement(s) with the highest potential to ensure sustainability in service delivery in the long term are then considered as base for a further process of design and project structuring.

The four main stages of analysis to design an implementation arrangement to follow are presented in Fig. 9.2 and Box 9.1. For more detailed guidance on the process to gradually advance the five business cases through the process of strategic planning for water security, please check the Handbook for the Implementation of NBS for water security (Altamirano et al. 2021).

It is important to clarify that while on the one hand the input to this first phase is expected to be a preferred strategy for water security, for which there is a clear strategic and economic case; on the other hand the further specification of a hierarchy of services to be provided by the strategy and/or specific green infrastructure

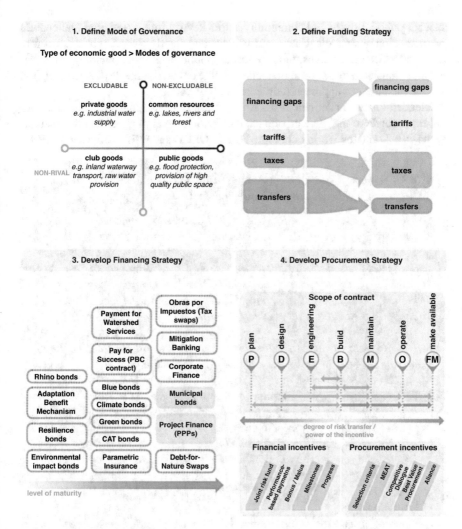

**Fig. 9.2** Main components of the implementation arrangement. (Source: Altamirano et al. 2021, p. 23)

investments and the potential sources of revenue helps to further shape the strategic case of the investment programme being considered and may even lead to significant changes in the solutions being thought as part of this preferred strategy. Box 9.1 presents the four more important steps considered in this process of designing an implementation arrangement.

Making use of system analysis, group model building and other collaborative techniques along with principles of New Institutional Economics, the FFWS enables a process of transdisciplinary collaboration to design fit for purpose implementation mechanism for water security projects and strategies. This process involves all relevant public, private and community actors key for implementation and enables the translation of strategic water security plans into clearly phased hybrid infrastructure clusters that can be absorbed by formal public investment planning processes and then translated into several financially viable or even bankable deals making use of a blended finance approach (Altamirano 2019, p. 7).

Blended finance is defined as the strategic use of development finance and philanthropic funds to mobilise private capital flows to emerging and frontier markets by the OECD and the World Economic Forum (OECD and WEF 2015; OECD 2018).

---

**Box 9.1 Steps to Design an Implementation Arrangement According to the Financing Framework for Water Security**

*Step 1: define the main services the project will create and categorize this in types of economic goods. It is important here to bear in mind that we categorize the services the asset created by the project delivers, not necessarily the asset itself. For example, a forest may provide services that can be considered private (such as reduction of sedimentation rate of hydropower plants), yet the forest itself may be a public good. This categorization enables the identification of which types of funding could be appropriate to ensure cost recovery.*

*Step 2: Funding strategy: the funding of a project could be either public or private. In general terms, the main sources of funding are what the OECD called the 3 T's: Taxes, Tariffs or Transfers. Once the sources of funding – who ultimately pay for the project- are determined the mechanisms to arrange capital upfront (financing) and how to place the project on the market (procurement) are selected.*

*Step 3: Financing strategy: depending on the type of project and whether the project sponsor is public or private, a variety of financing instruments could be used. In the graph below we show for example a variety of innovative financing instruments for Climate Adaptation and DRR (Altamirano 2019).*

*Step 4: Procurement strategy: which refers to how the government agency or private project sponsor responsible for the project can choose to make use of or to purchase the project. The graph shown here applies mainly to public infrastructures, while other sectors or types of transactions may need a different approach, such as the design of regulated markets or bottom-up community-based initiatives. At is shown in this graph in case of public procurement of infrastructures the government may choose to tender it as a fully integrated contract (e.g. involving the private sector from planning up to Operation and Maintenance) or choose for more traditional separate ones.*

(Source: Altamirano 2019, p. 13)

## 9.3    Green Versus Grey Infrastructure Projects: Structuring Investable NBS Propositions

Multiple factors slow down the rate of adoption of NbS for water security. Some of the more often cited are uncertain performance, higher (real and perceived) risk and an unattractive cash profile of NbS projects. However, the most fundamental challenge is that most public and private investment planning processes are geared towards grey infrastructure "projects" as investment units and do not fit the characteristics of natural infrastructure investments. This section presents how natural infrastructure is seen through the lens of the proponents of this approach versus the lens used by investors. The way hybrid infrastructure strategies are seen by eco-engineers and proponents in general versus financers and project developers create an important divide in language and interests. The criteria they both apply to judge the potential of green and hybrid versus grey-only infrastructure strategies are fundamentally different.

It is important to clarify that whether the project developer could be public or private, does not make a significant change in this divide; the only difference could be the capacity of the public project developer to carry more risks and financial losses than the private one. Our objective with the FFWS is to enable NbS proponents to engage in strategic planning and investment planning processes and work more effectively together with project developers, project sponsors and financiers.

### 9.3.1    Cost-Effectiveness of NBS Versus Grey Infrastructure

Here an important aspect to consider is context specificity. That is for the calculation of life cycle costs and comparison of NBS versus grey solutions and strategies "green infrastructure design and performance is generally more context-specific than grey infrastructure. NBS solutions for DRR need to be designed and built to fit the soil, terrain and hydrological conditions of each individual site" (American Rivers 2012, p. 9). For NBS projects this difference means, on the one hand, greater complexity and uncertainty in ex-ante cost estimations and cash profiles, and in the other hand often a greater value from addressing wider local concerns and values (Altamirano and de Rijke 2017), i.e. a wider set of co-benefits (see Chap. 6 Le Coent et al., this volume).

### 9.3.2    Cash Profiles of Green Versus Grey Infrastructure

Cost-benefits comparisons made of NBS solutions versus grey infrastructure for example for stormwater management; have found the following advantages of NBS versus grey infrastructure projects in terms of Total Costs of Ownership (American Rivers 2012, p. 9):

- Reduced built capital (equipment, installation) costs
- Reduced operation costs (e.g. energy costs)
- Reduced land acquisition costs
- Reduced repair and maintenance costs
- Reduced external costs (off-site costs imposed on others)
- Reduced infrastructure replacement costs (potential for longer life of investment)

Nonetheless, NBS have unique financing challenges inherent to their cashflow and risk profiles. Benefits are often unique, delayed, dispersed, non-guaranteed and non-financial, complicating the estimation of an internal rate of return (IRR). With respect to costs, capital expenditure is often spread over a longer-term, in comparison to grey solutions as construction time or time to reach functionality is often longer for green versus grey infrastructure.

While Total Costs of Ownership (TCO) are expected to be lower for NBS versus grey infrastructure in the long term, it is also important to consider the differences in the perceived risk profiles of green versus grey and the impact that will have on the cost of capital and on the "risk premium" to be charged by implementing parties to the procurement agency when opting for green versus grey. This will be especially the case in the earlier years of transition towards a hybrid infrastructure market, when risk perception will remain high and companies that engage in providing these NBS solutions will not have the required track record to prove to financiers that these companies have full overall control of construction and performance risks.

The multi-functional and innovative nature of green versus grey makes the financing of NBS solutions at scale significantly more challenging. Nevertheless, the specific characteristics of NBS also result in a net positive impact for on-site aesthetics and other co-benefits has often proven beneficial to generate new funding sources since these positive impacts and other co-benefits increase the willingness to pay from people in the immediate vicinity of these solutions. For example, in Portland, Oregon, residents were more willing to invest for on-site stormwater projects that brought scenic and other direct additional benefits (American Rivers 2012).

Our approach proposes a structure process to shape NbS projects and design fit for purpose implementation arrangements that improve the cashflow and risk profiles of NBS projects, enable the conversion of co-benefits into additional revenue sources and keep transactions costs and implementation risks inherent to multi-functional projects at a minimum.

### 9.3.3 Specifying Multiple Levels of Service: A Hierarchy of Functions to Guide Trade-Offs

A main advantage of NBS is that they can fulfil multiple functions. This also means that when NBS strategies are structured as investment projects, these may translate into projects that are contracted by multiple principals (public and/or private). As trade-offs between the functions provided by the NBS strategies may be expected, this could easily translate into significant contractual risks, during both the construction and operation of these projects.

To reduce these eventual contractual risks while increasing the possibility to monetize more co-benefits of NbS we propose a number of collaborative modelling protocols that help clarify:

- **Hierarchy of functions:** specifying which combinations of measures (green, grey and non-structural) ensure together 2–4 main functions; and then make clear how to prioritize in case of trade-offs between them. The final prioritization is a function not only of the physical processes, but ultimately a social construct that is influenced by how active different problem owners are and which function is valued more by public and/or private beneficiaries
- **Function curves, Life Cycle Costs (LCC), cashflow and risk profiles of natural infrastructure measures:** the function curves, risk matrixes and LCC of grey infrastructures are often well known, however that is not the case for green infrastructure. A wide variety of technical expertise (e.g. ecology, morphology, civil engineering, and so forth) and simulation models need to be considered to arrive to the definition of these variables which ultimately shape the cash and risk profile of these hybrid investment projects.

These two elements set basis for further in-depth analyses and will lead to the identification of alternative revenue generation strategies (funding strategy) and the choice of a family of implementation arrangements. Depending on whether the services provided – not the assets- can be considered public, toll, common resources or private goods different sources of funding would apply; tariffs can be applied to private and toll goods and taxes or transfers would be required to fund public services. Then depending whether taxes, tariffs or transfers are identified as the most important source of revenue as well as whether the public or the private sector will be the main project sponsor, different types of implementation arrangements will be considered for further development of the full business case.

More specifically investment in NbS for water security and watershed conservation could take any of the following four forms:

1. Public procurement contracts, which includes traditional Design-Bid-Build contracts but also Public-Private Partnerships and even unsolicited proposals made by the private sector but that require concession rights from the government authorities
2. Privately driven water stewardship investments,
3. Collective investment vehicles, and
4. Environmental and/or ecosystem markets

Although the design process will vary for different types of implementation arrangements, in most cases, investments will lead to investment projects and/or the delegation of operation and maintenance activities to third parties. Whenever a public or private entity needs to implement the envisioned activities, these entities will need to decide whether to implement themselves or to delegate implementation to another party: public, private or community. In that sense, independent of whether the choice for implementation arrangements is 1,3 or 4 (as above); the project sponsor will have to make financing and procurement choices. For doing so, this chapter presents the process to guide them in selecting the project delivery and finance mechanism that reduce transaction costs and ensure the right incentives are created for sustained service delivery (Altamirano 2019).

## 9.4 Spain, Medina Del Campo Aquifer Recovery as Illustration

The illustration presented here is a summarised version of the case presented in the Handbook for Implementation of NBS for Water Security (Altamirano et al. 2021). The NAS in question is the Medina del Campo aquifer, a groundwater body in Central Spain extending beneath Southern Valladolid and Northern Avila provinces. The area covering 3700 $km^2$ is highly impacted by droughts, groundwater exploitation, and degradation of the surface riverine ecosystems along the Zapardiel river. Climate projections indicate that these conditions will worsen in the future and probably threaten the economic wellbeing of the region, which is highly dependent on agriculture. A collaborative process with water users and related stakeholders has resulted in the identification and planning of 5 measures: aquifer recharge, technological transformation of fields, alternating crops, water abstractions control and other governance measures including the constitution of WUAs (water user associations). While the technological transformation of fields was not considered originally as part of the strategy within the NAIAD project, the analysis undertaken by Deltares, including the results of the first stakeholder engagement workshop found out this to be a critical component for the overall success of the NbS programme. Therefore, it was decided to include this measure as part of the preferred strategy in the design of the project preparation process.

The FFWS for the Medina del Campo case was implemented during the process of building commitment with water users, and during the later stage of strategy building for complying with the Water Framework Directive targets for groundwater. The assessment of existing data was a collaborative process between different NAIAD demo partners, the Duero River Basin Authority (CDH) and the research institute Deltares. Additionally, the findings from the NAIAD project and the FFWS application could be of use for the further design of the LIFE Integrated Project lead by the CHD. This LIFE-IP RBMP-Duero project aims to implement a river basin management plan in the central-south part of the Duero river basin, including the Medina del Campo area.

The most crucial success factor for successful implementation in the Medina del Campo case relates to behavioural change by agricultural water users, and how to effectively incentivize them to make significant changes in their agricultural practices. Existing traditional practices have compromised the sustainability of water resources in the long term. Given this key implementation challenge in this application of the FFWS relatively more attention was paid to the non-structural measures or soft components of the NbS strategy and the process included an in-depth institutional analysis.

### 9.4.1   Strategic Case: Theory of Change and Enabling Environment

Spain has been exposed to significant simultaneous changes, which have challenged water management efforts nationwide. On the one hand, European regulation requires from member parties the compliance with more demanding environmental goals. On the other hand, the lack of demographic retention in the rural areas and an aged farming sector affect this and other regions in Spain and set an important constraint for the implementation of the proposed measures.

Main drivers for implementation of an NbS strategy are to reduce water consumption by 25%, to restore ground-water-related ecosystem services, to improve water supply quality now affected by arsenic contamination, and to reduce flood and drought risk and other related risks such as landslide. The initiative stems from the strategic goals and responsibilities of the CHD to comply with European regulations and national water planning. The enabling environment is given by the structure of water rights, and the Water Framework Directive. Accordingly, the problem owner is the Duero River Basin Authority (CHD), as the authority in charge of water planning and the enforcement of the Water Framework Directive (FD).

In previous decades, the CHD granted water rights over the aquifer in a time when the knowledge on aquifer dynamics was rather scarce. Therefore, there was an overprovision of water rights on the aquifer. The situation as is now is presented in a Causal Loop Diagram (Fig. 9.3).[1]

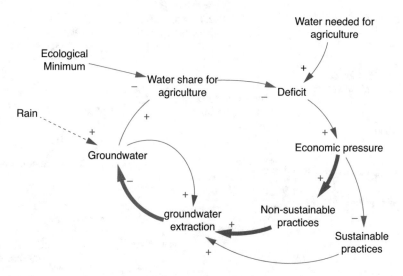

**Fig. 9.3** Business as Usual situation in Medina del Campo

---

[1] See Lane 2008 for getting familiarity with CLD representation of complex systems.

There is a balancing loop between water availability, the higher levels of ground-water the higher water extraction reducing the existing levels of water in the GWB. Water extraction is driven by agricultural production. A share of this produc-tion is the result of unsustainable water use practices, which is driven by economic pressure faced by farmers due to extremely low prices. Non-sustainable practices imply higher rates of water extraction, and consequently, it is represented with a thicker arrow. The economic pressure increases as there is a perception of the water deficit between water needs and availability, competing with the ecological mini-mum. As the positive contribution of rain to water levels is rather insufficient to balance water extractions, it is represented with a dotted line.

The NAS strategy proposed.

to introduce a change in the way water is managed towards a more sustainable water use regime includes:

1. Aquifer recharge (structural measure).
2. Formation of Water Users Association (non-structural, governance measure);
3. Control of abstractions; (non-structural measure to increase enforcement).
4. Transformation of the fields and.
5. Introducing alternating crops.

As presented in Fig. 9.4, these interventions aims at reinforcing sustainable prac-tices, reducing the needs of water and physically contributing to water stock in the GWB.

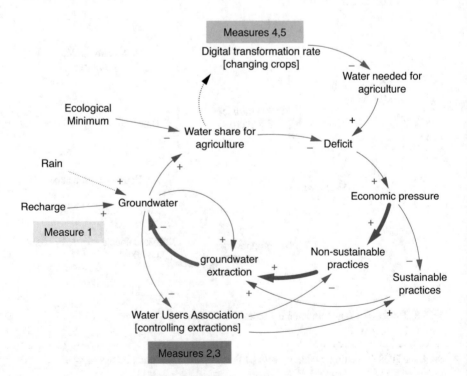

**Fig. 9.4** NAS strategy to achieve sustainable water use in Medina del Campo

## 9.4.2   Economic Case: Winners and Losers

The most important and direct benefit that results from the implementation of the NbS strategy in Medina del Campo is the reduction of drought risk and associated impacts for the agricultural sector. As agriculture is one of the main economic activities in the region, a reduction of this risk impacts directly economic resilience.

As previously explained the NbS strategy aims to reduce in the long-term water stress by conserving aquatic ecosystems, terrestrial ecosystems and wetlands protected under Natura 2000 policy. By balancing environmental and economic goals, the NbS investment programme is expected to contribute to the region goal of retaining youth and may also contribute to more young people becoming active in a new modern agricultural sector. In the medium to long term the program aims to avoid a potential future social conflict that could be triggered if aquifer condition worsens and is declared over-exploited.

The sector most impacted by the implementation of the measures in the short term is the agriculture sector, particularly the farmers, although it will also affect the whole agroindustry value chain. The paradox is that this is also the sector that will benefit the most in the long term with a more reliable and sustainable water provision model. Other interested groups include the environmentalist organisations, as well as business owners linked to the agriculture sector. The identification of pains and gains for different actors (see Table 9.1) was established upon the interpretation of interviews made to CHD officers, a representative of farmers and a representative of the Castilla y Leon Autonomous Community.

## 9.4.3   Commercial, Financial and Management Cases

Given the future scenarios of water scarcity, the focus service is reducing water consumption. Funding and governance have two main sources. The first one through centralized procurement and using the budget available from the Duero River Basin Authority. Another important source of funding emanates from the European Union level, where the environmental goals reached by the measures are the priority. This budget is also managed in a centralized manner and will be driven by the fulfilment of performance indicators in the aquifer, and effectiveness of the governance goals implementation, e.g. degree of parcels encompassed in a WUAS. Being that the service and benefits constitute a public good, the possibility of putting a tax scheme in effect is considered feasible and desirable. Some income has already been inputted by the water rights and their subsequent responsibilities. Figure 9.5 summarises the service hierarchy, funding and governance structures related to each of the three main functions the NAS strategy includes.

The implementation arrangement was structured according to the procurement practices of public commissioners: Medina de Campo municipality and the CHD. The delivery and proper operation of the aquifer recharge system is a responsibility of the Municipality, as such they will act as commissioner for this part of the NbS strategy. The **governance modes** that will be used for the implementation of

**Table 9.1** Pains and gains of existing value chains due to the implementation of NbS in Medina del Campo

| Sector | Target group | Winner/ loser | BAU-2050 | | Solution-2050 | |
|---|---|---|---|---|---|---|
| | | | Pain | Benefits | Pain | Benefits |
| Farming | Farmers | Loser short-term Winner long-term | Yes | No No extra costs or changes to their current exploitation schemes. Unrestricted exploitation in the short term. | Yes (short term) Reduction of water consumption and compliance with regulations | Yes (long term) Sustainable and reliable provision of water- avoiding abrupt discontinuation due aquifer overexploitation |
| Environmental | Environmental groups | Winner | Yes Acquire and ecosystems exploitation and loss | No | No | Yes Avoid ecosystems degradation and recovery of the aquifer |
| Commerce | Local business owners | Neutral | Yes | No | Yes (partially) | Yes |

Source: Altamirano et al. 2021 p. 199

this is part of the NbS strategy is therefore **public procurement contract**. Taking this into account, the CHD and the Municipality can develop further with support from EU innovation partners the specifics of the procurement strategy, including the **scope of contract**, financial incentives to consider in the payment mechanism as well as **procurement incentives** built in the awarding procedure (Fig. 9.5).

Given the innovative character of the solution, it is expected that the municipality will keep control over design and then delegate the responsibility for building the solution and possibly operate it to the winning private company or consortium.

Finally, both the municipality and the CHD oversee the management of water-related disaster risks such as droughts. Table 9.2 gives an overview of possible implementation arrangements for Medina del Campo NbS programme.

## 9.5   The Way Forward

As we advanced in the implementation of the FFWS and it further development to respond to the needs of our demonstration cases in NAIAD, we have observed that the demo leaders and the proposers of green infrastructure in at least half of our demonstration cases and therefore also the NbS they propose were not yet part of the formal public planning and investment programming process. In many cases the proponents of NbS are organisations active in advocacy and/or academic work and often with little familiarity with public and private investment planning processes. As a result, there is an implicit bias to shape these projects towards the creation of awareness, and less towards demonstration of their revenue generation potential. Our methodology has therefore supported demo leaders in considering how to move forward towards implementation and scale and restructure demonstration cases to create the investment case for public and private sectors alike.

### 9.5.1   The Missing Link: A Full Business Case

For plans and projects to access funding and financing is necessary to prepare a full business case for the entire investment programme and each of the projects that make part of it. Unfortunately, in most cases the proponents of NbS are organisations with an advocacy and/ or scientific background with limited involvement in public and private investment planning processes. As a result, often NbS pilots and demonstration projects are shaped more as awareness raising projects than as "investment projects" that could attract funds from either public authority aiming at reducing a risk, or private impact investors willing to accept lower returns in exchange for social and environmental impacts.

The criteria and level of detailing regarding implementation costs and risks differ greatly between the project descriptions of NbS proponents and the requirements for allocation of public funding or granting of loans by impact investors. In simple

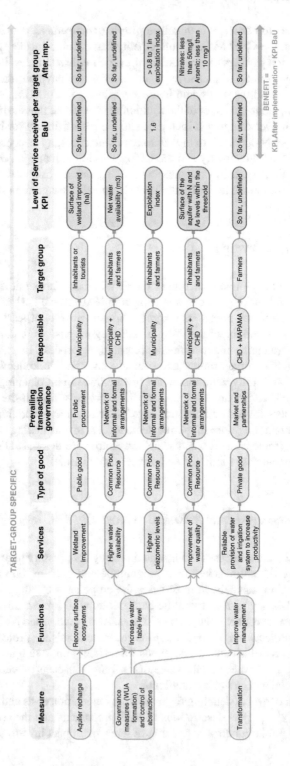

**Fig. 9.5** Service hierarchy, funding and governance structure for Medina del Campo. (Source: Altamirano et al. 2021 p. 200)

**Table 9.2** Implementation arrangement for NbS strategy in Medina del Campo

| Measure | Commissioner | Main tasks associated | Public-in-house/procure in the market/assumed by the market/assumed by a network | Degrees of private managerial freedom |
|---|---|---|---|---|
| Assessment of water quality (service: water quality) | CHD | Collection of information and inventory of water capture points | Assumed by a network | N/A |
| | | Geological mapping and survey, Piezometric -flow measurement and hydro-geochemical studies. | Procured in market | Low |
| | | Study of infrastructures and devices available for artificial recharge. | Public in-house | High |
| | | Tests of recharge devices and complementary construction works. | Assumed by a network | N/A |
| | CHD | Extraction inventory and checking correspondence between the authorized volumes and collected data | Public in-house | N/A |
| | CHD | Locate parcels whose crops are suspiciously been irrigated with groundwater extractions | Assumed by a network | N/A |
| Water recovery and resource efficiency (service: water availability and quality) | CHD | Creation of a recreational area near the Medina del Campo town | Procured in market | High |
| Artificial recharge of the groundwater body Medina del Campo (measure: Aquifer recharge) | CHD | Design: Civil engineering consulting company | Procured in market | Low |
| | | Construction works [CW] removing of waterproofing layer (soil and infiltration works) | Procured in market | High |

(continued)

**Table 9.2** (continued)

| Measure | Commissioner | Main tasks associated | Public-in-house/procure in the market/assumed by the market/assumed by a network | Degrees of private managerial freedom |
|---|---|---|---|---|
| Compensation payments for environmental services and climate change mitigation and adaptation services (measure: CUAs + governance measures) | CHD | Agroforestry programme: creation and management of new forested areas in currently cultivated plots | Assumed by a network | N/A |
| | | Assessing NbS impacts on reducing water risk for agriculture (minimal compensation, maximum compensation, assessment of damages) | Procured in market | N/A |

Source: Altamirano et al. 2021 p. 202

terms, what in the scientific and advocacy world could be considered a project, within investment cycles is considered a project idea. For this project idea to become an investment project that can be assessed for bankability and/or investability, many much more details and evidence needs to be gathered and more clarity needs to be achieved regarding the way NbS proposed will be implemented.

## 9.5.2   New Partnerships and Expertise Required

In order to ensure a successful implementation of NbS as well as to guarantee stable levels of service over time; it is key to consider not only lifecycle costs and their distribution over time but even more the skills and expertise required to undertake the activities. Based on an identification of key implementation resources hold by different actors, activities and risks can be assigned in such a way that the project can be delivered at the lowest costs, the highest quality while minimizing risks. By considering these aspects, the implementing agencies can be guided in their choices of who should take care of which life cycle phases of the project. In other words, this understanding of cost elements and cost drivers can guide the process of allocation of risks, responsibilities and rewards between the key implementing actors that could be either from the public sector, the private sector or the community.

An in-depth analysis of the strengths of Public, Private, People actors' is required to guide these allocations Given the differences in implementation arrangements and actors between NbS and grey infrastructure solutions up until recently, to find suitable implementing parties for large scale NbS projects may prove challenging.

Until recently NbS projects have been often undertaken by community volunteers coached by NGO's and/or environmental government authorities; and more often than not these projects have a piloting function and are of limited scale. In these projects often social objectives are equally important as those related to biophysical conditions or risk reduction; which influences significantly the design of NbS measures, the methods for their construction and the emphasis given to monitoring and data collection systems. This means all in all a very different project management style than the one normally applicable to grey infrastructure projects.

Meanwhile the provision and procurement of regular grey infrastructure is a relatively more formalized process where (large) construction companies and public infrastructure agencies are key players. In this sector risk-based asset management along the entire useful life of the asset is the new norm. Additionally, due to public procurement rules in this sector; risk allocation and the related liabilities carried per implementing party need to be clarified and agreed upon way in advance before project implementation.

### 9.5.3  Mosaic and the Need for Innovative Contracting Practices

The future is in mosaic projects, and their implementation requires innovative contracting practices, as concluded during the recent Environmental Market and Finance Summit.[2] Over and over, asset managers and market service providers told us that they redesigning projects that can responsively serve multiple markets, depending on where the demand is. This allows them to stack funding from multiple sources: carbon offsets, sustainable forestry, water quality credits, recreational use payments, wetland and habitat mitigation, and other revenue streams.[3] Additionally, in a recent market sounding research process undertaken by Deltares in Peru, in cooperation with the Natural Infrastructure for Water Security (NIWS)[4] project it was found that hybrid (green-grey) infrastructure projects are seen as more attractive to project developers than green infrastructure projects alone. According to the methodology proposed, a central building block is hybrid infrastructure clusters. These are after organised into hybrid and multipurpose infrastructure projects and formal performance-based contracts that can be funded by different revenue streams; depending on local institutional conditions and context specific preferences and the willingness to pay of beneficiaries"(Altamirano 2019, p. 5). However the contracting of multiple services by different authorities and blending of funds from the public and the private sector that benefit from these services requires the development of new public procurement and contracting practices that can deal with this complexity. In first instance this requires the clarification and agreement on a hierarchy of functions and associated levels of services that enable the making of trade-offs during the whole life cycle of green infrastructure: design, construction, operation and maintenance.

### 9.5.4  Policy Recommendations

Research and climate funds aim at the mainstreaming of NbS need to require a different mix of expertise and roles that ensure the applicability of the knowledge and evidence developed and increase their ability to influence public and private investment decisions.

---

[2] The summit was hosted by Forest Trends and AEMI. The summit main conclusions are summarized in the blog titled "Five Things We Can Do in the Next 24 Months to Mobilize Major Investments in Ecosystem Restoration and Climate Resilience, November 13, 2019. https://www.forest-trends.org/blog/five-things-we-can-do-in-the-next-24-months-to-mobilize-major-investments-in-ecosystem-restoration-and-climate-resilience/

[3] Idem 11.

[4] More information on the NIWS project lead by Forest Trends available here: https://www.forest-trends.org/who-we-are/initiatives/water-initiative/natural-infrastructure-for-water-security-in-peru/

Along with a different mix of expertise in the consortia, it is important that the right type of coaching is given to demonstration cases leaders to ensure they are able to achieve not only benefits in terms of awareness raising but also serve as pilots to demonstrate the investability and bankability of NbS projects.

Finally, a new type of mission-driven research programmes aimed at implementation of NbS at scale to deal with climate and water risks; needs to include additional mechanisms to increase accountability and impact of research efforts. These mechanisms could include the setting up of advisory boards or users board for clusters of projects where key representatives from public procurement authorities, banks, impact investors and companies are represented and have the opportunity to give feedback about the knowledge and evidence being developed from early on in the project.

# References

Altamirano MA (2017) A financing framework for Water Security. Oral presentation, Water Futures workshop, during the XVI World Water Congress, Cancun, Mexico

Altamirano MA (2019) Hybrid (green-gray) water security strategies: a blended finance approach for implementation at scale. Background paper Session 3. Roundtable on Financing Water, Regional Meeting Asia Manila, OECD

Altamirano M, de Rijke H (2017) Costs of infrastructures: elements of method for their estimation. Guide for the calculation of LCC of NBS. DELIVERABLE 4.2 : EU Horizon 2020 NAIAD Project, Grant Agreement N°730497 Dissemination

Altamirano MA, de Rijke H, Basco-Carrera L, Arellano B (2021) Handbook for the implementation of nature-based solutions for water security: guidelines for designing an implementation and financing arrangement, DELIVERABLE 7.3: EU Horizon 2020 NAIAD Project, Grant Agreement N°730497 Dissemination. 1st edition

American Rivers (2012) BANKING ON GREEN: a look at how green infrastructure can save municipalities money and provide economic benefits community-wide. Retrieved 15 Nov 2017, from   https://www.asla.org/uploadedFiles/CMS/Government_Affairs/Federal_Government_Affairs/Banking%20on%20Green%20HighRes.pdf

Global Commission on Adaptation (2019) Adapt now: the urgency of action. https://gca.org/global-commission-on-adaptation/report

Lane DC (2008) The emergence and use of diagramming in system dynamics: a critical account. Syst Res Behav Sci 25(1):3–23

Munich Re (2018) Natural disasters. Hurricanes cause record losses in 2017 – the years in figures. Published 04-01-2018. https://www.munichre.com/topics-online/en/climate-change-and-natural-disasters/natural-disasters/2017-year-in-figures.html

OECD (2013) Water and climate change adaptation: policies to navigate uncharted waters, OECD studies on water. OECD Publishing, Paris. https://www.oecd-ilibrary.org/environment/water-and-climate-change-adaptation_9789264200449-en

OECD (2018) Making blended finance work for the sustainable development goals. OECD Publishing, Paris. https://doi.org/10.1787/9789264288768-en

OECD, WEF (2015) Blended finance vol.1: a primer for development finance and philanthropic funders.   https://social-finance-academy.org/uploads/wef-oecd-blended_finance_a_primer_development_finance_philanthropic_funders_report-2015.pdf

UK HM Treasury (2018) Guide to developing the project business case: better business cases for better outcomes

World Bank Group (2019) Integrating green and Gray: creating next generation infrastructure. World Bank, Washington, DC. https://openknowledge.worldbank.org/handle/10986/23665

# Chapter 10
# Reducing Water Related Risks in the Lower Danube Through Nature Based Solution Design: A Stakeholder Participatory Process

**Albert Scrieciu, Sabin Rotaru, Bogdan Alexandrescu, Irina Catianis, Florentina Nanu, Roxane Marchal, Alessandro Pagano, and Raffaele Giordano**

## Highlights

- In the Lower Danube we identified that Nature Based Solutions co-benefits play a key role in enhancing social acceptance of the NBS considering the potential socio-economic benefits, even higher than in the reduction of flood damages.
- Contrarily to most of the existing approaches, in which the co-benefits are accounted for exclusively during the phase of NBS assessment, the Lower Danube taught us that the co-benefits need to be co-defined since the NBS design phase.
- The hydraulic models developed for the Lower Danube, based on the 2006 catastrophic flood, represented the base for envisaging the scenarios of the former floodplain restoration.
- A GIS infra-territorial indicator methodology to assess flood risk vulnerability was implemented in order to complement the current lack of available insurance data related to destroyed and affected dwellings.

A. Scrieciu (✉) · S. Rotaru · B. Alexandrescu · I. Catianis
National Institute of Marine Geology and Geoecology (GeoEcoMar), Bucharest, Romania
e-mail: albert.scrieciu@geoecomar.ro

F. Nanu
Business Development Group (BDG), Bucharest, Romania

R. Marchal
Caisse Centrale de Réassurance (CCR), Paris, France

A. Pagano · R. Giordano
Water Research Institute – National Council (CNR-IRSA), Bari, Italy

© The Author(s) 2023
E. López Gunn et al. (eds.), *Greening Water Risks*, Water Security in a New World, https://doi.org/10.1007/978-3-031-25308-9_10

## 10.1   Introduction

The Lower Danube wetlands, one of the most important European wetland ecosystem, lost nearly 80% of its surface over the last century due to river dredging, land reclamation and flood control (www.icpdr.org). Anthropic interventions along the Danube river water course, such as construction of the hydropower plants Iron Gates I and Iron Gates II and alterations along its banks through embankment, have generated high bank erosion processes as well as changes of the riverbed with negative impact on navigation. The negative effects induced by anthropic interventions coupled with climate change impact have intensified the flooding and drought events. Also there are problems with desertification, areas such as Dabuleni (called by the locals "Sahara of Oltenia") are in continuous expansion while others are emerging. In addition, catastrophic floods like the one which occurred in 2006 that devastated Rast town, had major impact along the whole Lower Danube Sector. The ecosystem resilience after these hazards is even more weaken by less destructive but more frequent floods which occur at high water discharges and have mainly a local impact.

After the catastrophic floods from 2006 the Romanian government approved "the Program for Ecologic and Economic Reshape of the Danube Floodplain within the Romanian sector". Among its priorities, it included a reconsideration of the defence lines against the flooding of settlements, by restoring some embanked areas to set up wetlands for biodiversity conservation, together with an evaluation of the economic potential within these embanked areas in the context of the wetlands restoration. Afterwards, a series of projects promoting Nature Based Solution (NBS) were implemented by National Administration Romanian Waters during 2006–2014 (Fig. 10.1).

In order to continue the abovementioned initiatives, NAIAD Project proposed the implementation of a wetland restoration project in the Lower Danube region (Romanian Case study) by designing NBS for dealing with water-related risk. For

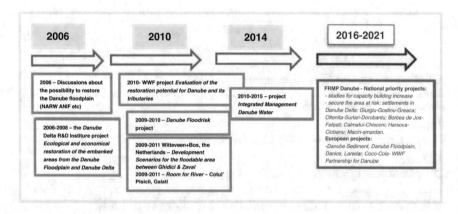

**Fig. 10.1**   Initiatives promoting NBS implementation in the Lower Danube

this objective, NAIAD implementation in the Lower Danube case study aimed at providing support to facilitate both local collaboration between the different stakeholders and the engagement of these local stakeholders in decision-making and policy setting, by integrating their (local) knowledge with the results of scientific models. This would contribute to build a strong cooperation and collaborative framework for the effective adoption of NBS for water risk management (as discussed in Chap. 6 this volume). Good governance of NBS requires creating local partnerships, facilitating cooperation, and identifying clear roles within those partnerships focusing on the local stakeholders knowledge of resources. In addition, local administrations need capacity building on implementation, management and monitoring interventions on the ground. Thus the process undertaken in the Lower Danube Case study is based on working with both communities at the local level and decision-makers at national level to align policy and practice.

The aim of this chapter is to describe the different steps for developing a Natural Assurance Scheme (NAS) for the Lower Danube Case Study (Zimnicea – Calafat sector) meant to improve the flood and drought management and capitalize the NBS co-benefits. This chapter is structured in the following way: first we present a physical risk assessment of the Lower Danube case study; second, we discuss the main results of the dedicated processes focusing on NBS design, with emphasis on active stakeholders engagement; and finally, we explore the possibilities to perform a damage assessment in areas vulnerable to flooding and without any access to insurance data.

## 10.2   Case Study Characterisation and Physical Risk Assessment

In terms of geology, the Lower Danube area, where our case study is located, is mainly characterized by newer formations, as expected, due to the high sediment influx from the Danube River. Most of the formations are young, ranging from the Upper Pleistocene fundament to the Holocene deposits that overlay it (Fig. 10.2). Since the construction of the Iron Gates dams, the sediment afflux has strongly decreased, although the main source of material is still being transported by the Danube. Geomorphologically, the case study area is part of the Oltenia terraced plain sector characterized by a sequence of terraced plains with sand dunes, bordered in the south by the large escarpment of the Pre-Balcanic Plateau (Constantinescu et al. 2015).

In hydrological terms, Danube River represents the most important component of the studied area, flowing for approximately 250 km between Ziminicea and Calafat. At Calafat, Danube has its largest meanders, starting from Cetate until Rast, with a change of direction up to 180°. The average slope of the Danube is only 0.043% due to a small difference in level, of only 7 m. The water flow in this area has been conditioned by both the reduced slope and some geological characteristics

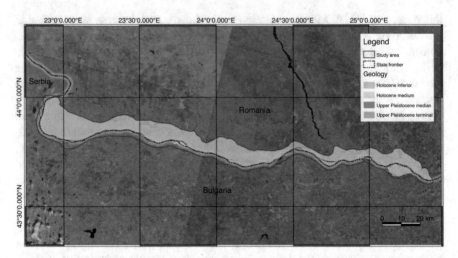

**Fig. 10.2** Map of the local geology in the study area

of the area, such as the slow subsidence in the Calafat – Rast sector and ascending movements towards East at the confluence with Jiu River. Therefore, the flow velocity decreases together with the transport capacity, resulting in the intensification of the sedimentation processes which has a series of consequences on the water course. These consequences are first, a decrease of the river bed depth (up to 3 m at average level compared to the usual depth of 5–8 m) concomitantly with its broadening of up to 1.3–1.6 km; and second, the formation of spits and sand banks which generate a large number of water course unplaiting. This means that the level fluctuations of the Danube in the area of Calafat reach amplitudes of 8–9 m in direct connection with the large variation of the flow. The multiannual average flow is approximately 5500 m³/s at Cetate and 5460 m³/s at Calafat hydrologic stations (h.s.). Exceptional flows have been recorded in 1940 when a maximum flow of about 15,000 m³/s was reached and in 1946 when a minimum flow of approximately 2000 m³/s was measured.

The last catastrophic flood in the Lower Danube area was recorded in 2006. The estimated return period for this flood event in this specific region (Danube River, Gruia h.s.) is 100 years (Liška et al. 2008). For a period of three months (from April 2006 until June 2006) the water volume recorded during this event was 116,000 × 10⁶ m³ (maximum values of 15,990 m³/s at Calafat h.s. and 15,970 m³/s at Bechet h.s.). The flood alert for the Lower Danube lasted more than 6 weeks and 13 dikes broke in Romania during the duration of the event from the 7th April to the 15th June 2006. Both overflow and infiltration caused dike breakage which led to the flooding of Rast – Bistret, Bechet – Potelu – Corabia, Tatina – Spantov – Manastirea, Calarasi – Raul, Oltina, Facaieni enclosures. In the studied area, one dike failure in Bechet has been recorded on the 24th April 2006 at 7:15 am.

Usually, during a year, the following variation can be observed: maximum flow in April–May-June and minimum flow in September–October; in between there is an autumn increase (November) and a winter decrease (January–February). The Danube River freezes across its entire width only in very frosty winters. The ice bridge forms usually at the half of January and its longest recorded duration was 54 days in 1954.

According to its chemical composition, the Danube water is carbonated-sulphated-chlorinated but its degree of mineralization is small, around 300 mg/l. Therefore, the water can be used for human consumption after treatment, for irrigations and for industry.

The Lower Danube experiences a temperate climate, with rains for the whole year, with hot and dry summers. Compared to the other areas of the country, this area is characterized by the highest temperatures, both in the summer and in the winter, with an annual average temperature over 11 °C. The studied area has some climatic peculiarities due to its location in the SW of the country being influenced by the Mediterranean climate, resulting in drought periods in the summer, with a maximum of two rain periods.

Average annual temperature during the 1991–2000 interval was 11.9 °C with the highest annual average temperature of 13.2 °C registered in 2000 while the lowest annual average of 11.0 °C was recorded in 1996. The highest monthly values are recorded in July and the lowest in January, therefore the highest monthly average value within this time interval (1991–2000) was recorded in July 1993 with 29.1 °C, and the lowest monthly average value was recorded in January 1996 with 27 °C.

The case study area is susceptible to severe droughts especially in the summer season, as mentioned above. Areas such as Dabuleni facing desertification, are in continuous expansion while others are emerging. Droughts result from a combination of meteorological, physical and human factors. Their primary cause is a deficiency in rainfall and the timing, distribution and intensity of this deficiency is in relation to existing storage, demand and water use. Temperature and evapotranspiration may act in combination with insufficient rainfall to magnify the severity and duration of droughts. Moreover, due to changes in land use, water demand and climate, the droughts may become more frequent and more severe in the future.

The Danube River is a major economic region reflected by the high percentage of cities located along its route. In our case study there are a number of major cities and rural areas starting from Zimnicea to Calafat. The distance between the Danube River and the settlements along its course varies between few hundred meters and one kilometre, with the exception of those settlements that have ports and where the infrastructure has been developed right up to the riverbank.

In between the populated urbanized areas, most of the land is being used for agriculture and pastures (Fig. 10.3) since this area is very fertile and produces very good yields. Initially this area corresponded to the former Danube floodplain, an area naturally flooded during high water levels, with a very important role in protecting the neighbouring settlements from catastrophic events by attenuating the flood peaks. The land use changes imposed by the agricultural and industrial evolution have altered the regulatory role of the floodplain resulting in the need for

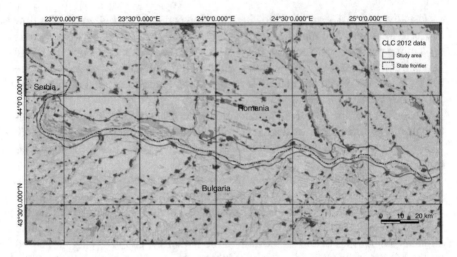

**Fig. 10.3** Map of the case study site overlapping with cities, towns and agricultural fields

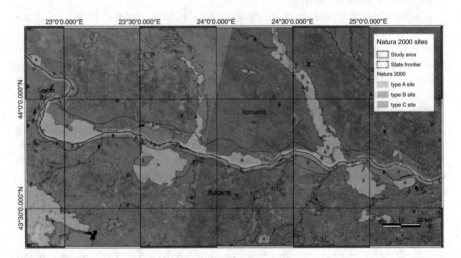

**Fig. 10.4** Map of the Natura 2000 sites in the case study region

implementing additional protection measures (grey infrastructure). The land along the riverbank is currently under the administration of private owners which therefore hinders the incentives that the local authorities could have regarding the maintenance and development of these areas.

However in the region, along the Danube riverbanks, some Natura 2000 sites listed under "Birds Directive" can be found (Fig. 10.4). These sites are strongly protected from human intervention and are strictly monitored by the National Agency for Environment Protection.

## 10.3 Nature Based Solution Design Process

Nature Based Solutions (NBS) have become not only a complementary but a valid alternative to grey infrastructures for coping with climate-related risks in urban and rural areas alike. As defined by the European Commission, NBS are solutions inspired and supported by nature which are cost-effective, and capable to simultaneously provide environmental, social and economic benefits and help build resilience. Several examples of NBS for dealing with climate-related risks are cited in the scientific literature - i.e. restoring wetland, restoring and protecting forests, renaturing watersheds, creating natural retention areas, creating groundwater recharge areas, etc. Moreover, NBS are increasingly recognized for their capacity to support ecosystems functions and to generate ancillary environmental, economic and social benefits considered as essential backbones of actions for climate-change mitigation and adaptation. However successful NBS seem to be, starting from design to implementation, a challenge due to several barriers. One of the key barriers that need to be addressed concerns the low level of stakeholders and local community engagement in the NBS design process.

To this aim we have developed a stakeholders engagement process in order to design intervention scenarios based on NBS implementation. The research performed in the Lower Danube case study, within NAIAD project, aimed at understanding the role of natural assurance schemes in complex natural, economic and social contexts. To increase the relevance and the potential for replicable results, the large scale case study approach was complemented with a focus on the analysis of the Dabuleni-Potelu-Corabia enclosure (area drained for land reclamation) specific NBS. The downscaling was performed in order to allow the assessment of this specific NBS effectiveness based on a combined bottom-up interest of communities for diversification of economic activities and a top-down concern for reducing the pressure on the grey infrastructure for flood protection by means of a cascade system of green solutions.

The different phases of the NBS co-design process are described in the following sections.

### 10.3.1 Main Beneficiaries and Regulatory Framework

The main beneficiaries from the process undertaken were the local communities, not only regarding the flood and drought protection, but also regarding the production of co-benefits related to the socio-economic and ecosystem dimensions. In contrast with grey flood prevention infrastructure that is already planned at river basin scale and has more solid financing sources, green infrastructure relies largely on local communities, first for acceptance, and second, for its implementation, monitoring and maintenance. Especially in Lower Danube case study, where NBS are in rural areas, and where farmers and local population are users, owners or administrators of land resources and have their own way of using these resources. Therefore, to ensure the acceptance of NBS at the local level, local communities, as well as the local administrations need to be engaged in decision-making processes for design and implementation.

In the context of the Water Framework Directive 2000/60/EC and of the Flood Directive 2007/60/EC on the principle of "more space for river", as well as the risk of climate change, specialized studies have been started since 2007 aimed at ensuring more favourable conditions for the drainage of flood events by repairing and restoring the ecological characteristics and regulatory functions of a part of the floodplain to the initial conditions that existed before the embankments construction, while at the same time ensuring the sustainable development of the adjacent areas in terms of income and revenue flows.

The principles of sustainable development focusing on green infrastructure aligned with the Water Framework Directive and the Flood Directive, have been applied in the general direction of the proposed scenarios developed for Dabuleni-Potelu-Corabia (DPC) enclosure, in order to reconnect this sector of the former Danube Floodplain with the Danube River to reduce water related risks (flood and drought) and to exploit the benefits and co-benefits generated by the implemented NBS. To this aim, local community knowledge and values were elicited and structured, to be used for the definition and assessment of the co-benefits to be produced.

## 10.3.2   Stakeholders Engagement Process

As already stated, a stakeholders process was implemented in the study area, with the scope of defining the key co-benefits to be produced through the NBS implementation and to support the assessment of its effectiveness. Three rounds of semi-structured interviews (approximate duration 1 h) with individual stakeholders (or group of stakeholders representing a single institution), one per stakeholder, were held first. The results of this activity showed that the whole area is increasingly affected by persistent droughts and, particularly in some locations, intense floods, both responsible for human and economic losses. Additional issues were also raised individually by the stakeholders, mainly related to the state of the environment and to the economic activities (e.g. agriculture, tourism, etc.), and identified as key elements to support the development of the area. Some problems were also discussed and highlighted by the stakeholders e.g. the negative effects associated to the lack of institutional cooperation and the limited stakeholders involvement in decision-making.

The first stakeholder workshop (approximate duration 3 h), was held in March 2018 at the headquarter of the River Basin Administration Jiu (Craiova), and oriented to the definition of a ranking among benefits and co-benefits. The main benefits were related to the reduction of the impacts of both floods and droughts, which were considered as equally important. Regarding the main co-benefits for the DPC enclosure, the highest-ranked ones were, in order of relevance: (i) the development of eco-tourism; (ii) the limitation of migration/depopulation; (iii) the increase in biodiversity; (iv) the development of fishing and aquaculture activities; (v) the increase of agricultural production. Furthermore, the following NBSs were identified as potentially relevant for the area: wetland restoration, retention areas, river renaturation, and reforestation.

The second stakeholder workshop was held, with the same stakeholders, in December 2018 (approximate duration 3 h) at the headquarters of the River Basin Administration Jiu (Craiova). A Casual Loop Diagram (CLD) was collectively built, using the Vensim® simulation software, to describe the current state of the system. The main benefits associated with the reduction of water related risks (i.e. 'drought' and 'flood magnitude') and the selected co-benefits ('biodiversity', 'eco-tourism', 'fish production', 'population' and 'agricultural production') are drawn in bold (Fig. 10.5).

The developed CLD was used to support the discussion about the expected impacts of the NBS, according to the stakeholders understanding. A performance assessment matrix was developed, allowing the stakeholders to provide qualitative weight to the capability of the NBSs to produce the selected benefits and co-benefits (Table 10.1).

The Table 10.1 shows the comparison among four potential NBS. As shown in this table, stakeholders perceived the wetland restoration as greatly effective in reducing water-related risks, but also in producing important socio-economic and ecosystem-based co-benefits, such as increasing biodiversity and fish production, enabling eco-tourism initiatives.

The stakeholders engagement process was supported by the physical assessment, the flood and drought modelling and NBS effectiveness, as described further in the text.

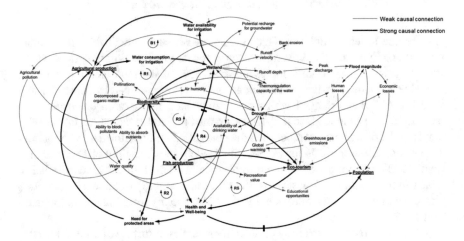

**Fig. 10.5** CLD to describe the dynamics related to the benefits ("drought" and "flood magnitude", in bold) and co-benefits (in bold and underlined) in current conditions in Dabuleni-Potelu-Corabia enclosure. (From Coletta et al. 2021)

**Table 10.1** The potential NBSs and the associated objectives/benefits and co-benefits

| Alternatives/ NBSs | Objectives/Benefits and co-benefits | | | | | |
|---|---|---|---|---|---|---|
| | Water-related risks reduction | Biodiversity increase | Agricultural production increase | Eco-tourism increase | Fish production increase | Population growth |
| Wetland restoration | +++ | +++ | −− | +++ | +++ | ++ |
| River renaturation | + | ++ | 0 | ++ | + | ++ |
| Retention areas | +++ | + | − | + | ++ | + |
| Forested areas | + | ++ | −− | +++ | + | + |

**Fig. 10.6** Dabuleni-Potelu-Corabia enclosure

### 10.3.3    Proposed NBS Scenarios for the Case Study Area of the Dabuleni-Potelu-Corabia Enclosure

The Danube River floodplain defence system was implemented between the years 1960–1966. Currently the existing structures for flood protection integrates a total area of approx. 4735.56 km², representing 92.15% of the total area registered as floodable Romanian territory. This defence system is represented by approximately 1200 km of dykes arranged within 50 distinct land reclamation enclosures. An "enclosure" is the name given to the areas drained for land reclamation. DPC enclosure (Fig. 10.6) was dammed between the years 1965–1966 at the degree of protection against catastrophic floods of 1%, with a safety reserve height of 1 m. The dike length is 32.4 km, the cross section is trapezoidal with the following characteristics: crest width – 5 m and slopes 1:3 to the water and 1:(4÷5) to the enclosure. As a protective measure against erosion, hybrid black poplars were planted in the dike-river bank area.

The DPC enclosure was fully drained, representing a surface of 14,665 hectares and the area irrigated has a surface of 2.86 hectares (approx. 20% of the total drained area). The draining of the DPC enclosure was achieved by opening the Celeiu channel that discharged the water with the help of five pumping stations: Corabia (3.2 m³/s), Stejarul (1.45 m³/s), Valcovia (10.75 m³/s), Racari (2.75 m³/s) and Celei (1.4 m³/s).

### Hydraulic Modelling of the Dabuleni-Potelu-Corabia Enclosure

Taking into account the particularities referring to land ownership and willingness to embrace the NBS implementation, we have developed two hydraulic models consisting on (i) partial flooding and (ii) total flooding of the enclosure surface. Further on, this two hydraulic models will represent the base for envisaging the scenarios for the restoration of the DPC enclosure.

The hydraulic modelling of the water flow in the DPC enclosure considered the embanked Danube river, as well as the inside enclosure, and was performed using the HEC-RAS software (developed by the US Army Corps of Engineers). The hydrological parameters used in the modelling were those corresponding to the recorded flood of 2006.

The hydraulic model of the Danube riverbed and DPC enclosure was developed using the topometric and topobathymetric cross sections, plotted throughout the embanked Danube river (Figs. 10.7 and 10.8) and the numerical model of the land in the enclosure area (Fig. 10.8). The equidistance of the cross sections is approximately 1 km and their location corresponds to the Danube kilometre landmarks (Figs. 10.7 and 10.8). The numerical model of the terrain consists of a grid with a resolution of 5 × 5 m.

The hydraulic model developed for the studied area consisted of a one-dimensional model corresponding to the embanked area of the Danube river and a two-dimensional model corresponding to the enclosure area behind the dikes. For the two-dimensional model area, the overall grid cell size was 30 × 30 m. The connection between the one-dimensional hydraulic model (1D) and the two-dimensional hydraulic model (2D) was achieved considering two spills, corresponding to the areas where the breaches appeared in the defence dikes during the 2006 flood.

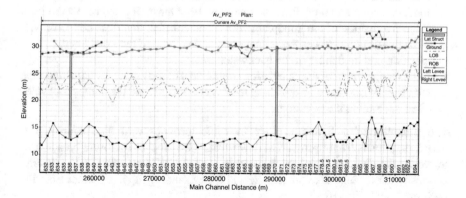

**Fig. 10.7** Longitudinal profile of the Danube River intersecting the plotted cross-sections within the Bechet – Corabia sector (Legend: Ground-river talweg, LOB-left overbank, ROB-right overbank, Left and Right Levees represent the defence dikes)

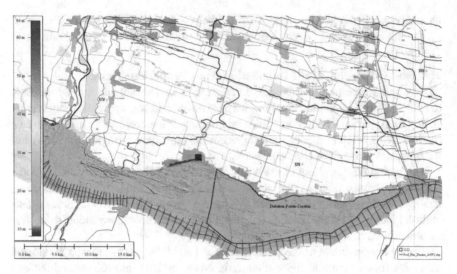

**Fig. 10.8** Cross sections locations

The calibration of the hydraulic model represents an important stage of the hydraulic modelling process and consists in validating the model results against the hydrological observations: flow rate and maximum level, total volume (increase and decrease) and rating curves (Fig. 10.9).

The main parameters that can be acted upon during the adjustment (calibration) process of the mathematical model that reproduces the flood waves propagation are:

- roughness coefficients (ni) (Fig. 10.10), which model the hydraulic resistance of the river channel;
- introduction of the accumulated mileage of the floodplain in the direction of the recorded flood propagation axis(1D);
- determining the higher areas and levels from where the floodplain begins to flood, detecting and modelling the local low level areas (located below the level of the riverbanks) of the floodplain with a polder effect that does not contribute to the flow, but influences the propagation and volume of recorded floods;
- detection and modelling of the backwater areas;
- optimal adjustment of the calculation coefficients of the model, by adjusting time and distance calculation steps along the river ($\Delta T$, $\Delta X$), and the number of cycles when integrating the equations.

The calibration of the hydraulic model took into account several dike breaches produced as follows, from upstream to downstream: the Bechet enclosure breach, followed by the division dike breach between Bechet enclosure and DPC enclosure, and lastly the Danube dike breach, created for evacuating the flood water from the DPC enclosure (Fig. 10.11).

The roughness coefficients resulted from the calibration process have values of 0.03 for the riverbed and 0.07 for the floodplain (Fig. 10.10).

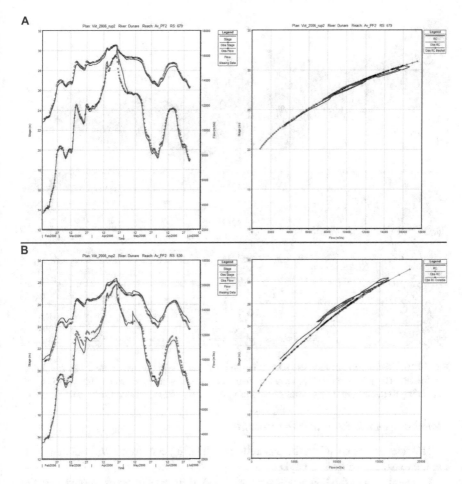

**Fig. 10.9** Calibration results of the hydraulic model (flow and level hydrographs, left side and the rating curves, right side) in the sections of Bechet (**a**) and Corabia (**b**) hydrometric stations

**Planning Scenarios for the Restoration of Dabuleni-Potelu-Corabia Enclosure**
The proposed restauration scenarios for the DPC enclosure integrated the conclusions of the meetings held with the stakeholders from our case study area. Based on the extreme flood from 2006 that recreated the Potelu Lake, the majority of our stakeholders have been in favour of restoring the former floodplain, mainly being driven by the potential benefits and co-benefits that the implementation of the NBS can bring. The new formed Potelu Lake was drained one year after, in 2007, causing the loss of the benefits and co-benefits associated with the wetland. The proposed planning scenarios were focused on the total flooding (optimistic planning scenario) and partial flooding (realistic planning scenario) of the enclosure, in order to recreate the former Potelu Lake that will act as a natural retention area to reduce the flood peak and promote additional uses such as fish farming and recreation.

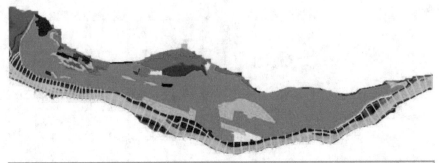

| Color | Value | Name | Default Manning's n |
|---|---|---|---|
| | 0 | no data | |
| | 1 | 112 - Artificial surfaces/ Urban fabric/ Discontinous urban fabric | 0.03 |
| | 10 | 324 - Forest and semi-natural areas/ Shrubs and/or herbaceous vegetation associations/ Sclerophyllous vegetation | 0.2 |
| | 11 | 411 - Wetlands/ Inland wetlands/ Inland marsh | 0.03 |
| | 12 | 511 - Water bodies/ Inland waters/ Water courses | 0.03 |
| | 2 | 121 - Artificial surfaces/ Industrial, commercial and transport units/ Industrial or commercial units | 0.03 |
| | 3 | 211 - Agricultural areas/ Arable land/ Non-irrigated arable land | 0.05 |
| | 4 | 221 - Agricultural areas/ Permanent crops/ Vineyards | 0.035 |
| | 5 | 231 - Agricultural areas/ Pastures/ Pastures | 0.035 |
| | 6 | 242 - Agricultural areas/ Heterogeneous agricultural areas/ Complex cultivation patterns | 0.06 |
| | 7 | 243 - Agricultural areas/ Heterogeneous agricultural areas/ Agriculture mixed with areas of natural vegetation | 0.06 |
| | 8 | 311 - Forest and semi-natural areas/ Forests/ Broad-leaved forest | 0.08 |
| | 9 | 321 - Forest and semi-natural areas/ Shrubs and/or herbaceous vegetation associations/ Natural grasslands | 0.2 |

**Fig. 10.10** Distribution of land use categories and roughness coefficient adopted. The roughness coefficients (Default Manning's n) used for the enclosure area were taken from Chow (1959), corresponding to the different categories of land use presented in Corrine Land Cover 2018

*Planning scenario 1 (optimistic scenario)*

This scenario is based on the modelling of flooding conditions without considering the existence of defence dykes.

The total area of the DPC enclosure is 14,665 ha its maximum estimated capacity would be $75 \times 10^7$ m$^3$ (at 28 m ground level). In normal retention conditions (at 24.5 m ground level) the estimated stored water volume will be $24 \times 10^7$ m$^3$ (Fig. 10.12).

The flooding simulation of the DPC enclosure was performed based on the recorded flood wave from 2006 (Figs. 10.13 and 10.14). The connection between the one-dimensional (1D) hydraulic model corresponding to the Danube river and the two-dimensional (2D) hydraulic model corresponding to the DPC enclosure was made through side spillways. The spillways crest level was extracted from the numerical terrain model of the DPC enclosure representing the lowest values identified at the base of the dykes.

The hydraulic modelling results of scenario 1, indicates a decrease of 430 m$^3$/s of the maximum water flow downstream of the DPC enclosure as well as a lowering with 36 cm of the maximum water level in the upstream part (Bechet hydrometric station).

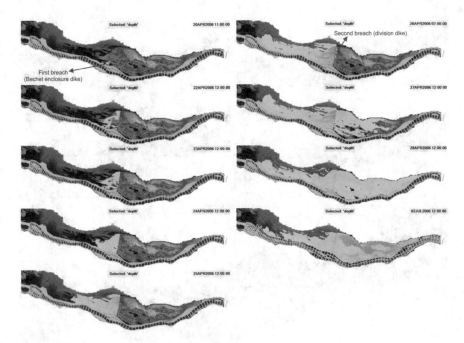

**Fig. 10.11** The results of the flood wave propagation in the Bechet and DPC enclosures generated by the breaches in the Danube dike and the interior division dike, at different times

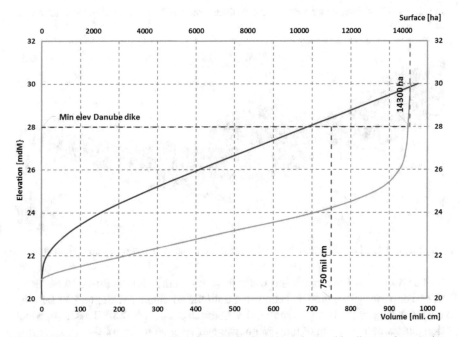

**Fig. 10.12** Characteristic curves of Dabuleni-Potelu-Corabia enclosure (blue line – volume variation, brown line – water surface variation)

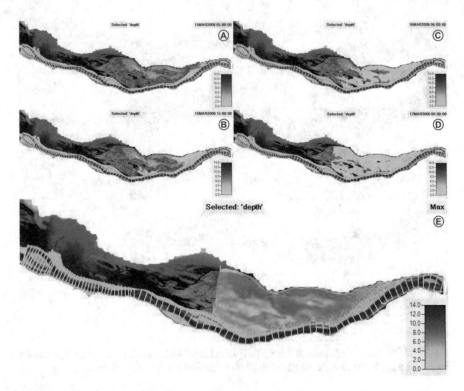

**Fig. 10.13** Flood wave propagation in the DPC enclosure without considering the existence of defence dykes

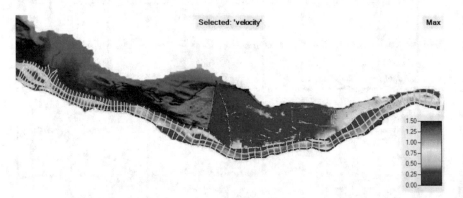

**Fig. 10.14** Water flow velocities in the DPC enclosure based on the recorded flood from 2006

The decrease of the maximum water flow of 430 m³/s is considered to be maintained downstream of the DPC enclosure until the confluence with the river Olt (the main tributary of the Lower Danube) (i.e. a stretch of about 60 km). This flow reduction generates a lowering of the maximum water level with approx. 20 cm according to the rating curve from the section of Corabia h.s.

The main effect generated by the water level decrease is represented by the reduction of the pressure on the existing defence dykes.

The estimated value of the engineering works proposed to be carried out within the framework of the planning scenario 1 was calculated taking into account the costs of similar investments. This calculated value is approx. M€ 8 and represents the total investment needed for implementing the engineering works without considering the costs of expropriation of land areas, permits and authorizations, as well as other commissions and fees; which means that the total costs would be higher.

*Planning scenario 2 (realistic scenario)*

This scenario is based on restauration of the former Potelu Lake (Fig. 10.15) which implied readapting the main irrigation and drainage channels.

The limits of Potelu Lake will be represented by contour levees with the crest level corresponding to the level that will ensure a protection against catastrophic floods with the probability of 1%.

The total length of the contour levees surrounding the Potelu Lake will be 47 km, however, if the northern limit will be extended up to the base of the existing terrace, the total length will be reduced to 27 km.

Based on our workshops and the inputs elicited from the local stakeholders, the main uses of Potelu Lake would be fish farming and recreation.

The total area of the proposed Potelu Lake would be 6230 hectares and its maximum estimated capacity will be $36 \times 10^7$ m$^3$ (at 28 m ground level). In normal retention conditions (at 24.5 m ground level) the estimated stored water volume will be $14.5 \times 10^7$ m$^3$ (Fig. 10.16).

The reduced transport capacity of the Potelu Lake supply/drainage channels, together with its small volume (approx. $21 \times 107$ m$^3$) reserved for the attenuation of the flood waves from the Danube River ($116{,}000 \times 10^6$ m$^3$) would have a negligible impact in reducing the water flow and flood peak.

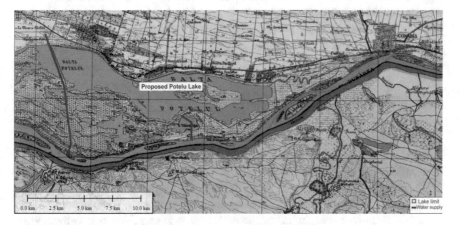

**Fig. 10.15**  The extension of the proposed Potelu lake

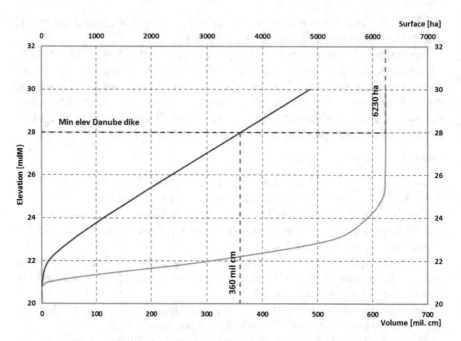

**Fig. 10.16** The characteristic curves of the proposed Potelu Lake (blue line – volume variation, brown line – water surface variation)

The value of the construction works proposed for the planning scenario 2 was estimated taking into account the costs of similar investments. The calculated costs would be approximately 47,000,000 Euros in the case of delineating the entire Lake Potelu with contour dykes (47 km in total length). However, if the northern limit of the lake would be extended up to the base of the existing terrace, reducing the length of the estimated dyke construction (down to 27 km), the calculated costs would be approximately 29,000,000 Euros. These costs refer only to the engineering works without considering the costs of expropriation of land areas, permits and authorizations, as well as other commissions and fees; which means that the final costs would be higher.

## 10.4   Economic Assessment

DPC enclosure has been largely damaged during the 2006 flood event. High population density and geographical constraints make the area extremely vulnerable to devastating consequences of flood events.

Considering the current lack of insurance data available to assess the insured damage and to calibrate damage curves specific for the Lower Danube area, we opt for a GIS infra-territorial indicator methodology to assess flood risk vulnerability.

In order to gather information on the economic assessment we performed a literature review to collect quantitative information about the 2006 flood damage. Consistent post-event research on the flood damage has been performed (Liška et al. 2008; Schwarz et al. 2006). These researches are the unique source of information concerning the damage, in terms of number of damaged houses, most damaged communities and global amount of losses.

## 10.4.1 Literature Review on 2006 Flood Damage

The estimation of the amount of damage for the 2006 flood is approximately M€ 400 for the Lower Danube area with 14,000 people displaced in Romania and 63 displaced in Bulgaria (Liška et al. 2008). For Romania the estimated damage are approximately M€ 200–300 with no human losses recorded. When focusing on the counties of the studied area, we obtained detailed information on the tangible direct damage (Table 10.2).

When comparing the affected Romanian counties, we observe that both Dolj and Calarasi (Calarasi is not located in the studied area) counties were amongst the most damaged ones. Dolj county represents the largest proportion of the total number of affected constructions (54.6% of the total destroyed dwellings) and railroads (87.8%). Olt county has not been strongly affected. Concerning agricultural damage, two of the four areas heavily impacted are located in the studied area: Potelu and Calarasi villages (not to be mistaken with Calarasi county and Calarasi town). It represents 41.184 ha of productive land flooded and 10.802 people lost their livelihoods. In terms of loss of productivity the studied area represents 54.9% of the total loss of hectare of productive land (Table 10.3).

Then, it is possible to define the average costs of a claim. We consider the total number of damaged infrastructures (10522) for the overall Romania for a total cost of M€ 250. The average costs of a claim is € 23,759. In this study we focus only on assessing the direct damage to the dwellings (destroyed and affected), for the overall Romania it represents 31% of the total damage, thus M€ 77.9. For the studied area, it represents 12%; so we consider the damage to the dwellings destroyed and affected for approx. M€ 30.8. So the average costs of claims for dwellings are € 9300 (total number in the area: 3279).

## 10.4.2 GIS-Based Indicators

The second step is to apply the GIS-based indicators methodology developed by CCR to perform a damage estimation in an area without any access to insurance data. The objective of this GIS-based infra-territorial indicator is to propose a vulnerability mapping method for disaster assessment in the NAIAD case studies. This indicator should be seen as a tool for decision-making process to assess the areas

**Table 10.2** Affected and destroyed infrastructure in Romania, focus on the counties in the studied area (Liška et al. 2008)

| County | Affected constructions | | | | Railroads and roads | | | Other constructions |
| | Destroyed dwellings | Affected dwellings | Dependencies | Wells | Railroads (km) | Local, county and national roads (km) | Bridges and footbridges (km) | |
|---|---|---|---|---|---|---|---|---|
| Olt (ROU) | 3 | 5 | 87 | – | – | 4 | 11 | 4 |
| Dolj (ROU) | 372 | 919 | 596 | 2133 | 21,0 | 3 | 169 | 48 |
| Total of the ROU counties | 681 | 2598 | 2763 | 3881 | 23,9 | 22 | 487 | 67 |
| Dolj proportion in the total (%) | 54,6 | 35,3 | 21,5 | 54,9 | 87,8 | 13,6 | 34,7 | 71,65 |

**Table 10.3** Damage done by floods on productive land and on human well-being; (Schwarz et al. 2006); focus on the studied area

|  | Evacuation | Loss of productivity |
|---|---|---|
| Balta Potelu | – | 8.200 ha arable land flooded |
|  |  | 5.900 ha forest flooded |
|  |  | 2.600 ha pastures flooded |
| Balta Calarasi | 2.480 displaced | 1.222 ha arable land lost |
|  |  | 4.686 ha arable land lost |
| Total | 10.802 | 41.184 ha |
| Studied areas proportion in the total | 23% | 54.9% |

**Fig. 10.17** Method to implement GIS-based risk indicator

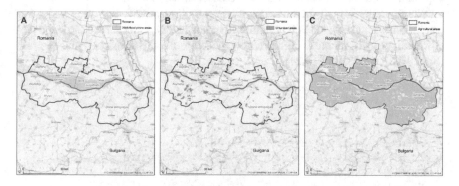

**Fig. 10.18** Indicators used for the GIS-based methodology

vulnerability to water-related hazards and to select the more appropriate area for preventive measures implementation. The proposed methodology (Fig. 10.17) combines a range of geographically based input layers from physical (flood extent) to socio-economic (land-use) ones within DPC enclosure.

Indicator-based approach participates in the development of standard methods to asses flood damage potential for assets at risks (Eleutério et al. 2010). Thus, three types of input layers have been overlaid in ArcGIS, through the raster calculator (Fig. 10.18). The layers have not been weighted; the location is the common key between the different layers. The flood prone area (hydraulic models developed by Geoecomar – Fig. 10.18a) and the urbanized areas (Corine Land Cover 2018 – Fig. 10.18b) have been overlaid for the vulnerability exposure of urban areas and in a second step, flood prone area and agricultural areas (Fig. 10.18c) have been overlaid for the exposure of agricultural areas.

Following the development of the spatial analysis, the obtained exposure indicator is aggregated at the community scale using meshes of 250 m, 500 m and 1 km resolution.

The figures below show the result of the overlaying process with urbanized and with agricultural areas. We observed that the central area of Bechet and Calarasi are the most at-risk areas (Fig. 10.19). The orange areas considered as "moderate" risk are agricultural areas largely damaged during the 2006 flood event, as visible in the Fig. 10.20. The GIS-based indicator highlights the large exposure of agricultural areas to flood risk. This is consistent with the findings in the literature review on agricultural damage.

Furthermore, it is possible to define the percentage of the communities at risk of flooding taking into account their urban or agricultural exposure. When comparing the flood exposure, the results indicate that all of the agricultural areas are heavily at risk (from 80% to 98% exposure), the most exposed communities from Romania being Ostroveni, Bechet and Grojdibodu while regarding the urbanized exposure, the most exposed ones are Ostroveni, Ianca and Grojdibodu (Fig. 10.21).

The exposure of agricultural areas is largely more significant than the urbanized exposure, which is consistent with the elements found in the literature review.

When we performed the assessment using meshes of 250 m, 500 m and 1 km resolution, with urbanized and agricultural inputs layers, we observed the same pattern with agricultural areas being more exposed to flood risk than urbanized areas (Figs. 10.22, 10.23, and 10.24).

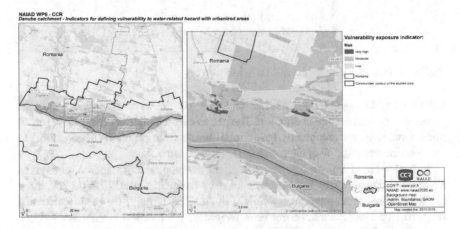

**Fig. 10.19** GIS-based vulnerability exposure indicators of urbanized areas

**NAIAD WP6 - CCR**
*Danube catchment - Indicators for defining vulnerability to water-related hazard with agricultural areas*

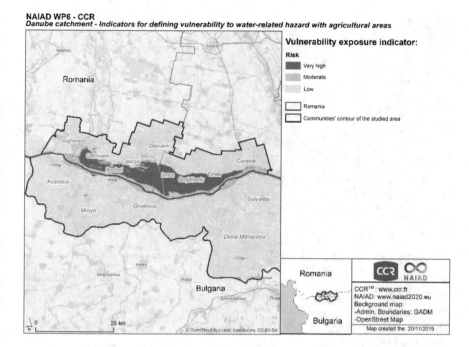

**Fig. 10.20** GIS-based vulnerability exposure indicators of agricultural areas

## 10.5  Conclusions and Lessons Learned

Besides the physical assessment of the NBS effectiveness in reducing the water-related risks at local level, the NAIAD implementation contributed to develop an integrated assessment framework, capable to account for the stakeholders preferences over the co-benefits to be produced. For a more detailed definition of co-benefits, the reader might refer to Chap. 5 in this book. For a detailed description of the different phases of the implemented process, a reader might refer to Giordano et al. 2020. In this section, we would like to describe the key lessons learned during the whole process.

Firstly, we learned that rich and diverse knowledge is required for an effective NBS design and assessment process. To this aim, the adopted approach allowed to collect individual risk perceptions, to detect and keep track of the main differences among risk perceptions and problem framings, and to avoid pursuing immediate and unanimous consensus. Secondly, the experiences carried out in the Lower Danube demonstrated the key role that the co-benefits play in enhancing social acceptance of the NBS. Most of the stakeholders involved in the process expressed high interest in the potential socio-economic benefits – i.e. eco-tourism, fishery production, reduce de-population (Giordano et al. 2020) – even higher than in the reduction of flood damages. Therefore, we learned that contrarily to most of the existing approaches, in which the co-benefits are accounted for exclusively during

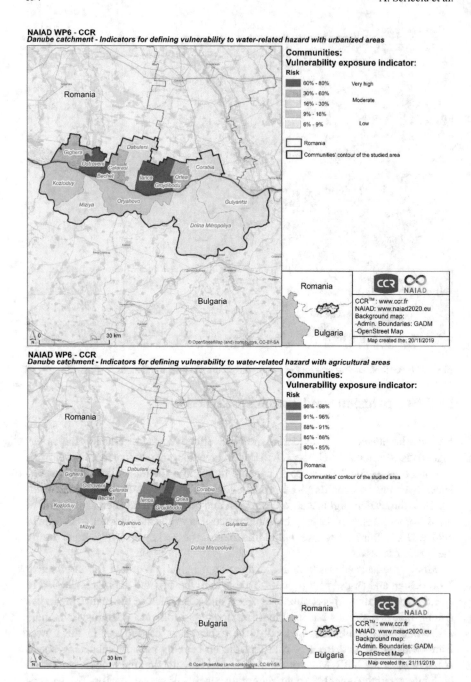

**Fig. 10.21** Percentage of the communities areas at risk of flooding: vulnerability exposure of urban (up) and agricultural areas (down)

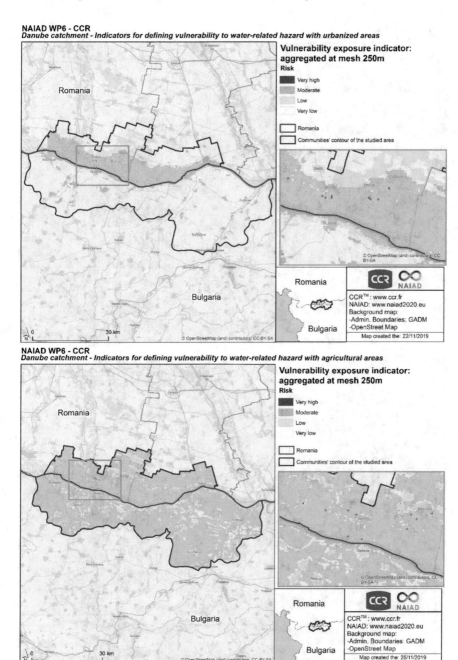

**Fig. 10.22** GIS-based indicators with urbanized areas (up) and agricultural areas (down) input layers – mesh of 250 m

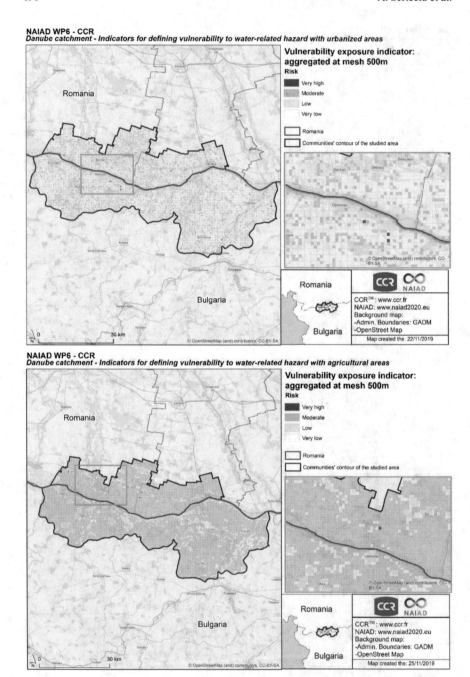

**Fig. 10.23** GIS-based indicators with urbanized areas (up) and agricultural areas (down) input layers – mesh of 500 m

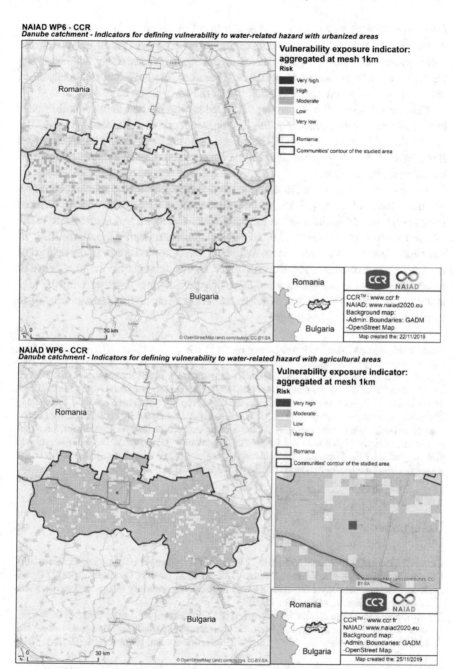

**Fig. 10.24** GIS-based indicators with urbanized areas (up) and agricultural areas (down) input layers – mesh of 1 km

the phase of NBS assessment, co-benefits need to be co-defined since the NBS design phase. Thirdly, given the importance of the co-benefits in enhancing the social acceptance of the NBS, guaranteeing the equality in accessing and benefiting the co-benefits is of utmost importance. Therefore, the detection and analysis of potential trade-offs and conflicts should be at the basis of the NBS design.

Moreover NAIAD Project attempted to assess the insured damage and to calibrate damage curves specific for the Lower Danube area associated with specific water related risks. For this, a GIS infra-territorial indicator methodology to assess flood risk vulnerability was implemented in order to complement the current lack of available insurance data.

The literature review on the 2006 flood in Romania provides relevant elements on the number of damaged assets. Nevertheless, the lack of data on the insured losses generates difficulties in the local damage assessment and furthermore to the avoided damage assessment. The GIS-based indicators can support the lack of insurance data, with the objective to define the most exposed areas. In the context of the Lower Danube case study, the GIS-based results demonstrate the prevalence of agricultural areas exposure to flood. The literature review together with the GIS analysis are able to provide a reliable cost estimation of the damage related to destroyed and affected dwellings.

**Acknowledgements** This work was funded by the H2020 project NAIAD (Grant 730497) from the European Union's Horizon 2020 research and innovation program. We acknowledge the support of our colleagues Iulia Puiu and Camelia Ionescu from WWF Romania. We thank Elena López Gunn for reviewing this chapter and providing many useful comments. Part of the data used within NAIAD Project was obtained through the Core Program (Program Nucleu) - "Developing integrated management for pilot areas of the Romanian Danube sector, influenced by climate change and anthropic interventions, by applying complex research methodologies", funded by the Romanian Ministry of Research and Innovation, contract no. 13N/08.02.2019. Sabin Rotaru was also supported by the Romanian Young Academy, which is funded by Stiftung Mercator and the Alexander von Humboldt Foundation for the period 2020–2022.

# References

Chow VT (1959) Open-channel hydraulics, McGraw-Hill civil engineering series. McGraw-Hill
Coletta VR, Pagano A, Pluchinotta I, Fratino U, Scrieciu A, Nanu F, Giordano R (2021) Causal loop diagrams for supporting nature based solutions participatory design and performance assessment. J Environ Manag 280:111668
Constantinescu Ş, Achim D, Rus I, Giosan L (2015) Embanking the Lower Danube: from natural to engineered floodplains and back. In: Geomorphic approaches to integrated floodplain management of lowland fluvial systems in North America and Europe. Springer, New York, pp 265–288
Eleutério J, Martinez D, Rozan A (2010) Developing a gis tool to assess potential damage of future floods. WIT Trans Inf Commun Technol 43:381–392
Giordano R, Pluchinotta I, Pagano A, Scrieciu A, Nanu F (2020) Enhancing nature-based solutions acceptance through stakeholders' engagement in co-benefits identification and trade-offs analysis. Sci Total Environ 713:136552

Liška I, Wagner F, Slobodnik J (2008) Joint Danube survey 2–final scientific report. ICPDR–
    International Commission for the Protection of the Danube River, Vienna
Schwarz U, Bratrich C, Hulea O, Moroz S, Pumputyte N, Rast G, Bern MR, Siposs V (2006)
    Floods in the Danube River Basin. Flood risk mitigation for people living along the Danube:
    the potential for floodplain protection and restoration. WWF, Vienna

# Chapter 11
# Multidisciplinary Assessment of Nature-Based Strategies to Address Groundwater Overexploitation and Drought Risk in Medina Del Campo Groundwater Body

Beatriz Mayor, África de la Hera-Portillo, Miguel Llorente, Javier Heredia, Javier Calatrava, David Martínez, Marisol Manzano, María del Mar García-Alcaraz, Virginia Robles-Arenas, Gosia Borowiecka, Rosa Mediavilla, José Antonio de la Orden, Julio López-Gutiérrez, Héctor Aguilera-Alonso, Laura Basco-Carrera, Marta Faneca, Patricia Trambauer, Tiaravani Hermawan, Raffaele Giordano, Eulalia Gómez, Pedro Zorrilla-Miras, Marta Rica, Laura Vay, Félix Rubio, Carlos Marín-Lechado, Ana Ruíz-Constán, Fernando Bohoyo-Muñoz, Carlos Marcos, and Elena López Gunn

## Highlights

- Noticeable improvement in knowledge of the MCGWB geological, geophysical and hydrogeological aspects: aquifer geometry, behaviour and ecosystems-aquifer interactions.
- Assessment of MAR effectiveness in the Trabancos sub-basin.
- Participatory Modelling activities have been used to develop a comprehensive understanding of the scope of the system.

B. Mayor (✉) · P. Zorrilla-Miras · M. Rica · L. Vay · E. López Gunn
ICATALIST S.L., Las Rozas, Madrid, Spain
e-mail: beatriz.mayor.rodri@gmail.com

Á. de la Hera-Portillo · M. Llorente · J. Heredia · R. Mediavilla · J. A. de la Orden
J. López-Gutiérrez · H. Aguilera-Alonso · F. Rubio · C. Marín-Lechado · A. Ruíz-Constán
F. Bohoyo-Muñoz
Instituto Geológico Minero de España (IGME), Madrid, Spain

J. Calatrava · D. Martínez · M. Manzano · M. del Mar García-Alcaraz · V. Robles-Arenas
G. Borowiecka
Universidad Politécnica de Cartagena (UPCT), Región de Murcia, Spain

L. Basco-Carrera · M. Faneca · P. Trambauer · T. Hermawan
Deltares, Delft, The Netherlands

E. López Gunn et al. (eds.), *Greening Water Risks*, Water Security in a New World, https://doi.org/10.1007/978-3-031-25308-9_11

- The main variable controlling risk in Medina del Campo is human activity.
- Reduction of groundwater abstractions would be the most critical necessary measure to recover GRES and aquifer storage.
- NAS strategy 1 with crop changes and water management measures showed the highest cost-benefit ratio.

## 11.1   Introduction

The Medina del Campo Groundwater Body (MCGWB) is the biggest aquifer system located in the Duero River Basin, in Central Spain. Being a drought prone area, the MCGWB has experienced intense exploitation in the last decades mainly for irrigation (which represents 96% of the total annual extractions), and for drinking water supply and other (industrial) uses. As a result, the sharp decrease in groundwater levels (Fig. 11.1) has induced water quality degradation, severe deterioration of dependent wetlands and streams, and eventual reduction in the delivery of ecosystem services in the area.

The Groundwater Directive 2006/118/EC and the Water Framework Directive 2000/60/EC impose the obligation for the Duero River Basin Authority (DRBA) to assess the impact and damages from existing pressures, and to take measures to restore the good quality status by 2027. In the case of the MCGWB, the main threats identified include lowering piezometric levels, diffuse agricultural pollution ($NO_3$ contents up to 190 mg/L), and elevated arsenic contents of lithological origin (up to 240 µg/L). A first measure established by the DRBA to address these pressures was a water transfer from the neighbouring Adaja River and the Cogotas reservoir to substitute groundwater by surface water for irrigation in the Adaja irrigation district (6000 ha). As a result, a localised recovery on piezometric levels was detected in the surrounding area due to the double effect of stopping groundwater extractions and increased replenishment from surface return flows.

The NAIAD framework was applied in the case study with the aim to contribute in finding and evaluating different nature-based strategies to address these challenges. The study pursued to identify and assess possible Natural Assurance Schemes (NAS) that could help reduce water related risks while restoring the aquifers system status and functions. With this objective, a series of NAS strategies were co-designed with local stakeholders combining NBS and soft measures (Table 11.1). The process followed the iterative steps set by the NAIAD stakeholder protocol described in López-Gunn et al. (2022 – Chap. 2 in this volume). The collaborative approach was combined with an analysis of their legal and technical feasibility by the DRBA.

R. Giordano
CNR-IRSA, Bari, Italy

E. Gómez
Climate Service Center Germany (GERICS), Helmholtz Center Geesthacht, Hamburg, Germany

C. Marcos
Confederación Hidrográfica del Duero, Valladolid, Spain

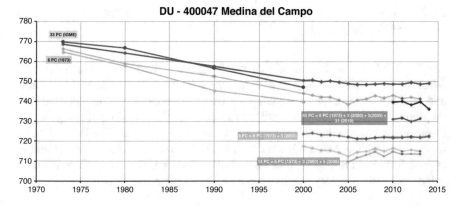

**Fig. 11.1** Groundwater level evolution in Medina del Campo Groundwater body (1975–2012) according to river basin management plan. (Source: CHD 2013)

**Table 11.1** NAS strategies considered for reducing vulnerability against drought risk while restoring the aquifer system status and functions in the Medina del Campo Groundwater Body

| Business as usual | Strategy 1 | Strategy 2 |
|---|---|---|
| **Measures** | | |
| No measures are applied | **NBS:** | **NBS:** |
| | Crop change towards drought resilient species | Aquifer recharge through the Zapardiel river using the Adaja irrigation infrastructure to restore riverine ecosystems |
| | Soil and water conservation practices | |
| | **Soft:** | **Soft:** |
| | Water user associations (WUAs) formation | WUAs formation |
| | Abstractions monitoring and control | Abstractions monitoring and control |
| | Environmental awareness raising | Environmental awareness raising |

The NAS strategy considered for reducing flood risks was based on an intervention project designed by the DRBA consisting on the removal of dikes, the re-naturalisation of the Zapardiel River banks and the enhancement of floodplains upstream to prevent floods in the town of Medina del Campo. However, the assessment of this strategy could not be performed because it was not technically possible to simulate the potential effects of the measures on reducing the flooding impacts. Therefore, only the drought related NAS strategies were analysed and are discussed hereafter.

This chapter presents the different methods and tools developed to assess the impacts and effectiveness of the selected NAS strategies in biophysical, economic and social terms, following the approaches presented in Chaps. 4, 5 and 6 of this volume. It also summarizes the approaches for integrating all these assessments, as well as the main conclusions and lessons learnt with views to the design and implementation on NAS for adaptation to droughts in the study area.

## 11.2 Biophysical Characterization and Assessment of the MCGWB

The MCGWB is the central part of the former so-called Los Arenales Aquifer System (Fig. 11.2), currently divided into three Groundwater Bodies. It is a semi-arid region, with average precipitation around 450 mm/year and high dependency on groundwater for agriculture and urban demands. The MCGWB belongs to the Duero river basin, administratively encompassed within the Castilla and Leon region (Spain). The MCGWB covers an area of 3700 km$^2$ (see Fig. 11.2) and is crossed from West to East by four main rivers – Guareña, Trabancos, Zapardiel and Adaja – and one tributary of the Adaja river by the left margin – Arevalillo.

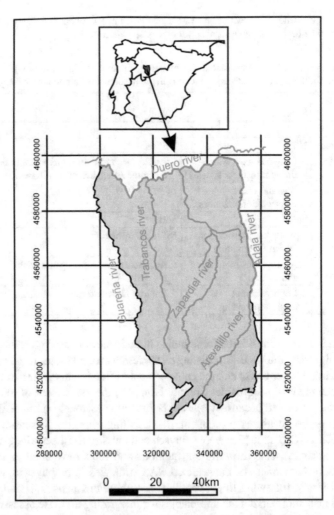

**Fig. 11.2** Location of Medina del Campo groundwater body within the Duero basin. (Source: modified from de la Hera-Portillo 2020a)

Intensive groundwater exploitation since the early 1970s, mainly for agriculture, induced a decline of groundwater levels up to 30 m in 40 years between 1972 and 2012. The sustainability indicator defined by the WFD, the Exploitation Index (ratio between groundwater recharge and abstractions), has passed the threshold value of 0.80 for the mentioned period (1972–2012) and is currently established in 1.65, according to the DRBA (DRBA, oral communication).

The overall aim of the analyses included in this section was to improve the understanding of the aquifer system structure and functioning, including the aquifer-rivers-wetlands relationships, as a basis to support simulation of the potential physical effects of the NAS strategies on the evolution trends of groundwater quantity and quality under different climate scenarios.

## 11.2.1 Improvement of Geological, Geophysical and Hydrogeological Knowledge

The first step was to check that most part of the hydrogeological knowledge of the MCGWB came from the former recognition of Los Arenales Aquifer System. The previously existing geological, geophysical and hydrogeological knowledge was compiled and reviewed as a baseline to build a conceptual model of the MCGWB.

The study has detected an important knowledge gap with regard to the geometry of the aquifer, which was estimated around 500 m deep based on the level of the existing wells. The existing geological and geophysical information has been integrated into a 3D geological model that allowed to reproduce the geometry of the whole aquifer system, showing that the real depth can go down to 2000 m in the area of Arevalo (De la Hera-Portillo et al. 2020b). This finding is key to understanding the response of the aquifer to groundwater management actions. The long period of high water pumping rates has caused the upper unconfined aquifer to act as a perched aquifer which recharges the deep regional aquifer, being the latter the most intensively exploited over the last 50 years. Under natural conditions, the rivers and wetlands played as discharge zones for the entire aquifer system, whereas in the current conditions they are solely connected to the upper layer (perched aquifer), receiving discharges from the unconfined aquifer only in wet periods. Under natural conditions (i.e. before the increase in water pumping that started in the 1970s), there were over two thousand wetlands and surface water bodies associated to the MCGWB, of which only a few remain today (de la Hera-Portillo et al. 2020a).

A second analysis consisted in a review of historical information about one of the most representative wetlands of the MCGWB: Lagunas Reales wetland. They are located 5 km southwest of Medina del Campo town (Fig. 11.2). There are references reporting the use of these ponds for fishing of tench and eels along 1500–1800, although water transfers from the Zapardiel River were also a frequent practice in order to keep the appropriate water level. The sedimentological analysis shows that these ponds should have been permanent before 1700, seasonal between 1700 and 1983, and fed by groundwater after 1983. From that date onwards, the system became disconnected from the regional aquifer (De la Hera-Portillo et al. 2020a), as a result of the piezometric drop caused by increasing pumping for irrigation. The sediments also show that the salinity of the waters has changed over time and there are markers of human activity (Mediavilla et al. 2020).

A set of facts called our attention when studying the aquifer-rivers historical relationship: concerning the four rivers crossing the MCGWB: (1) All these rivers are currently dry; (2) the existence of gauging stations in these rivers was anecdotal; (3) the available information on gauging measurements points out that some base-flow occurred mainly in the middle and lower sections of these rivers some decades ago (De la Hera-Portillo et al. 2020a); (4) The historical data available on the IGME databases show that the groundwater levels of the regional aquifer in the 1970s (considered the reference for natural conditions) were slightly lower than the water table of the upper unconfined aquifer.

The previous results allow us to conclude that the current conceptual model of the MCGWB presents a perched unconfined aquifer recharging the regional deep aquifer, which has been intensively exploited. The consequences are seen in the damage caused to the surface water ecosystems (rivers and wetlands) and terrestrial ecosystems that relied on groundwater (woods and riverine vegetation) today disappeared.

## 11.2.2   Assessment of Groundwater-Related Ecosystem Services

Due to the above mentioned pressures, the flow to humans of groundwater-related ecosystem services (GRES) in MCGWB has been reduced. To improve understanding of the aquifer-rivers-wetlands system and support simulation of the impact of particular NBS on the system, the current and potential future trends of the following set of Provisioning Services (PS) and Regulating & Maintenance Services (RM) related to GRES have been assessed: (i) Provision of fresh groundwater for irrigation (PS1); (ii) Provision of good-quality groundwater for human supply (PS2); (iii) Capacity of maintaining base-flow to streams (RM1); (iv) Capacity of maintaining riverine forests (RM2); (v) Capacity of maintaining wetlands (RM3).

The assessment of PS1 and PS2 current trends faced a main problem: the lack of data representativity with respect to particular aquifer depths. Many monitoring boreholes have long screened intervals, thus providing averaged data from large aquifer thicknesses. With respect to PS1, the statistical analysis of 169 piezometers with data records from 1985 to 2018 (and with at least ten continuous years of data) showed two periods with different behaviour (Borowiecka et al. 2019a, b): (1) between 1985 and 2001 there was a generalised tendency to lowering levels across the MCGWB pointing to intensive groundwater exploitation. (2) Between 2002 and 2018, only 25% of the observation points, mostly to the SE and E of MCGWB, showed clear increasing piezometric trends, while the rest showed weak and unclear tendencies. This points to diverse and localised causes, most probably abandonment of dried and expensive exploitations and replacement of groundwater for surface water for irrigation from the Cogotas reservoir (see above). Thus, the regional piezometric drawdown since 1985 reduced discharge from groundwater bodies to surface ecosystems and induced fast decrease of the Provisioning and Regulating & Maintenance Services PS1, RM1, RM2 and RM3 flows. Meanwhile, local reduction of groundwater exploitation just induced slow and localised recovery of groundwater levels over the last two decades.

With respect to PS2 assessment, elevated ionic balance errors in a number of analysis, and mixing of water from different flowlines in long-screened boreholes led to uncertain chemical representativity of $NO_3$ data and non-conclusive statistical results. To gain some insight about the current trends of PS2, two maps of $NO_3$ were drawn for two different moments, 1978–85 (99 data points) and 2018 (62 data points). Three zones were mapped: (A) with $NO_3 < 10$ mg/L, representing unpolluted groundwater; (B) with $NO_3 = 10$–49 mg/L, representing polluted groundwater but complying with the legal requirements on water quality, and (C) with $NO_3 \geq 50$ mg/L, representing polluted groundwater not complying with legal requirements. Comparing both maps, it seems that the surface area of zone A could have decreased around 3–4%, the surface of zone B could have decreased around 22%, and the surface of zone C could have increased some 25% between both periods. This would mean that a general improvement of PS2 is being produced. However, the assessment is weak for many reasons, most notably the mixing of data from different depths.

With the aim of understanding the present trends of Regulating & Maintenance Services RM1, RM2 and RM3, a geographical analysis was performed by comparing orthophotos of the study area from 1956 to 1957 (close to natural conditions) and 2017 (strongly modified regimes) years combined with spatial data from the Copernicus programme, with a shapefile of riparian zones, and a DRBA shapefile on wetlands and lakes (Poirée 2019). Concerning RM1, the length of streams receiving base-flow has decreased dramatically in three of the four main rivers: Guareña, Zapardiel and Trabancos, while it seems not to have changed significantly in the Adaja river. Thus, the partial piezometric recovery detected is not yet having an impact on the base-flow to rivers. Concerning RM2, the surface of riparian vegetation has decreased between 9% and 42%, most significantly for the Trabancos (36%) and Zapardiel (42%) rivers. Together with the lowering of the water table, land use changes, mostly for agricultural and urban uses, have also played an important role in the lowering of piezometric levels. With respect to RM3, the proportion of wetlands surface from 1956–57 to 2017 is just around 0.5%.

The hypothetical future trends of PS1, RM1, and RM3 GRES have been assessed through groundwater flow modelling. The numerical model, (using MODFLOW software) covers the MCGWB and the bordering groundwater bodies to the East (Los Arenales) and West (Tierra del Vino). The geometry, hydraulics, recharge, and exploitation functions were based on a sound review of all the available geological and hydrogeological data. The PS2 (potential evolution of $NO_3$ trends) is underway. After an acceptable calibration, groundwater flow was simulated for three different climate scenarios: (a) no change in current precipitation, (b) 3% increase in precipitation, (c) 8% decrease in precipitation (CEDEX 2012); and three different groundwater management scenarios: (1) Business as usual (BAU) (current exploitation index –EI- 1.65), (2) a reduction of EI to 0.85 by year 2050 and beyond, and (3) a reduction of EI to 0.8 by 2050 (DRBA goal) and beyond. The last two models aimed to provide sensibility on the impact of small groundwater management changes. The main results are presented in García-Alcaraz et al. (2019) and summarized hereafter: no changes in the piezometric level with EI = 0.8 by 2050, and 1 m lower (−1 m) with EI = 0.85, for whatever climate scenario. By year 2350, the recovery of the piezometric level would be 2 m higher with EI = 0.8 than with EI = 0.85 for whatever climate scenario. For EI = 0.8, the piezometric recovery rate between years 2018 and 2350 would be 0.012 m/year if the precipitation decreases 8%, 0.021 m/year if the precipitation increases 3%, and 0.018 m/year if there are no changes. Thus, groundwater management (through reducing the EI) has a much larger impact on piezometric recovery than climate change (through modified aquifer recharge).

Given the small impact and high uncertainty of climate projections, the assessment of wetland surface evolution relied mainly on the effect of groundwater management. Assuming a wetland area of 7000 ha in year 1950 as the reference for natural conditions, the best recovery trend is produced by management scenario 2 above: wetland surface recovery will be around 33% by year 2050 and 51% by year 2350 (Fig. 11.3).

### 11.2.3   Assessment of Managed Artificial Recharge (MAR)

One of the potential NBS considered as identified by stakeholders was Managed Artificial Recharge (MAR) through river-bed infiltration using the Trabancos River channel as infrastructure (Fig. 11.5). Its potential impact was assessed through modelling and simulation of the recharge conditions (see location in Fig. 11.4).

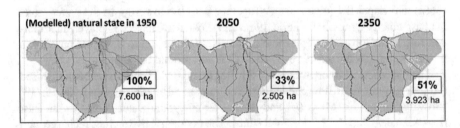

**Fig. 11.3** Modelled evolution of wetlands surface area with no precipitation changes and Exploitation Index = 0.8. (Modified from García-Alcaraz et al. 2019)

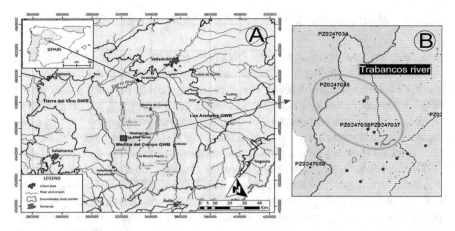

**Fig. 11.4** (**a**) Location of the MCGWB in the Duero Basin (Spain) (modified from De la Hera-Portillo 2020a, b). (**b**) Detail of the Managed Artificial Recharge area in the Trabancos River

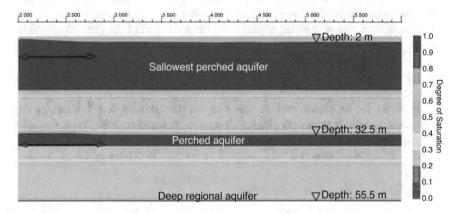

**Fig. 11.5** MAR impact in MCGWB considering an average-type decade: 3 wet years +5 average years +2 dry years, t = 3650 days. The x axis represents the distance from the river where the recharge would take place (Aguilera et al. 2019)

MAR as an action to increase the renewable water resources in the system has been analysed by an unsaturated-saturated flow model (Valle et al. 2018; Aguilera et al. 2019) using VS2DTi code (Hsieh et al. 2000). The area selected to simulate this MAR was the Trabancos River (Fig. 11.4). The results show that simulated MAR, with a recharge volume estimated in 3 Mm$^3$ in an average year, has no impact on the deep regional aquifer after 10 years of artificial recharge. It only increases the level of the perched isolated aquifers in the river surroundings (Fig. 11.5). A new groundwater model is under construction to test the potential effects on the Zapardiel basin. A simulation of the impact of the MAR on the deep aquifer was also performed, showing only a slight impact in the long term (after 10 years).

### 11.2.4 Hydrological and Water Allocation Assessment

Understanding the hydrology of the MCGWB (and its associated sub-basins) is essential for understanding the water demanded by the existing water users in comparison with the water availability. This section assessed the potential impact on the MCGWB water balance of the NBS in Medina NAS strategy 2, i.e. crop changes towards more drought resilient species as well as crop rotation. The hydrological and water allocation assessment was carried out using the RIBASIM model. The analysis conducted shows significant impacts when changing the status quo (Fig. 11.6).

The system is under pressure in the present situation. Today most irrigated areas are dependent on groundwater. However, over abstraction of groundwater for irrigation purposes is depleting groundwater dramatically as aforementioned. Agricultural practices are not being carried out considering this water stress situation. Therefore, a change of public subsidies towards support of the use of drought-resilient crops and more efficient technologies, in combination with technical and capacity support, would be advisable. According to the hydrological and water allocation assessment, the water demand is drastically reduced when farmers change their cropping plan. The model shows that this results in an improvement of the present situation, with high increases in the water level. A more sustainable system that is also more resilient to uncertain possible futures is possible when changing the existing farming practices with the new proposed measures. Moreover, it is recommended to conduct a detailed study of possible measures that improve and expand the current infrastructure for supplying surface water beyond the existing intervention, as the hydrological analysis shows considerable opportunities and co-benefits particularly in those sub-basins linked to Adaja and Duero rivers.

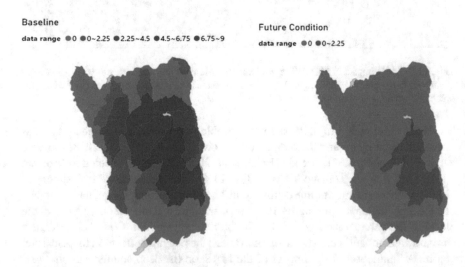

**Baseline**
data range ●0 ●0~2.25 ●2.25~4.5 ●4.5~6.75 ●6.75~9

**Future Condition**
data range ●0 ●0~2.25

**Fig. 11.6** Results for the transient regime for the baseline current situation and the future condition with the application of NBS in NAS strategy 2. (Source: own elaboration)

## 11.3   Risk Assessment of Natural Hazards in MCGWB

The assessment of hazards has proven that the MCGWB is located in an inflexion point with regard to global and regional climate models. This means no significant changes in the main studied variables, i.e. temperature, rainfall, floods and droughts, can be expected due to climate change. In technical terms, this is proven by the fact that the maximum average variations in such parameters are still less than a quarter of the standard deviation (approx). Hence, the current situation with regard to risk should be considered as the starting point towards any societal, economic or land use planning changes, and these changes will bear responsibility for any increase or decrease in risk.

With regard to floods, a rainfall-runoff to flood model was followed as a first approach for the Zapardiel River, aiming at providing an estimate of floods in Medina del Campo Town. The first step was to analyse the precipitation, which showed extremely variable with no significant trends. The model was calibrated at the entry point of the Zapardiel into Medina del Campo town. A coupled 1D and 2D approach would have been ideal to fully model the Zapardiel flow through Medina del Campo, given the river has been channelized. The river was modelled without any obstacles due to specific characteristics of the study. Nevertheless, an approach to test the impact of those elements is highly advised. Therefore, the most similar to natural conditions was modelled, excluding the channel, which is in fact designed with legal bindings regarding floods. The concluding remark is that, under natural conditions, the Zapardiel River can overflow within a return period of 50 years (that is a 2% chance each year) with very limited impact, if any. An event with a 1% chance a year will already cause several areas to become flooded, and an event with a 0.2% chance a year (matching the legal limit to consider flood prone zones) will provide a flood that could reach the Town Hall. Meanwhile, any action upstream focused on retaining flood waters could reduce the peak of the flow, hence it would contribute to lessen the impact to a degree still to be estimated. Given the low slope of the area, a very large amount of water should be retained in the basin in order to minimize the impact of floods in Medina del Campo. River restoration measures could help in order to attain this goal. Nevertheless, further analyses considering the obstruction structures should be performed. In fact, a coupled groundwater model and a distributed rainfall-runoff model with a mixture of 1D/2D approaches would be the best way to proceed.

Overall, the main variable controlling risk in Medina del Campo is human activity. It is on the choice of the inhabitants to give the river its natural space and carry out activities compatible with the scarce rainfall, allowing for manageable risk; or to continue a model that adds pressure to the ongoing physical situation, which would build in increased risk.

## 11.4    Social Assessment: Risk Perception and Selection of NAS Strategies

The methodologies developed by Giordano et al. (2022 – Chap. 5 in this volume) were applied in Medina case study to undertake a social analysis of risk perception and support the selection of NAS strategies, while detecting implementation barriers and levering actions. This section focuses on the barriers due to lack of effective interactions among different decision-makers and stakeholders. The work assumes that effective NAS require cooperative actions from stakeholders, since individuals do not make decisions in a vacuum.

The process involved three main steps. First, a participatory mapping exercise with local stakeholders was carried out to identify the network of stakeholder interactions, the information flows between them and the tasks developed by each of them to address risk reduction. This exercise was part of the second stakeholders workshop aimed to co-design the NAS strategies, in line with the stakeholder engagement protocol described in Chap. 2 (López-Gunn et al. 2022 – this volume). Second, the research team analysed the map in order to detect the main vulnerabilities and barriers, as well as key elements in the network. The obtained maps detected four key barriers, as well as several interventions that could be implemented to overcome them. The identified barriers were the following:

First, the **role of water user associations (or WUAs)**, which is strongly affected by the farmers' willingness to associate, depending on the farmers' risk awareness. The lack of WUAs involvement in the NAS process would very importantly reduce the effectiveness of controlling individual water abstractions.

Second, the **low level of acceptance of the water rights allocation**. Most of the involved farmers considered the process unfair and not fully transparent. More effective implementation of the water rights would have led to a reduction in aquifer exploitation.

Third, the **technical support to farmers for enabling crop change toward less water-demanding crops**. The lack of WUA involvement affected the sharing of this technical information and, thus, the farmers' behavioural change towards more sustainable water consumption.

Fourth, the **low level of accessibility of information concerning the groundwater state**, mainly due to the low level of availability and reliability of data on the groundwater level (i.e. groundwater metering). This information would have an impact on the farmers' risk awareness and, consequently, the formation of WUAs and the implementation of the crop change strategy (NAS strategy 2).

The exercise showed that, in order to facilitate the cooperative work of decision-makers and stakeholders for NAS co-design and implementation, actors need to be better involved in the process. Information should flow within the network and tasks should be carried out cooperatively. In the third and final step, a simulation model was used to identify the most effective interventions, as described in Giordano et al. (2022 – Chap. 5 in this volume). The list of identified interventions to address these barriers and a comparison of their effectiveness at tackling those barriers is shown in Fig. 11.7.

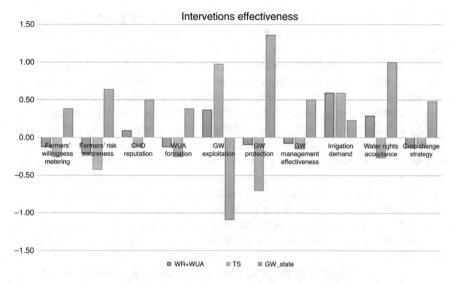

**Fig. 11.7** Comparison among the three intervention scenarios. (Source: own elaboration)

Figure 11.7 shows that the most effective interventions are those aiming at enhancing the availability, reliability and accessibility of the information related to the groundwater state. These interventions could have a twofold positive impact. On the one hand, they could lead to an increase in the farmers' risk awareness and, in doing so, they could contribute to enable the shift towards less water-demanding crops. On the other hand, the availability of reliable information on the groundwater state could allow the DRBA to enhance the effectiveness of the territory control and the groundwater management. This, in turn, could have a positive impact on the farmers' perception of the DRBA role and, consequently, could lead to a higher acceptance of the water rights management process.

## 11.5  Economic Assessment of NAS Effectiveness

The two NAS strategies proposed for coping with drought hazard risks in the MCGWB (see Table 11.1) were economically evaluated under different scenarios following the economic cost-benefit analysis framework presented in LeCoen et al. (2022 – Chap. 6 in this volume). The aim was to assess whether the NAS strategies can help to increase the provision level of groundwater-related ecosystem services while reducing the economic impacts of hydrological risks.

The soft measures in both strategies would reduce water availability for farmers by restricting groundwater abstractions. In addition, the MAR in strategy 2 would not result in increased water availability for irrigation and thus would not reduce drought damages, whereas strategy 1 would reduce drought economic risk through the introduction of alternative crops and soil conservation practices. It must be noted that reducing groundwater extractions is a must rather than an option. Therefore, the "Business as usual" (BAU) strategy was redefined as the BAU with the restriction of groundwater abstractions, and both strategies were compared with this redefined BAU scenario. The climatic and socio-economic/regulatory scenarios considered were also selected using a combination of expert-based and participatory co-development approaches. The evolution of the Common Agricultural Policy (CAP) subsidies was identified by stakeholders as a critical driver of land use change. Consequently, both a current CAP scenario and a more environmentally-oriented CAP scenario were considered. Regarding climatic scenarios, a preliminary assessment of trends in average and maximum precipitation using the RPC 4.5 and RPC 8.5 IPCC projections (CEDEX 2012) showed that no significant trends existed in any of the rainfall series to the project time horizon (2050), in line with the risk assessment results previously mentioned (Llorente et al. 2018). Consequently, only one climate scenario, based on the historical trends, was simulated (Calatrava et al. 2019).

The economic impacts of the NAS strategies in terms of reducing drought risk have been assessed using an agro-economic model calibrated to the technical, economical and hydrological characteristics of the study area. This model simulates land and water allocation among cropping alternatives to improve the aquifer's conditions and reduce drought risk in irrigated agriculture, under the different climatic and socio-economic scenarios considered, and computes several economic, social and resources use indicators. The method and results of the economic assessment of the impact of NAS strategies on the avoided damages is detailed in Calatrava et al. (2019). The impact of the CAP scenarios on the results for both strategies is negligible. The insurance value of the NAS strategies, calculated as the difference in the mean annual avoided damage between each strategy and the BAU + control of groundwater abstractions case, is 12 M€/year for strategy 1 and 0 for strategy 2, as the recharge has no effect on the agricultural irrigation capacity (see Chap. 6). The largest share of the costs for both strategies corresponds to the soft measures, especially to the creation of Water Users' Associations.

Although the strategies' co-benefits have not been monetarily evaluated, the combination of the different analysis performed allowed for the qualitative and/or quantitative assessment of major expected co-benefits, as summarised in Table 11.2.

Identified co-benefits include an increase in water productivity, job generation and the profitability of agricultural employments, suggesting a greater potential for higher agricultural wages. Regarding the environmental co-benefits, both strategies would have similar impacts on the improvement of the aquifer's quantitative and qualitative status, although the artificial recharge in strategy 2 would accelerate the aquifer's improvement and would positively impact on some riverine ecosystems. Lastly, strategy 1 implies a less intensive farming than strategy 2, as crop rotations and less water-demanding crops are fostered, resulting in an environmental improvement of agricultural systems.

The results of the economic assessment show that the second strategy does not reduce drought risks but improves local riverine ecosystems, while the benefits of

**Table 11.2** Co-benefits assessment for each strategy with respect to the BAU situation

| Item | CO-benefit | BAU + GW control | Strategy 1 | Strategy 2 |
|---|---|---|---|---|
| Aquifer quantitative status | Reduced extractions | −50.50% | −50.50% | −50.50% |
| | Aquifer exploitation index | 1.00 | 1.00 | 1.00 |
| | Changes in piezometric levels | No effect on the deep aquifer | No effect on the deep aquifer | No effect on the deep aquifer |
| Aquifer qualitative status | Changes in water quality | Compliance with standards | Compliance with standards | Compliance with standards |
| Groundwater-dependent ecosystems | Generation of, and support to, ecosystems | – | – | Local ecosystems improvement |
| Environmental status of agricultural systems | Irrigated area with permanent crops (%) | 19.49 | 23.95 | 19.49 |
| | Irrigated area under crop rotation (%) | – | 78.50 | – |
| Water use | Water use efficiency (m³/ha) | 5629 | 5303 | 5629 |
| | Average water productivity (€/m³) | 0.88 | 0.97 | 0.88 |
| | Social profitability of water (jobs/Mm³) | 15.76 | 16.18 | 15.76 |
| Agricultural employment | Employments in agriculture (jobs/ha) | 0.0887 | 0.0858 | 0.0887 |
| | Profitability per employment (€/job) | €28,240 | €32,687 | €28,240 |
| | Minimum area to grant an average salary (ha) | 11.95 | 10.65 | 11.95 |

Source: Calatrava et al. (2019). Average values for the two CAP scenarios considered

strategy 1 largely outweigh its costs, with a 3.17 benefit/cost ratio, even if the existing co-benefits were not considered (see Calatrava et al. 2019). However, strategies 1 and 2 are not conflictive but highly complementary and should be ideally combined to accelerate the aquifer's recovery and increase other environmental co-benefits.

An exercise to develop business models for the analysed strategies was carried out collaboratively with the stakeholders during workshop 3 of the stakeholders engagement protocol (López-Gunn 2022 – Chap. 2 in this volume), following the methodology of the NAS canvas framework (Mayor et al. 2022, Chap. 8 in this volume). The results showed the required actors, roles and elements required for the implementation of the strategies, and the possible funding instruments that could help the implementors face the upfront and maintenance costs. The resulting NAS canvases for Medina strategies can be found in Mayor et al. (2019, 2021).

## 11.6 Integration of Results from the Biophysical, Social and Economic Assessments

### 11.6.1 Qualitative Integration of Results in Medina Del Campo-Drought

The first approach for integration of results was the construction of a Systems Dynamic Model to understand the qualitative relations, in line with the methodological approach described by Basco et al. (2022 – Chap. 7 in this volume). The results of the qualitative system dynamic model developed for Medina case study can be seen in Fig. 11.8. The conceptual map describes the complex interconnections of the MCGWB according to stakeholder's perception. The system dynamics qualitative model (SDQM) aggregates knowledge coming from different stakeholders with different degrees of expertise. It represents the dynamic hypotheses and participant's views on the cause-effect relationships underlying social, economic and environmental factors. It reveals the complexity of the interactions composing the system. In this SDQM, 82 feedback loops are responsible for the underlying behaviour of the system. In order to facilitate the comprehension of the diagram and

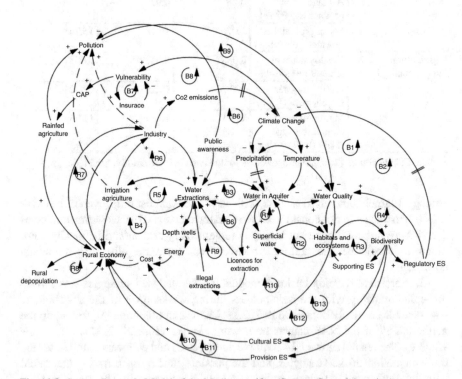

**Fig. 11.8** System Dynamic Model of the Medina aquifer. (Source: Own elaboration)

to easily find strong leverage points, the most important loops are highlighted. We identified 9 positive loops indicating self-reinforcement of a given change within the loop and 13 balancing loops indicating self-regulation of a given change within the loop (Fig. 11.8). The stakeholders selected the variable 'water in the aquifer' as the core variable of the system.

According to stakeholder's view, the underlying causes of the aquifer water level decrease are driven by the dominance of reinforcing loop R5 and R9, which are reinforced by loops R6 and R7. This means that the main factors affecting the aquifer are water extractions that are induced by a need of perpetuating an economic system based on agriculture and agriculture-based industry. A dominance of loopR5 would weaken loops R1, R2, R3, R4, which are responsible for the delivery of the main ecosystem services associated with the aquifer.

The dominance of balancing loop B3 regulates the system. This means that when the water level is reduced to a certain point, the strength of R1 is also reduced (water extractions decrease).

## 11.6.2 Quantitative Integration of Results: Effectiveness of NBS for Strategic Adaptive Planning in Medina Del Campo

The effectiveness of NAS strategies for MCGWB requires the integration of the results from the previously described studies in a manner that is understandable, acceptable and sufficiently accurate for decision-makers and end users. The meta-model developed in the final phase of the project allows assisting the Medina case study in the multi-criteria decision making and evaluation of the NAS strategies considering possible futures. Overall, the meta-model allows the combination of biophysical, economic and social indicators resulting from the groundwater and water allocation models and carrying out a type of multi-criteria analysis. With this meta-model, users can access the results of different models in a fast-integrated systems model. Meta modelling enables relevant stakeholders to be involved in the whole modelling chain using a collaborative modelling approach. The interactive visualization of the integrated results and the resulting integrated indicators enables a rapid and comprehensive assessment of the effectiveness of NBS and soft measures in terms of direct benefits and co-benefits. An overview of the indicators used for assessing the effectiveness of NBS for Medina del Campo are presented in Fig. 11.9.

The current prototype, already including the stakeholder's views and understandings, can be accessed at: https://app.powerbi.com/view?r=eyJrIjoiMDYzOThmY-zAtZWYxZi00NWFiLTk4OGUtYmU2MmVhZDAzZWFhIiwidCI6IjE1ZjNmZT-BlLWQ3MTItNDk4MS1iYzdjLWZlOTQ5YWYyMTViYiIsImMiOjh9

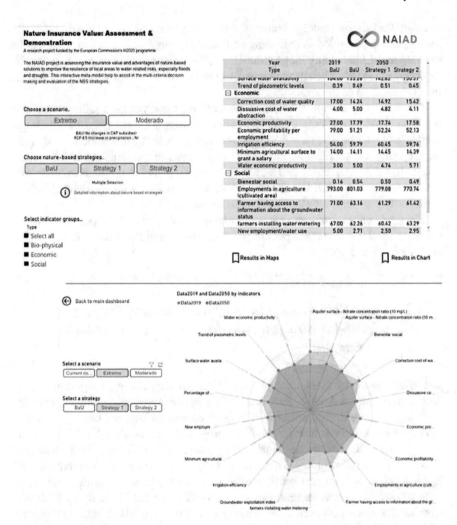

**Fig. 11.9** Visualization of meta-model dashboard. (Source: NAIAD website. https://app.powerbi.
com/view?r=eyJrIjoiMDYzOThmYzAtZWYxZi00NWFiLTk4OGUtYmU2MmVhZDAzZWFhIi
widCI6IjE1ZjNmZTBlLWQ3MTItNDk4MS1iYzdjLWZlOTQ5YWYyMTViYiIsImMiOjh9)

The main elements of the meta-model are:

- *Main dashboard* which summarizes and qualitatively compares the valuation of
  indicators based on the selection of two scenarios (extreme and moderate) and three
  strategies (Business as usual, strategy 1 and strategy 2). The user can select what
  scenario, strategy, and indicators to evaluate by clicking on the appropriate boxes;
- *Spatially distributed results* for each indicator can be accessed through the indicator table in the main dashboard;
- *Description* of strategies that are included in the models. This also includes the
  strategies that are monitored and modelled separately (groundwater recharge and
  floods).

The meta-model should not only be considered as a tool, rather as a collaborative modelling process that facilitates the conveyance of ideas, working outputs and mental models. The development of trust, acceptance of other's perspectives and shared sense of ownership is indispensable for creating partnerships that allow participatory adaptive planning and management.

## 11.7  Conclusions and Lessons Learnt

The work performed in the Medina del Campo case study has entailed important progress in the understanding and description of the MCGWB from many disciplines. Furthermore, several models and tools with high performance levels have been generated and made available for use by the DRBA, which will inform the current and upcoming River Basin Planning stages.

Some of the main conclusions and lessons derived from the biophysical, risk, social and economic assessments above presented, as well as from their integration to assess the effectiveness of the co-developed NAS strategies, are summarised below.

- Identification of the main changes (stages) that occurred in the MCGWB at space-time scale: a previous unperturbed stage (before 1970) with groundwater levels close to ground surface and discharge of groundwater into the rivers and wetlands; a transition period characterized by an increasing and intensive groundwater pumping (between 1970 and 2010); and a stage with shy and slow signs of recovery at the local level from 2010.
- Between years 1956 and 2017, the surface of riparian vegetation in the streams has decreased between 9% and 42%, most significantly in the Trabancos (36%) and Zapardiel (42%) rivers. The proportion of wetlands surface from 1956–57 to 2017 is just around 0.5% of the 7600 ha identified. Even though the piezometric levels are recovering in part of the MCGWB, this is not yet having impact neither on the capacity of the aquifer to maintain riparian forest, nor to generate baseflow to rivers and discharge to wetlands.
- Reducing groundwater exploitation at a regional scale is the most effective measure to produce a widespread recovery of the groundwater levels, and thus increase water supply and resilience against droughts. Given that the rivers, wetlands and riverine forests are linked to the water table, recovering GRES related to regulation and maintenance of surface ecosystems would require to increase the groundwater stored in the aquifer system, and thus a reduction of the exploitation index.
- Improving the representativity of the monitoring networks is a main and paramount need to allow improved assessments of the current and future trends of groundwater quantity and quality in the MCGWB.
- The implementation of MAR to the unconfined aquifer could improve to some extent the riparian conditions of the Trabancos river water in the short term. Nevertheless, the simulated impacts on the deep aquifer are almost imperceptible in the long term.

- The most effective NAS strategy in terms of economic impacts and co-benefits would be strategy 1, although a combination of both would provide the highest impact. However, these measures would need involvement of the farmers. They are aware of the aquifer value and recognise their reliance on groundwater for the future, but they would also need support to address the implementation and opportunity costs in a joint effort between public and private investments.
- Open communication and awareness raising for farmers about the situation, as well as the collaboration between the DRBA and all affected stakeholders in an iterative process to find suitable solutions, can contribute to tear down some of the barriers for the acceptance and adoption of the proposed nature based strategies chosen by the stakeholders. An open and direct provision of high quality information about the evolution of groundwater quantity and quality would improve confidence and awareness among stakeholders.

# References

Aguilera H, Heredia J, De la Orden JA, De la Hera A, Del Barrio Beato V (2019) Assessing the feasibility of managed aquifer recharge through unsaturated zone modeling with V"2DTI in a multilayer system. In: 46th IAH Congress, Málaga, Spain, pp 450

Borowiecka M, García-Alcaraz MM, Manzano M. (2019a) Analysis of piezometric trends in the Medina del Campo Groundwater Body to understand the status and drivers of changes of groundwater-related ecosystem services. NAIAD EU Project. In: Proceedings of IAH2019, the 46th Annual Congress of the International Association of Hydrogeologists, Málaga (Spain). J. Jaime Gómez Hernández and Bartolomé Andreo Navarro (eds.) Electronic Ed. www.aih-ge. org; pp 296

Borowiecka M, García-Alcaraz MM, Manzano M. (2019b) Analysis of piezometric trends in the Medina del Campo Groundwater Body to understand the status and drivers of changes of groundwater-related ecosystem services. NAIAD EU Project. Proceedings of the 46th International Association of Hydrogeologists, Málaga, Spain, 22–27 September 2019

Calatrava J, Mayor B, Martínez-Granados D (2019) DELIVERABLE 6.3 DEMO Insurance value assessment report- Part 1: Spain – Medina del Campo. EU Horizon 2020 NAIAD Project, Grant Agreement N°730497

CEDEX (2012) Evaluación del impacto del cambio climático en los recursos hídricos en régimen natural [Assessment of the impact of climate change on natural regime water resources]. Centro de Estudios y Experimentación de Obras Públicas. 268 pp

CHD (2013) Informe sobre zonificación de las masas de agua subterránea Tordesillas, Los Arenales, Medina del Campo y Tierra del Vino y propuesta de otorgamiento de concesiones y autorizaciones en dichas masas. 46 pp

De la Hera-Portillo A, López-Gutiérrez J, Llorente M, Mayor B, Mediavilla R, Heredia J, López-Gunn E (2020a) Reconstruction of the river-aquifer interaction in Medina del Campo groundwater body (Duero basin, Spain) as base knowledge to restore aquatic ecosystems. In: 6th IAHR Europe Congress. 14th–17th September 2020. Warsaw (Poland)

De la Hera-Portillo A, Marín-Lechado C, López-Gutiérrez J, Ruíz-Constán A, Rubio F, Bohoyo F, Orozco T, Llorente M (2020b) First geophysical-based assessment of the groundwater storage in an intensively exploited multilayer aquifer. Results obtained in NAIAD Project for the Medina del Campo groundwater body. 10 Asamblea Hispano-Portuguesa de Geodesia y Geofísica. Toledo

García-Alcaraz M, Manzano M, Faneca M, Trambauer P, Pescimoro E, Altamirano M (2019) Understanding the potential of nature based solutions to recover the natural ecosystem services of the Medina del Campo Groundwater Body in the NAIAD EU project. In: Proceedings of IAH2019, the 46th annual congress of the international association of hydrogeologists, Málaga (Spain). J. Jaime Gómez Hernández and Bartolomé Andreo Navarro (eds.); pp 156

Hsieh P, Wingle W, Healy RW (2000) VS2DI—A Graphical Software package for simulating fluid flow and solute or energy transport in variably saturated porous media, Water-Resources Investigations Report 9-4130. U.S. Geological Survey, Lakewood

Llorente M, De la Hera A, Manzano M (2018) "DELIVERABLE 6.2 From hazard to risk: models for the DEMOs. Report Part 1: Spain – Medina del Campo". EU Horizon 2020 NAIAD Project, Grant Agreement N°730497

Mayor B, López Gunn E, Zorrilla P, Nanu F, Groza I, Schrieu A, Marchal R, Le Coent P, Graveline N, Piton G, Biffin T, Daartée K, Pengal P, Krauze K, Douglas C, van der Keur P, Jørgensen ME, Calatrava J, Manzano M (2019) Deliverable 7.2: From Bankability to Suitability report: value capture and business models to catalyse implementation of NAIAD demo's NAS strategies. H2020 NAIAD project. Grant Agreement No 730497

Mayor B, Zorrilla-Miras P, Le Coent P, Biffin T, Dartée K, Peña K, Graveline N, Marchal R, Nanu F, Scrieu A, Calatrava J, Manzano M, Lopez-Gunn E (2021) Natural assurance schemes canvas: a framework to develop business models for nature-based solutions aimed at disaster risk reduction. Sustainability 13:1291. https://doi.org/10.3390/su13031291

Mediavilla R, Santisteban JI, López-Cilla I, Galán de Frutos L, de la Hera-Portillo Á (2020) Climate-dependent groundwater discharge on semi-arid inland ephemeral wetlands: lessons from Holocene sediments of Lagunas Reales in Central Spain. Water 12(7):1911

Poirée L (2019) Evolución de las zonas riparias de los ríos Trabancos, Zapardiel, Adaja y Guareña (cuenca del Duero) entre 1956 y la actualidad. (Evolution of the Trabancos, Zapardiel, Adaja and Guareña rivers riparian areas (Duero River basin) between 1956 and present). Master Thesis. Water and Terrain Science and Technology Master, Universidad Politécnica de Cartagena, Spain. Supervisers: Marisol Manzano and Virginia Robles

Valle P, Aguilera H, Heredia J, De la Orden JA, De la Hera A, Del Barrio Beato V (2018) Evaluación del impacto de la recarga artificial en la zona no saturada (ZNS) mediante modelización numérica: masa de agua subterránea de Medina del Campo. Congreso Ibérico AIH-GE, 2018, pp 145–154. 015

# Chapter 12
# Natural Flood Management in the Thames Basin: Building Evidence for What Will and Will Not Work

Mark Mulligan, Arnout van Soesbergen, Caitlin Douglas, and Sophia Burke

**Highlights**

- We write for those planning to implement or evaluate the effectiveness of, natural flood management (NFM) interventions.
- We describe application of spatial Policy Support Systems and the spreadsheet-based Eco:Actuary Investment Planner to help understand the investment costs and benefits associated with different types and magnitudes of NFM.
- We also outline example deployments of low-cost FreeStation, //Smart: monitoring equipment to evaluate existing NFM interventions particularly leaky debris dams, retention ponds and regenerative agriculture as deployed in the Thames Basin of the United Kingdom.
- Our guidance is focused on organisations lacking technical and/or financial capacity for NFM effectiveness evaluation, since we document applications of open-access, user-friendly, novel, low-cost decision support tools, monitoring equipment and protocols, using examples from the Thames Basin.

## 12.1 Introduction

This chapter is written to support those considering the use of natural flood management (NFM) interventions or wanting to monitor the performance of existing interventions. We focus on three types of NFM interventions – leaky debris dams, retention ponds and/or regenerative agricultural techniques such as low tillage (Box 12.1) – within the Thames basin. In 2017 the UK's Department of Environment, Food and ural

M. Mulligan (✉) · A. van Soesbergen · C. Douglas
Department of Geography, King's College London, London, UK
e-mail: mark.mulligan@kcl.ac.uk

S. Burke
AmbioTEK Community Interest Company, Essex, UK

© The Author(s) 2023
E. López Gunn et al. (eds.), *Greening Water Risks*, Water Security in a New World, https://doi.org/10.1007/978-3-031-25308-9_12

**Box 12.1: Natural Flood Management (NFM) Interventions Considered in this Chapter (Source: Authors' Own)**
NFM are landscape management activities that aim to retain more water in the landscape and reduce the speed at which water travels downstream thus reducing the concentration of runoff during a rainfall event and thus flood risk to downstream assets. NFM therefore has the potential to reduce flooding.

**Leaky Dams** are a flood mitigation technique in which logs are anchored across a stream to slow river flow. They store floodwaters upstream and encourage floodwaters to spill out of bank in low impact areas (such as forests and farmland) upstream of the assets at risk. There are many designs, most of which represent only a partial barrier.

**Retention Ponds** are bespoke topographic depressions that provide additional storage capacity for floodwater and slow it's flow into agricultural drainage systems and thus rivers. They are particularly common in the lower parts of agricultural fields

(continued)

**Regenerative Agriculture** is a land management technique that involves no or low tillage, the use of cover crops and diverse crop rotations to help restore soil structure to a more natural state, encouraging infiltration and reducing runoff generation. This management technique has the potential to increase the water storage capacity of the soil, thereby reducing downstream runoff generation and flood risk.

Affairs (DEFRA) released £1 million for local governments and community groups to apply for grants up to £50,000 to fund the building of NFM measures in pilot projects to help build the evidence base. Most projects opted to invest in leaky debris dams. Subsequently, the government released a further £14 million funding for larger scale projects in the north of the UK. An obligation of the DEFRA funding was that organisations monitor how well the interventions worked and thus help build the evidence base for NFM. However many organisations lacked the technical and/or financial capacity to monitor NFM effectiveness within the £50,000 budget limit. We therefore developed and deployed monitoring equipment and evaluation protocols to help these organisations fulfill their reporting obligations, since this equipment was also needed for our own investigations of the effectiveness of NFM. We also developed model-based decision support tools to assist organisations whilst planning scales and types of NFM interventions they might make. The methods for these tools are described in Chap. 4. In this chapter we show the results of two of these tools: the Eco:Actuary Investment Planner and the FreeStation //Smart: monitoring system.

## 12.2  Study Site: Thames Basin Context

The Thames River is in the south of the United Kingdom and runs about 350 kilo-
metres through the counties of Gloucestershire, Wiltshire, Oxfordshire, Berkshire,
Surrey, London, Essex and Kent before reaching the North Sea. The Thames basin
(Fig. 12.1) is home to around 13 million people, the majority of which live in
Greater London. The catchment is approximately 16,000 $km^2$ in size and is predominantly peri-urban with urban centres (i.e. Swindon, Oxford, Reading, Slough,
London) surrounded by suburbs interspersed with pockets of rural land. The 2013/14
winter storms lead to widespread flooding in the basin with clean-up costs of over
£1 billion (Thompson et al. 2017). London has been identified as one of the most
at-risk cities globally for flood hazard due to high levels of economic activity and
asset value at risk (EA 2009). Part of this risk derives from coastal inundation and
part of it from fluvial flooding.

Within the UK there has been a general move away from traditional flood
defences towards a flood risk management framework based upon more holistic
approaches to flood management including use of natural solutions (DEFRA
2012) – so called natural flood management (NFM). Most of the risk however is still
managed through traditional engineered defences, with £930 million invested during 2014–2015, and a further £180 million spent on maintenance in the same period

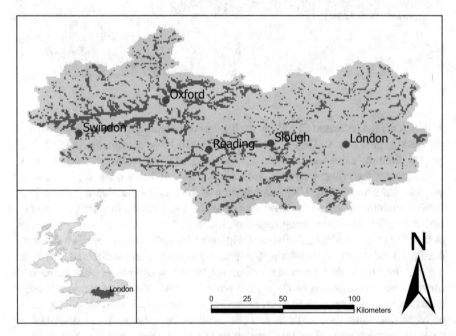

**Fig. 12.1** *Main:* Thames non-tidal catchment and floodplain areas, outside London (Mulligan
2019), showing the position of the five largest cities and inset the position of Thames basin in the
UK. The following map from the Eco:Actuary Policy Support System was used in the figure: topographic floodplain

(EA 2014). Despite the interest in NFM within the UK, barriers exist towards implementation of NFM interventions within the Thames basin. These include lack of space, high cost of land, potential resistance from landowners and lack of funding. NFM are also considered an unproven technology and may be more suited to mitigation on smaller streams and rivers (see Chap. 4).

## 12.3   Risk Assessment

Here we apply the Co$tingNature, WaterWorld and EcoActuary Policy Support Systems to provide an assessment of flood risks and natural flood mitigation in the Thames basin. We start by using the Co$tingNature Policy Support System (Mulligan 2015; methods: www.policysupport.org/costingnature) at 1 km spatial resolution to assess the Thames basin for its potential for, and exposure to, ecosystem relevant natural hazards (Co$tingNature is a globally applicable model that includes the following hazards: cyclones, tsunamis, landslides and flood)s. Hydrological flood is the greatest hazard for most of the Thames, with only landslides having a greater risk in some areas (Fig. 12.2). The greatest hazard potential is located along the main channel of the Thames, followed by its major tributaries. With regards to socio-economic exposure, high levels of exposure exist all along the main channels of the Thames with the greatest levels where the Thames runs through the cities of Oxford, Reading, Greater London and the Thames Estuary developments (Fig. 12.2). Within towns and cities, population and GDP have the greatest exposure, whereas for the rest of the basin it is agriculture; although, there are some isolated pockets throughout the basin where infrastructure has the most exposure (particularly roads and bridges). The urban areas of London, Windsor, Reading, Oxford and parts of the Thames Valley all have high hazard and exposure (Fig. 12.2).

Flood hazard is determined not just by precipitation events but also by the configuration of green infrastructure (water stores within the landscape) such as: floodplains, canopies, soil, water bodies and wetlands. Urbanisation removes these stores as cities spread into the surrounding landscape removing canopies and wetlands, channelizing rivers and concreting over soil stores. According to the natural flood storage module of WaterWorld (Mulligan 2013; methods: www.policysupport.org/waterworld), within the Thames basin, every year some 8.5 $km^3$ of water is stored by green infrastructure (Fig. 12.3). At the catchment scale most of this storage occurs in the form of floodplains, canopies and soils, only negligible amounts of water is stored in water bodies and wetlands, since these are sparsely distributed. However, only one-third of this green infrastructure within the Thames is protected, and thus much of this infrastructure is at risk from land use change and/or intensification. Damage to this green infrastructure will increase flood risk of downstream assets.

However, it is not just the presence of green infrastructure which is important but also its location within the catchment relative to rainfall and runoff generation and assets at risk. The main river channel of the Thames and many of its major tributaries have insufficient water storage capacity to accommodate the water generated

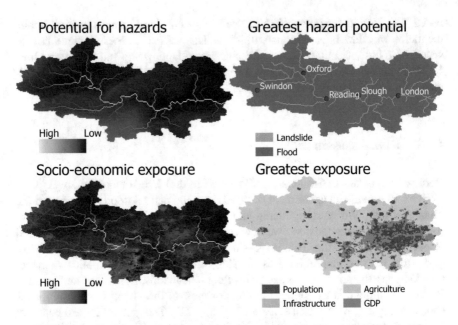

**Fig. 12.2** Spatial distribution of hazard potential and socio-economic exposure within the Thames basin (Mulligan 2020a) Hazard and exposure are both expressed in relative terms from low (0) to high (1) within the Thames basin. In this context, hazards refer only to ecosystem relevant hazards i.e. inland floods, cyclones, coastal inundation, landslides and soil erosion. Clockwise from top left: Potential for hazards; highest potential along the main channel of the Thames; Greatest hazard potential are erosion (orange) and hydrological flood (blue); Socio-economic exposure, from low (purple) to high (yellow); with greatest levels along the lower Thames; Principal socio-exposure to natural hazards is agriculture, GDP, population, and infrastructure. The following maps from Co$ting Nature were used in the figure; *clockwise from top left*: Relative potential for ES (ecosystem service) relevant hazard, Dominant hazard potential, Relative socio-economic exposure to ES relevant hazard, Dominant exposure. (Source: own elaboration)

during precipitation events as shown in Fig. 12.4 by application of the EcoActuary Policy Support System (methods: www.policysupport.org/ecoactuary). With insufficient stores in the landscape, flood risk increases and encourages dependency on protection from grey infrastructure. As both flood hazard and exposure is anticipated to increase in the future, action needs to be taken to reduce the fluvial flood risk within the Thames basin. Hazard is expected to increase due to changes in timing and magnitude of rainfall events and changes in land-use (Garner et al. 2017; Wheater and Evans 2009). Exposure is also expected to increase in the future due to increases in population and assets (Sayers et al. 2018). Without suitable mitigation this will likely lead to greater flooding and flood damage in the future (Wheater and Evans 2009; Ashley et al. 2005). The Thames basin has several physical and social characteristics which leave it vulnerable to flooding: notably, low elevation, high population density, high value assets, limited natural flood storage/drainage coupled with many commuters and pressured transport infrastructure and systems. Whilst our focus is fluvial flooding throughout the basin, key commercial and residential areas in central London are also at risk of coastal inundation.

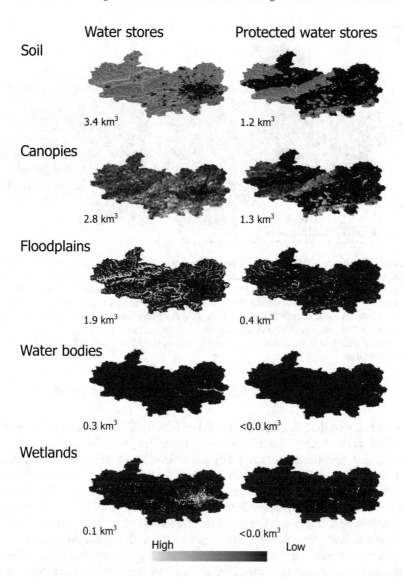

**Fig. 12.3** Spatial distribution and annual total water storage of the green infrastructure within the Thames basin; all water stores (*left*) versus protected water stores (*right*) (Mulligan 2020b). Purple areas have no storage and yellow areas have high storage. Numbers provided relate to the amount of water stored per year in km³. Every year 8.5 km³ of water is stored within the green infrastructure of the Thames, with soil and canopies acting as the greatest stores. Only one-third (2.9 km³/year) of this storage is protected. The following maps from WaterWorld were used in the figure, top to bottom and left to right: soil storage capacity, soil protected storage capacity, canopy storage capacity, canopy protected storage capacity, floodplains total storage capacity, floodplains protected storage capacity, water body storage capacity, water body protected storage capacity, wetland storage capacity, wetland protected storage capacity. Note that the wetland areas is overestimated in the urban area since the road surfaces pond water during rainfall and store it temporarily. (Source: own elaboration)

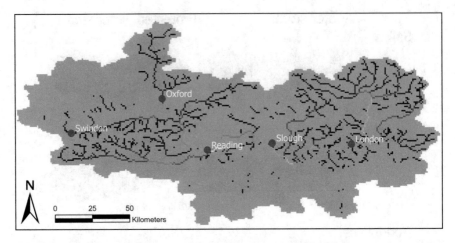

**Fig. 12.4** The river non-tidal network of the Thames basin showing locations where there is insufficient water storage upstream of the landscape to accommodate runoff events (Mulligan 2020c). As the colour of the river lightens the greater the excess water surplus; Reading, Slough and east London are particularly vulnerable to fluvial flooding. The Thames through London is tidal and thus not considered here. Potential water stores within the landscape are canopy, soil, wetlands and water bodies but see Fig. 12.3 for differences in storage capacity. The following map from Eco:Actuary was used in the figure: accumulated realised excess over permanent stores. (Source: own elaboration)

## 12.4 Natural Flood Management (NFM) Effectiveness

In order to assess the impact of NFM interventions within the Thames basin we created and applied novel monitoring and modelling techniques – namely the Eco:Actuary Investment Planner (EIP) and //Smart: monitoring sensors (methodological detail provided in Chap. 4). First, using the Eco:Actuary Investment Planner we identified the scale of investment required for different types of NFM intervention to achieve a given reduction in peak flow at key river sites. For this we focused our analysis on four sub-catchments of the Thames: the Lee, Stort, Mole, Coln and one adjacent catchment: the Medway (Fig. 12.5, Table 12.1). For the second part of our analysis we used //Smart sensors to monitor the ability of existing NFM interventions to store or slow the flow of water at sites in the basin. We primarily present results of NFM within the Thames basin but also present results from interventions in neighbouring catchments where these results add value. The monitoring and modelling methods complement each other and allow stakeholders to both plan for new interventions and assess the effectiveness of existing ones.

In order to evaluate the effectiveness of the NFM interventions, we (1) applied FreeStation //Smart: to assess the ability of the interventions to hold water within the landscape, (2) applied the EIP to evaluate the financial cost associated with the interventions to achieve a 15% reduction in flood peak downstream, and (3) applied Co$tingNature to assess the co-benefits associated with each intervention. We consider that the 'best' intervention is one which offers good value for money from a flood mitigation perspective.

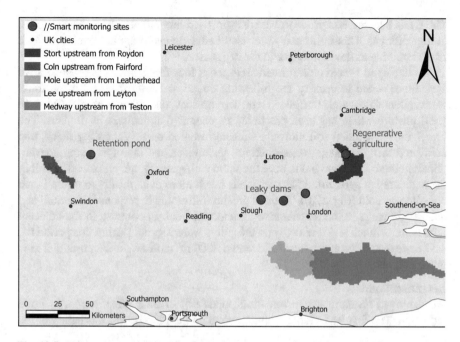

**Fig. 12.5** Sub-catchments of the Thames Basin with Locations of EIA simulation (Table 12.1) and locations of //Smart monitoring. Thames basin outline in dark grey

**Table 12.1** Sub-basins of the Thames Basin and adjacent (Medway) used as study sites for application of the Eco:Actuary Investment Planner

| Study sites | Upstream catchment area (km²) | Predominant geology |
| --- | --- | --- |
| Lee River | 663 | Chalk |
| Stort River | 175 | Chalk |
| Mole River[a] | 54 | Weald Clay |
| Coln River | 80 | Limestone |
| Medway River | 793 | Weald Clay |

[a] Complex geomorphology and steep river gradient contribute to very high flood risk

## 12.4.1   Ability of NFM Interventions to Hold Water in the Landscape

**Regenerative Agriculture**

Regenerative agriculture is a system of principles and practice that increases biodiversity, enriches soils, improves watersheds and enhances ecosystem services (TGI 2017). There are three main tenets: minimising/avoiding tillage, eliminating bare soil and encouraging plant diversity (Burgess et al. 2019). It is thought that these practices improve the structure of the soil which increases the infiltration of rainwater rather than the water flowing as surface runoff. Therefore land managed this way will store and slow more water and thus cope better with extreme weather events.

The fields become less prone to waterlogging and less prone to becoming parched during drought. This infiltrated water also better drains to aquifers, meaning that rivers continue to flow for longer in the dry season.

We deployed FreeStation //Smart: sensors at four farms practicing regenerative agriculture – one in each of the following counties: Hertfordshire, Lincolnshire, Buckinghamshire and Suffolk. Here we present the results of the farm in Hertfordshire which has been practicing regenerative agriculture for 9 years. The FreeStations included soil moisture sensors installed on neighbouring fields, one cultivated using regenerative agricultural techniques and the other using conventional methods. The two fields have the same soil, geology and climate. The soil in the regenerative agriculture field responds much more dynamically to rainfall than the adjacent field (Fig. 12.6). There is greater infiltration in response to rainfall and quicker drainage in the regenerative agricultural field. In contrast, in the adjacent field there is much less drainage and the soil is waterlogged. During this period the 2.57 ha regenerative agriculture field stored 2000 m$^3$ more water than a similar area in the neighbouring field (Fig. 12.7).

### Retention Ponds

We deployed //Smart: at two retention ponds to test their effectiveness in flood reduction. The first was a large retention pond on the Pipp Brook, a tributary of the

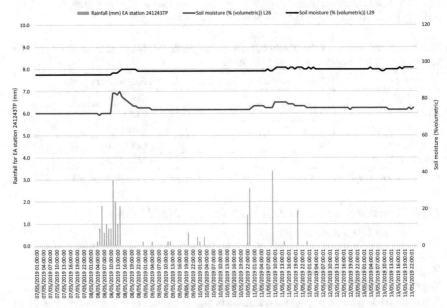

**Fig. 12.6** Volumetric soil moisture in fields practicing regenerative (*red*) and conventional (*black*) agriculture in Hertfordshire, UK, and nearby rainfall for a week in May 2019. As the conventionally managed soils (black) are water logged (100% soil moisture), rainfall is unable to infiltrate and thus flows over the surface as runoff. In comparison the soil managed using regenerative agricultural practices (red) is able to respond to rainfall events due to the lower soil moisture and quicker drainage, meaning less runoff and slower flow of water to the river system

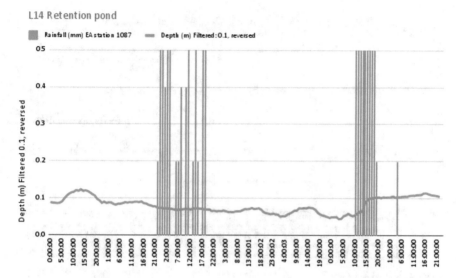

**Fig. 12.7** Cumulative water storage in fields practicing regenerative (*red*) and conventional (*black*) agriculture in Hertfordshire, UK for a week in May 2019. The soil managed using regenerative agricultural practices stored 2000 m³ more than a similar area under conventional agriculture at the end of this week. Storage value calculated using the 'Mitigation of flood risk by this soil management intervention' function in //Smart:River online interface for this FreeStation

River Mole in Surrey. The second pond was a small retention pond on agricultural land within the Stour River catchment in Warwickshire.

We present the results of the small retention pond on agricultural land for a week in July 2019. Both water level and soil moisture responded to the rainfall event (Fig. 12.8). About a day after the rainfall event the soil moisture returned to its previous level but water level in the retention pond remained high. Over that same period of time, the pond stored an additional 6 m³ of water over the pond surface area of 2500 m² (Fig. 12.9). The amount of water stored in the soil was much higher owing to the large increase in soil moisture (from a dry start) and larger wetted area (10,000 m²) than the reservoir. The soil stored approximately an extra 6000 m³ during that event, assuming the rainfall event wetted an area of 1 ha (10,000 m²) around the storage pond.

**Leaky Dams**
We deployed //Smart: at four leaky dams within the Thames catchment: two located on Silk Stream, a tributary of Brent River; one on Merryhills Brook, a tributary of Salmons Brook; and one on a tributary of Blackwater River.

We present here the results of a run of 31 leaky dams on the River Stour at Paddle Brook for a week in January 2021. A FreeStation stage logger is set upstream of the 31 dams and another downstream of them. The downstream logger shows discharge with lower peaks and higher troughs (longer recession) than the upstream and this indicates correct functioning of the dams. Over the course of the week the level of the river increased in height by up to 0.7 m (1.5 m³/s) upstream but only 0.6 m³/s

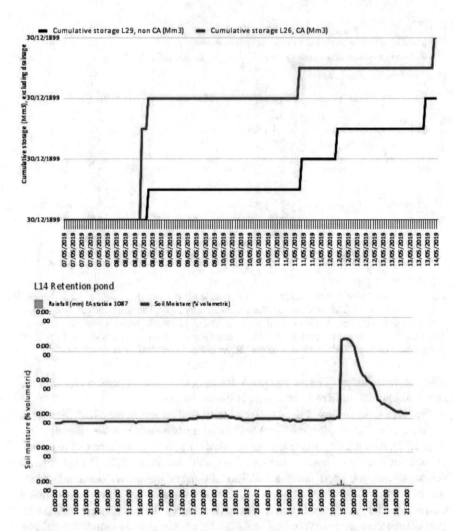

**Fig. 12.8** (*upper*): Water level and nearby rainfall in the farm retention pond for a week in July 2019. (*lower*): Volumetric soil moisture as measured by a soil moisture probe immediately next to the retention pond for the same week. Both the water level of the retention pond and the adjacent soil moisture increased in response to the rainfall event. Soil moisture quickly returned to pre-rainfall levels but the water level in the retention pond remained elevated and drained slowly through the narrow outlet

downstream (Fig. 12.10) so around 25% of peak flow is mitigated. The following month the impact is even greater (1.0 m³/s) for some events. 1.0 m³/s represents 0.09 m³/s (3.6%) reduction of peak flow per dam. Estimating the cumulative difference in discharge between the upstream and downstream dam for the same week in January (Fig. 12.11) shows the total slowed volume over the period represents 1.75 m³ or 0.06 m³ per dam over the period.

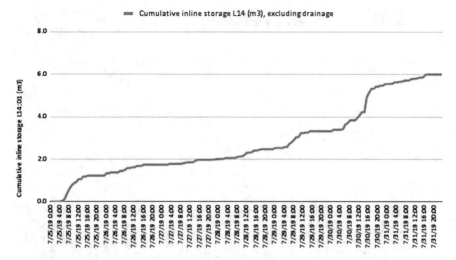

**Fig. 12.9** Cumulative water storage in farm retention pond in Stour catchment for a week in July 2019. The retention pond stored an additional 6 m³. Storage value calculated using the 'mitigation of flood peak by inline storage intervention' function in //Smart:River online interface

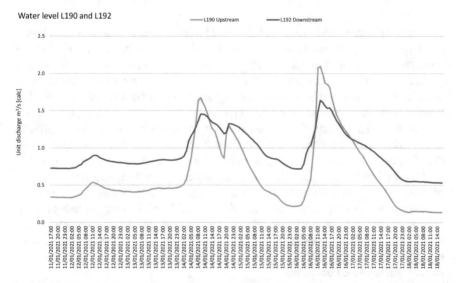

**Fig. 12.10** Water level upstream and downstream of 31 leaky dams on the River Stour at Paddle Brook for a week in January 2021 showing a series of peaks associated with rainfall events. The discharge upstream of the dams is flashier than downstream with the peaks mitigated by almost 0.5 m³/s or 25%

During a 7 day period, we calculated that the small retention pond stored 6 m³, leaky dam reduced flow by a maximum of 0.09 m³/s (totalling 0.16 m³/s over the week) and regenerative agriculture field 2000 m³. These values should not be directly

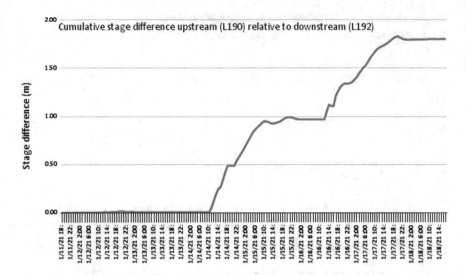

**Fig. 12.11** Cumulative difference in discharge between gauges upstream and downstream of 11 dams. The total slowed storage over the period represents 1.75 m³ or 0.06 m³ per dam. Calculated using the Difference in stage change either side of dam in FreeStation //Smart

compared because of differences in intensity of rainfall events, antecedent conditions and size of interventions. However rough magnitudes of differences can easily be observed: large area based interventions such as regenerative agriculture have a greater potential to reduce flood risk than point based interventions such as leaky dams and retention ponds. Regenerative agriculture is more scalable as well. The additional 2000 m³ stored by using regenerative agriculture is based on a single 2.57 ha field. Therefore a farmer implementing regenerative agriculture on a 86 ha farm (UK average farm size), with similar soil conditions as described here, would therefore have the capacity to store an extra 67,000 m³ relative to the counterfactual.

This section has provided illustrative examples of how the //Smart: system can be used to investigate the effectiveness of natural flood management interventions. We presented snapshot data to give an indication of the possible magnitude of water storage of each intervention type. With wider deployment of these monitoring stations we will be able to create a stronger evidence base for the relative water storage and flow-slowing capabilities of these interventions under a wide variety of environmental and weather conditions.

## 12.5 Cost Effectiveness of NFM

NFM interventions are often cheaper to install than traditional grey infrastructure (e.g. Vineyard et al. 2015). This is particularly the case if being used in areas of low land prices upstream of urban centres. However NFM interventions can still be costly to build and to maintain at scale. We have estimated the investment costs of three types of NFM per unit of storage for the next 20 years based on literature (for example, Burgess-Gamble et al. 2018) and discussions with stakeholders implementing NFM interventions in the Thames basin. *Leaky dams:* There is great variability in costs of installing leaky dams but reported capital costs per dam range from £100 to £3000 with median cost range of £175–£500. The reported lifetimes of leaky dams are also variable, although most estimates are around 10 years. There is very little information on maintenance costs of leaky dams. They need to be inspected frequently (Quinn et al., 2013; Dodd et al., 2016). Labour costs for maintenance are estimated at £50/h but hours needed are dependent upon type and location of dam.

**Retention Ponds** Construction of retention ponds typically requires heavy machinery meaning costs range from £2000–£6000 per pond with storage capacity of around 1500 m$^3$, £2.3 to £4 per m$^3$ (Shipston Area Flood Action Group, *pers. comm*). Costs of maintenance are mainly for dredging and these are estimated at £0.1–1 per m$^3$ of per year. The opportunity cost of installing a retention pond on farmland is high variable. This is estimated here as the potential annual value of producing a wheat crop in the area occupied, around £100/year for a retention pond of 1500 m$^3$, i.e. £0.07 per m$^3$ per year. *Regenerative Agriculture:* The cost of adjusting soil and crop management practices is generally quite low or negligible. However, purchase of specialist soil management and cultivation machinery may be required. Posthumus et al. (2015) describe the investment cost of reduced tillage or zero tillage as zero. However, equipment maintenance costs are estimated at £50 and £67 per hectare respectively. The loss of production is estimated to be £32 per hectare based on cereal crops. In order to convert maintenance cost per hectare (an average of £58.50 per year) into the cost per cubic metre of extra soil storage for floodwater provided by Regenerative Agriculture (RA), we used the soil moisture data measured in one of our paired sites, where soil moisture was measured in adjacent RA and non-RA plots under the same soil type, slope, and meteorological conditions and the difference in storage calculated. A hectare that costs £58.50 per year in maintenance is equivalent to £0.08/m$^3$ additional water stored. Opportunity costs (loss of production) are estimated to be £32 per hectare per year, which is £0.05 per m$^3$/year, based on the estimated 700 m$^3$ of extra storage per hectare.

The trade-offs between the cost of building and maintaining NFM interventions in comparison to their benefits in flood mitigation have yet to be evaluated for the Thames catchment. We use the Eco:Actuary Investment Tool to undertake this comparison for five sites within the Thames (Table 12.1). We calculated the approximate cost of investment required to achieve the mitigation goal (enough storage for a 15% reduction in downstream flood peak) for each of the three NFM interventions. For the cost per m$^3$ of mitigation (GBP/m$^3$) for each intervention we used the following values for the Thames: £61.63/m$^3$ for leaky dams, £16.82/m$^3$ for retention ponds and £2.85/m$^3$ for regenerative agriculture.

Each river requires a different volume of flood storage in order to mitigate peak flows by the required 15%, depending on the characteristics of the flow regime (Fig. 12.12). The costs of achieving the 15% reduction varies by intervention and

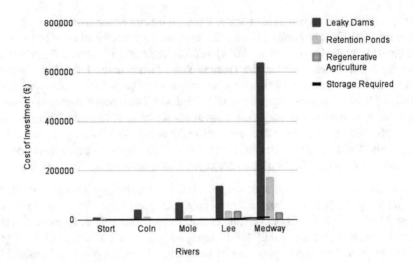

**Fig. 12.12** Investment required to achieve 15% flow mitigation for each of the five study rivers (Table 12.1). Consistently, regenerative agriculture is the cheapest flood mitigation option. Whereas leaky dams are the most expensive option, with retention ponds being intermediate. The locations of modelled at-risk-assets are Roydon for the Stort River, Fairford for the Coln River, Leatherhead for the Mole River, Lea Bridge for the Lower Lee and Teston Bridge for the Medway River

the flow conditions of each river. For the same volume of flood storage, the most cost efficient method in all cases is regenerative agriculture. The most expensive in all cases were leaky dams, as they have such a small impact per dam. So, although leaky dams are individually low cost, they are not as cost effective as other measures since they are challenging to scale. In regards to the differences between rivers in the amount of storage required to mitigate flood risk, the historic flow condition data used by the EIP tend to be a more important predictor than catchment size. For example, the River Coln at Fairford has one of the largest catchments, but the volume of flood storage needed to reduce the flood peak by 15% is low as much of the catchment overlies chalk (Table 12.1).

## 12.6   Co-benefits of NFM

Co-benefits represent the additional benefits to people provided by NFM interventions, beyond flood mitigation. In order to evaluate these, we consulted previous research, literature and reports. Leaky dams provide no co-benefits. Although The Flood Hub (2019) assessed leaky dams as having the co-benefits of improving water quality and habitat provision, this benefit does not extend to the leaky dams we surveyed, which are installed on ephemeral/winterbourne rivers and with a gap between the base of the 'dam' and the riverbed. This means that these dams do not permanently hold water back and are therefore not creating habitat. The gaps are sufficiently large that there is also no filtering function for water. Retention ponds have the potential to provide many more co-benefits including: water quality enhancement through settling of water-borne sediment and biological degradation of pollutants; biodiversity benefits through the generation of new aquatic habitat; amenity and aesthetic benefits and in some cases recreation and health benefits (Susdrain 2019; Berry et al. 2017; Brown et al. 2012; Alves et al. 2018). The degree to which these benefits are provided depend on the size of the pond, the geophysical context as well as the at risk asset distribution downstream.

Regenerative agriculture has many co-benefits including reduction in soil erosion, improved water quality, improved soil health and increased biodiversity (Mitchell et al. 2019; Palm et al. 2014). Because of the large geographic area that regenerative agriculture can be applied to, it has the greatest ability of the three NFM interventions studied to provide co-benefits in addition to the primary benefit of flood mitigation. Also, because regenerative agriculture can lead to the storage of more carbon in the soil (Mitchell et al. 2019) its co-benefits have national and global importance.

## 12.7 Lessons Learned and Advice

We find that leaky dams, which have attracted recent funding in the UK due to their low capital cost and relatively simple implementation, are generally not cost-effective natural flood management interventions. Our study shows that they provide little flood storage ability in many cases, they provide no co-benefits and require building at scale – and thus significant investment – to achieve a meaningful reduction in river flow for large rivers. They also have significant maintenance costs and can be a flood risk liability as they can be undermined by extreme events and break up, blocking culverts and other grey infrastructure and thus can even lead to more serious flooding (Woodland Trust 2016). However, where leaky debris dams raise flows and encourage activation of flood plains on low value land they can have much greater impact on downstream flows, particularly for smaller rivers or where the assets at risk are nearby downstream. Similarly building the dams from Willow or other species that can bind together and continue to grow instream as a 'living dam' provides the potential for self-maintenance, reducing maintenance costs and downstream risks. Thus careful planning and professional installation of leaky debris dams can maximise their benefits and minimise associated risks. Retention ponds provide a variety of co-benefits but are expensive to build for the volumes of water stored and take up valuable farmland. In comparison, regenerative agriculture provides flood storage for considerably less investment. In addition, regenerative agriculture can provide notable local and global co-benefits such as increased soil biodiversity (Kertész, and Madarász 2014), reduced soil erosion (Hobbs et al. 2008) and enhanced carbon sequestration (West and Post 2002). The challenge with land-based NFM such as regenerative agriculture is the requirement for landowner/ farmer investment and buy-in. Encouraging farmers to change long-held agricultural management practices can be challenging and requires incentives and risk management, both to ensure that farmers incomes are protected against adverse impacts on productivity and to meet the capital costs of transitioning machinery. Building leaky log dams in rivers is easier since there are no opportunity costs since the streams are not farmed. Land owners still require reassurance that they will not be held liable for any negative downstream impacts of the dams.

The specific results discussed in this chapter are relevant only to these study sites but the open-access Eco:Actuary Investment Planner and //Smart: monitoring equipment can be readily applied elsewhere. The Eco:Actuary Investment Planner uses Environment Agency data as standard so can easily be applied to other parts of the UK. However, the tool can still be used outside of the UK if long term flow and stage records are available. //Smart: technology has been designed for use globally in a wide variety of contexts. As the design and build instructions are freely available, the component parts can be purchased inexpensively and the stations require no coding capabilities, these sensors remove many of the existing barriers to low cost environmental monitoring. The decision as to whether to use the Investment Planner or //Smart: system depends on whether the NFM intervention is being planned or already in place (Fig. 12.13). If the NFM intervention is already in place then the //Smart: system is the best choice for assessing the effectiveness of the intervention. If not, the Eco:Actuary Investment Planner is the most suitable for understanding cost benefit and likely effectiveness ahead of deployment. Figure 12.13 supports users through the process of identifying whether the tools are appropriate for their situation.

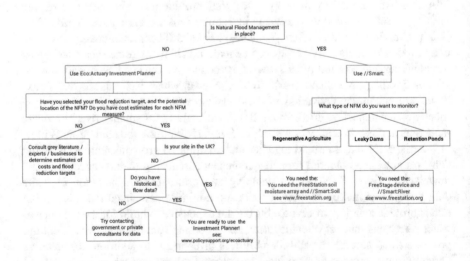

**Fig. 12.13** Decision tree to support the application of the //Smart: monitoring system and the Eco:Actuary Investment Planner. As //Smart: sensor technology and user documents are continually being updated please consult http://www.freestation.org/ for the most up-to-date guidance on materials as well as build and installation instructions. The Eco:Actuary Investment Planner is available at http://www.policysupport.org/ecoactuary

## 12.8 Practical Advice for those Intending to Deploy // Smart: Sensors

Field monitoring in a range of rural, peri-urban and urban environments is not always easy to do well, especially at low cost. Adhering to the following recommendations will ensure accurate and long-term data:

- To reduce theft and vandalism we recommend providing an explanation on the outside of the sensor about the purpose of the device and stating that it does not have any monetary value (Fig. 12.14). We found this beneficial as it allows interested people to find out about the device whilst also decreasing people's curiosity to open it to see what is inside. In urban settings, where the risk of vandalism or theft is particularly high, install the station out of reach, out of sight, and with steel cable ties. Ideas for urban camouflage include bat/bird boxes and using coloured camouflage tape / paint. Devices should always be installed with landowner permission.
- Have a local steward or champion who lives or works in relatively close proximity to the device(s). With someone nearby the device can be quickly reset if it develops problems and/or can be easily checked on for signs of vandalism, treefall or other issues (see points 3 and 4).
- When building stations be mindful of what wildlife may interact with the sensors or equipment. Rabbits can chew through low-lying cables which connect the soil moisture sensors to the main station, or sometimes even the cables higher up on

**Fig. 12.14** Signage on FreeStations. Communicating the purpose of the station helps to reduce theft and vandalism

### What is this?

This is a low-cost, open source **flood** monitoring system installed to help manage flooding from this river by providing a real-time river level monitoring system and by testing the effectiveness of natural flood defenses upstream. It is of no commercial value since it uses open source hardware and is built by volunteers.

### What does it do?

It measures the distance between the instrument and the water to detect when the water is rising rapidly or when defenses are storing water. It sends this information to a freely accessible database at www.freestation.org.

*Contact us at: www.freestation.org*

the station itself (e.g. from solar panel to logger housing). This problem is solved by encasing vulnerable wires in steel conduit tubing and by placing all other cables high out of reach.

- The stations need good mobile reception which is affected by topography between the station and the proximity of the local phone mast. We found that raised external aerials mounted on a 3 m fibreglass extendable, windproof pole helped to increase mobile reception. If signal is poor, then: i) select a different location (such as further upslope – the phone signal can vary significantly over a short distance); or ii) change to a FreeStationLocal that does not connect to the internet and relies on a local steward to download the data every month using the Freelay system or mobile phone tethering (www.policysupport.org).

- When siting the station, keep in mind the orientation and positioning of the device so that the solar panels receive maximum light, avoiding large trees and structures to the south (for locations north of the equator). If locating a soil moisture sensor in an agricultural field, be mindful of the maximum height of the crop so that the solar panel is mounted high enough to be unaffected by the crop. We found maize and mustard cover crops particularly problematic in this regard. If necessary, the logger and solar panel can be separated from the sensor by long wires. This is useful for when a thick tree canopy overshadows a stream, in this instance the logger with the solar panel can be separated from the river level sensor with an extended cable so that the logger/solar panel can be installed outside the canopy on the bank with a clear view south. The solar panel also needs to be kept clean.

- It is very important that the area under the footprint of the sonar water depth sensors used on FreeStage devices is kept clear of vegetation and debris to allow the sensor a clear view of the water surface. Regular vegetation clearance of reeds, brambles and trash from upstream may be necessary. Always install the sensor facing downstream of any instream mounting post.

- It is best to fix stations to permanent infrastructure, particularly in a flood risk zone, but only if this provides a clear view of the water. For example, the sensor can be affixed to a bridge or wall but only if cantilevered out to avoid the wall interfering with the sensor. The simplest installation is to mount from a bridge where there is a clear view downwards. This is because the structure is secure and there is little chance of vegetation growing under the bridge. If there is no structure, make sure to place a post firmly into the stream bed with a post driver and mount the station to the post.

- When deploying sensors in agricultural fields, good communication with the farmer is imperative so that the monitoring equipment is compatible with their farming equipment and/or can be retrieved, if necessary, while the soil is being tilled or seeded or the crop harvested. With regards to compatibility, ensure sprayer arms are able to operate over top of the stations, and only use an extended aerial pole in agricultural fields if necessary or if no spraying will take place.

Ideally locate stations between tractor "tram lines" so that the stations line up with the end of the spray arms and are as far as possible from the tractor.

## 12.9 Achievements and Remaining Barriers to NFM in the Thames

Progress has been made in the evaluation and use of natural flood management interventions in the Thames Basin. We have developed low-cost, open-source tools for monitoring NFM interventions *in situ* and a modelling tool for planning the scale of deployment necessary to achieve a defined mitigation target in a gauged basin. Through the deployment of this monitoring and modelling capability across a range of interventions in the Thames Basin, in partnership with the UK Environmental Agency, landowners, river and wildlife trusts and water companies we have increased knowledge on the effectiveness of three different types of NFM intervention in a variety of different contexts. Though the evidence base to evaluate effectiveness of NFM investments now, barriers to the use of NFM still exist.

These include: deployment, monitoring and maintenance barriers. With regards to deployment, within the Thames there is a lack of public land to install NFM interventions which makes implementation more difficult. This problem is further compounded because private landowners may be hesitant to host NFM measures, due to: i) the potential for litigation (i.e. if a leaky dam fails and causes damage downstream, or someone drowns in a retention pond), and ii) the opportunity cost of losing land to an NFM intervention and uncertainty around longevity of any incentive or compensation scheme as well as liability for long term maintenance or replacement costs.

With regards to the remaining barriers for monitoring, areas of low sunlight and low mobile phone signal are difficult to monitor with live systems. Non-networked monitoring stations require monthly visits to download data and check for problems and this substantially increasing the operation costs for monitoring systems distributed, like NFM measures, across large landscapes.. This is one example of why a local champion is particularly beneficial.

A key barrier remaining for NFM interventions long term is maintenance as this is pivotal to whether NFM interventions actually maintain their utility within the landscape. The best interventions are ones that are income generating such as retention ponds that also have another purpose (i.e. marinas or fishing ponds) and regenerative agriculture (which can reduce the costs of crop production). Leaky dams can require substantial maintenance, as the soil around the dam can erode during high flows thus weakening the structure and making the logs vulnerable to being swept downstream. These maintenance costs need to be factored into cost:benefit calculations for different NFM interventions, but they are difficult to estimate and are often overlooked.

## 12.10   Conclusion

Our results show that regenerative agriculture has a greater ability to reduce flooding than leaky dams or retention ponds in situations like the Thames basin, since RA can be applied cheaply over large areas. This result is particularly interesting considering the United Kingdom's departure from the European Union and its replacement of the EU's Common Agricultural Policy with a new subsidy scheme. The government's proposed plans centre around the idea of landowners receiving public money for the provision of public goods (i.e. ecosystem services), whilst private goods (crop production) are supported by the market. Our research has implications for this newly developed policy in two notable ways. First, our results show the positive and scalable flood alleviation effect of regenerative agriculture. Second, our development of a monitoring system that can quantify the amount of water stored (and thus benefit provided by agricultural land) serves as a starting point for developing parametric subsidies based directly on ecosystem service provided by land.

The UK government has invested into small-scale NFM projects, many of which funded pilot studies with leaky debris dams. We find that these can be an ineffective method of flood reduction because they are difficult to scale. Only in rare cases where many dams are built closely upstream of assets at risk of flood on small rivers, or when these dams activate large floodplains on low cost land do we find these effective. Our results illustrate the value of providing an evidence-base for decision making. The DEFRA pilot project scheme, combined with our technology and the efforts of project partners, has been instrumental in providing that evidence based. Natural flood management interventions clearly have a role to play in ecosystem based adaptation to climate change and flood risk reduction, but investments need to be carefully planned with respect to the type of NFM used, its placement and scale, and the evidence base for its efficacy. Scaling and maintenance considerations are paramount. Open access, practical and bespoke NFM tools now exist in Eco:Actuary and //Smart: to support organisations make such evidence-based decisions.

**Acknowledgements**  *This chapter is dedicated to our dear friend and colleague at King's College London, the late Professor John Anthony (Tony) Allan, whose recognition that we should "listen to farmers as they are the ones who manage the water" inspired and facilitated the work on regenerative agriculture presented below. Tony was an accomplished and highly influential academic whose virtual water concept has generated impact throughout the world. He was also very humble, gentle and selfless and was always in support of those around him, especially his students and early career colleagues.*

Many thanks to all our supporters during the NAIAD project. Our //Smart:River supporters included Matt Butcher (Environment Agency), Archie Ruggles-Brise (Spains Hall Estate), Mike McCarthy, Phil Wragg and Geoff Smith (Shipston Area Flood Action Group), Rob Dejean (Hallingbury Marina), Mark Baker (Unum Ltd), Nigel Brunning (Johnstons Sweepers Ltd), Joanna Ludlow (Essex County Council), Dean Morrison and Ed Byers (South East Rivers Trust), and Stephen Haywood and Lucy Shuker (Thames21). Our //Smart:Soil supporters were John and Paul Cherry, Ian Waller, Tony Reynolds and Andrew Maddever, and their neighbouring farmers.

# References

Alves A, Fomex JP, Vojinov Z et al (2018) Combining co-benefits and stakeholder perceptions into green infrastructure selection for flood risk reduction. Environments 5(29). https://doi.org/10.3390/environments5020029

Ashley RM, Balmforth DJ, Saul AJ et al (2005) Flooding in the future – predicting climate change, risks and responses in urban areas. Water Sci Technol 52(5):265–273

Berry P, Yassin F, Belcher K et al (2017) An economic assessment of local farm multi-purpose surface water retention systems in a Canadian prairie setting. Appl Water Sci 7:4461–4478

Brown C, Arthur S, Roe J (2012) Blue Health: Water, health and well-being – sustainable drainage systems. Research Summary. CREW Available via CREW. https://www.crew.ac.uk/sites/www.crew.ac.uk/files/sites/default/files/publication/Blue%20health%20suds%20summary.pdf. Accesssed 7 Nov 2019

Burgess PJ, Harris J, Graves AR, et al (2019) Regenerative agriculture: identifying the impact; enabling the potential. Report for SYSTEMIQ by Cranfield University. Available via Food and Land Use Coalition. https://www.foodandlandusecoalition.org/wp-content/uploads/2019/09/Regenerative-Agriculture-final.pdf. Accessed 6 Mar 2020

Lydia Burgess-Gamble, Rachelle Ngai, Mark Wilkinson, Tom Nisbet, Nigel Pontee, Robert Harvey, Kate Kipling, Stephen Addy, Steve Rose, Steve Maslen, Helen Jay, Alex Nicholson, Trevor Page, Jennine Jonczyk, Paul Quinn (2018) Working with Natural Processes – Evidence Directory. Environment Agency, UK https://assets.publishing.service.gov.uk/media/6036c5468fa8f5480a5386e9/Working_with_natural_processes_evidence_directory.pdf. Accessed 7 Nov 2019

DEFRA (2012) Tackling water pollution from the urban environment. https://www.gov.uk/government/consultations/tackling-water-pollution-from-the-urban-environment. Accessed 12 May 2018

Dodd J, Newton M, Adams C (2016) The effect of natural flood management in-stream wood placements on fish movement in Scotland, CD2015_02. Centre for Expertise in Water, Aberdeen

EA (2009) Thames estuary 2100: managing flood risk through London and the Thames estuary. TE2100, consultation document'. Available via gov.uk. https://www.gov.uk/government/publications/thames-estuary-2100-te2100. Accesssed 1 May 2018

EA (2014) Flood and Coastal Erosion Risk Management: Long-Term Investment Scenarios (LTIS) 2014. Available via gov.uk. https://assets.publishing.service.gov.uk/government/uploads/system/uploads/attachment_data/file/381939/FCRM_Long_term_investment_scenarios.pdf. Accesssed 1 May 2018

Garner G, Hannah DM, Watts G (2017) Climate change and water in the UK: recent scientific evidence for past and future change. Prog Phys Geogr 41(2):154–170

Hobbs PR, Sayre K, Gupta R (2008) The role of conservation agriculture in sustainable agriculture. Philos Trans R Soc B Biol Sci 363(1491):543–555

Kertész Á, Madarász B (2014) Conservation agriculture in Europe. Int Soil Water Conserv Res 2(1):91–96

Mitchell JP, Reicosky DC, Kueneman EA et al (2019) Conservation agriculture systems. CAB Rev 14(001)

Mulligan M (2013) WaterWorld: a self-parameterising, physically-based model for application in data-poor but problem-rich environments globally. Hydrol Res 44(5):748–769

Mulligan M (2015) Trading off agriculture with nature's other benefits, spatially. In: Zolin CA, Rodrigues R de AR (eds) Impact of climate change on water resources in agriculture. CRC Press

Mulligan M (2019) Floodplain. Model results from the Ecoactuary version 2 policy support system (non commercial-use). http://www.policysupport.org/ecoactuary [prepared by user caitlin.douglas_kcl.ac.uk]

Mulligan M (2020a) Model results from the Costingnature version 3 policy support system (non commercial-use). http://www.policysupport.org/costingnature [prepared by user caitlin.douglas_kcl.ac.uk]

Mulligan M (2020b) Model results from the Waterworld version 2 policy support system (non commercial-use). http://www.policysupport.org/waterworld [prepared by user caitlin.doug-las_kcl.ac.uk]

Mulligan M (2020c) Accumulated realised excess over permanent stores masked where Thames area is >0. Model results from the Ecoactuary version 2 policy support system (non commercial-use). http://www.policysupport.org/ecoactuaryPopout [prepared by user caitlin. douglas_kcl.ac.uk]

Palm C, Blanco-Canqui H, DeClerck F et al (2014) Conservation agriculture and ecosystems: an overview. Agric Ecosyst Environ 187(1):87–105

Posthumus H, Deeks LK, Rickson RJ, Quinton JN (2015) Costs and benefits of erosion control measures in the UK. Soil Use Manag 31:16–33

Quinn P, O'Donnel G, Nicholson A, Wilkinson M, Owen G, Jonczyk J, Barber N, Hardwick N, Davies G (2013) Potential use of runoff attenuation features in small rural catchments for flood mitigation: evidence from Belford, Powburn and Hepscott, Royal Haskoning and environment agency report. Joint Newcastle University

Sayers P, Penning-Rowsel EC, Horritt M (2018) Flood vulnerability, risk, and social disadvantage: current and future patterns in the UK. Reg Environ Chang 18(2):339–352

Susdrain (2019) Retention Ponds. Susdrain https://www.susdrain.org/delivering-suds/using-suds/ suds-components/retention_and_detention/retention_ponds.html. Accessed: 7 Nov 2019

TGI (2017) Regenerative agriculture: a definition. Terra Genesis International. http://www.terra-genesis.com/wp-content/uploads/2017/03/Regenerative-Agriculture-Definition.pdf. Accessed: 06 May 2020

The Flood Hub (2019) Natural Flood Management, Leaky Wood Dams. https://thefloodhub.co.uk/ wp-content/uploads/2019/03/Leaky-Woody-Dams-Natural-Flood-Management.pdf. Accessed 7 Nov 2019

Thompson V, Dunstone NJ, Scaife AA et al (2017) High risk of unprecedented UK rainfall in the current climate. Nat Commun 8(1):107

Vineyard D, Ingwersen WW, Hawkins TR et al (2015) Comparing green and grey infrastructure using life cycle cost and environmental impact: a rain garden case study in Cincinnati, OH. JAWRA 51(5):1342–1360

West TO, Post WM (2002) Soil organic carbon sequestration rates by tillage and crop rotation. Soil Sci Soc Am J 66(6):1930–1946

Wheater H, Evans E (2009) Land use, water management and future flood risk. Land Use Policy 26(1):S251–S264

Woodland Trust (2016) Natural flood management guidance: woody dams, deflectors and diverters. Sussex Flow Initiative. Available: https://www.woodlandtrust.org.uk/media/1764/natural-flood-management-guidance.pdf. Accessed 12 May 2020

# Chapter 13
# Giving Room to the River: A Nature-Based Solution for Flash Flood Hazards? The Brague River Case Study (France)

**Guillaume Piton, Nabila Arfaoui, Amandine Gnonlonfin, Roxane Marchal, David Moncoulon, Ali Douai, and Jean-Marc Tacnet**

**Highlights**

- Flash floods were studied with a multiscale hazard modelling approach including effects of Nature-based protection measures
- Relevant strategies are based on combinations of headwater natural water retention measures and of giving-room-to-the-river in the lowlands to increase ecological status and decrease flood levels
- Benefits were assessed with top-down approaches interesting for national and regional supporting stakeholders and also locally by survey to benefit from local dwellers' vision and willingness to contribute

## 13.1 Introduction

The Brague River is located in the region between Nice and Cannes along the French Mediterranean coast (Fig. 13.1). In the region, a quite dense urban belt is located aside the shore, where many hilly catchments finish their way to the sea. On

G. Piton (✉) · J.-M. Tacnet
Université Grenoble Alpes, INRAE, CNRS, IRD, Grenoble INP, IGE, Grenoble, France
e-mail: guillaume.piton@inrae.fr

N. Arfaoui
Catholic University of Lyon, Lyon, France

A. Gnonlonfin
University Côte d'Azur, IMREDD, Nice, France

R. Marchal · D. Moncoulon
CCR Caisse Centrale de Réassurance, Department R&D, Cat & Agriculture Modelling, Paris, France

A. Douai
University Côte d'Azur, GREDEG, CNRS, Valbonne, France

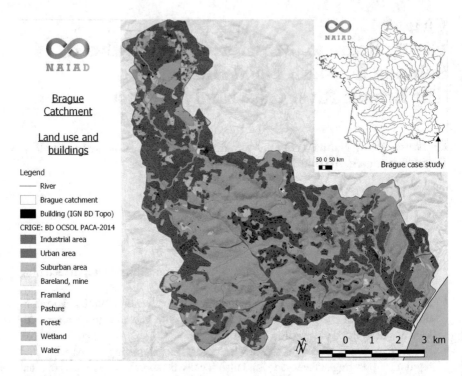

**Fig. 13.1** Brague River catchment land use and location in France

third October 2015, severe rainfalls triggered extreme flash floods in three catchments: Brague (area: 68 km²), Grande Frayère (22 km²) and Riou de l'Argentière (47 km²) as well as generalized urban runoff in the cities of Cannes and Nice and in the villages in between. Twenty people died, about €M 550–650 of losses were observed all over the area, as well as cascading hazards and failures on transportation, communication and energy networks (Préfécture des Alpes-Maritimes 2016).

The Brague River catchment was selected to perform an in-depth analysis. Its catchment has rural headwaters, a forested central part where is located both the Sofia-Antipolis activity area and natural parks, and finally lowlands occupied by urbanized areas, though pastures prevail in the lowlands' central area (Fig. 13.1).

On Oct. 2015, the basin was severely hit by an extreme flash flood (time return period >100 years). Numerous data on the event features were later collected by the local authorities (Préfécture des Alpes-Maritimes 2016). The disaster resulted in four fatalities within the catchment. More than 50 million € of damage were recorded in both the Biot and Antibes municipalities. Several campsites were flooded and later closed by the administration for safety reasons. Access roads to the Sophia-Antipolis activity area were cut. Building hosting business activities were damaged in number. It is worth mentioning that huge volumes of large wood (i.e., logs longer than 1 m) were released from the natural park forests causing bridges obstruction and, thus aggravating flooding (Fig. 13.2). This event was used as a calibration case to study flash flood hazards and risks and their effects on ecosystems.

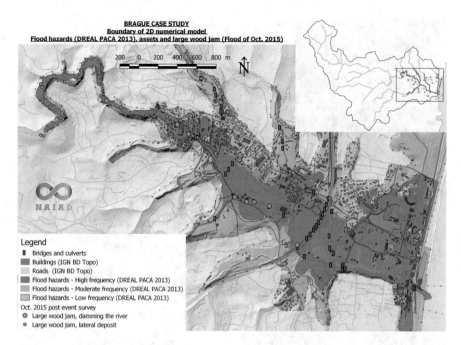

**Fig. 13.2** Map of the Brague lowlands with location of bridges and large wood jams inventoried after the oct. 2015 disaster and flood hazard map (DREAL PACA 2013)

In essence, flood risk is still very high in the area and social demand for new infrastructure works aiming at protecting the remaining people living in residential areas and working in industrial areas is extremely high. The Brague River was selected to perform a comprehensive flood risk analysis and to study the performance of flood protection measures either grey (i.e., civil engineering) and/or green and blue (i.e., Nature-based solutions – NBS). This chapter synthesizes the key lessons learnt after 3 years of research by five research teams working together. More details can be found in Pengal et al. (2018) on the main catchment features, Piton et al. (2018a) on hazards, Gnonlonfin et al. (2019) on the economical assessment and Tacnet et al. (2019) on integration of methods. The chapter first provides a short description of the methods used to assess flood hazards. It secondly describes how the protection strategies were defined, valuated (costs and benefits) and how there protection efficacy was assessed. It finally discussed the lessons learnt from this case-study and their possible replication and transfer to other sites.

## 13.2   Risk Assessment

Two modelling approaches were implemented in the Brague case study to address different scales relevant with risk assessment needs and types of planned NBS (Fig. 13.3):

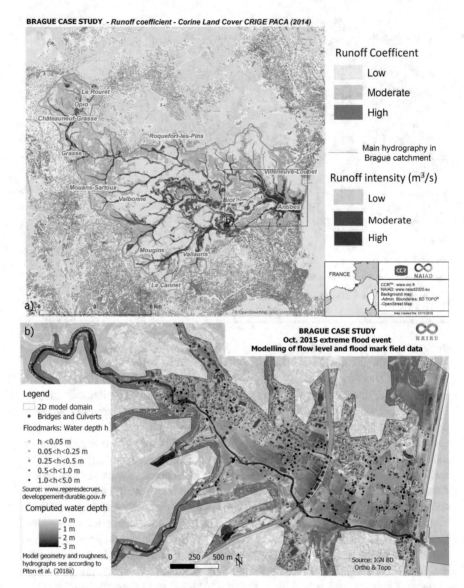

**Fig. 13.3** Multi-scales Hazard modelling: calibration results of (**a**) the CCR surface runoff hazard at catchment scale and (**b**) of the Iber flood hazard in the lowlands

- Surface runoff hazards were modelled by CCR (Caisse Centrale de Réassurance) at the catchment scale with their own models working on regular grids with pixel size of 25 m (Moncoulon et al. 2014),
- Flood hazards were modelled only on the lowlands with IBER (Bladé et al. 2014). This open-source software solves the shallow water equations in 2D on an irregular mesh. In our case study, elements were typically of size 1–3 m in urban

area or channels, and up to ten time bigger in pastures and forests, thus much finer than the CCR's model but providing results on a smaller area.

The catchment-scale modelling enabled to compute damages in the upper part of the catchment and eventually efficacy of small natural water retention measures (Strosser et al. 2015) spread in the catchment. However, a 25 m pixel size was too coarse to capture precisely hazards at house scale in the catchment lowlands, or effects of large wood jams or the protection efficacy of the measures planned to be studied in the lowlands. That is why the focus on the lowlands was performed with another more accurate tool, i.e., the 2D model. A satisfying calibration of both models was performed on the Oct. 2015 event (Fig. 13.3, details in Gnonlonfin et al. 2019). For instance the computed value of runoff-related damage at catchment scale in current situation is only 2% higher compared to the actual damage observed on Oct. 2015.

Both models computed the spatial distribution of hazard intensity of either historical events (e.g., Oct. 2015 flood) or theoretical events of known time return (e.g., 20 years, 100 years and 500 years). They were then coupled with methods to compute the related damages. In a second stage, models were tuned to include protection measures (see later) and re-run to assess the related avoided damages.

## 13.3    Evaluation of Protection Strategies

### 13.3.1    Tailoring Protection Strategies

The design of strategies was inclusive to integrate different knowledge from stakeholders. Stakeholders represent all the actors whose interests can be affected, positively or negatively, by the strategies. They include public actors, representatives of civil society (institutional stakeholders) and the population (individual stakeholders). In a first step, the design of strategies has aimed for the integration of experts and institutional stakeholders' knowledge. The strategies were defined by INRAE (previously Irstea) during years 2017 and 2018 based on:

(i)   Past engineering reports, previously planned measures and actually implemented measures, e.g., on the tributaries of the Brague called Vallon des Combes and Vallon des Horts;

(ii)  Feedbacks from Oct. 2015 and Nov. 2011 events on suitability of existing works and problems related to large wood;

(iii) Five stakeholders workshops that were organized by Univ. Côte d'Azur where needs, possible protection measures and policies were debated; and

(iv)  Expert knowledge trying to tailor to the catchment peculiarities these local knowledge with the know-how and state-of-the-art in design of NBS and flash flood protection.

In a second step and in the perspective to integrate individual stakeholders' knowledge and preferences, alternative protection strategies were proposed by 405 citizens during a survey performed by Univ. Côte d'Azur on 2019 (see later).

It was first envisioned to rely on water retention measures, i.e., flood retention basins and small natural water retention measures. The total rainfall volume related to the Oct. 2015 event on the Brague catchment was about 8.6 Mm³, more than half of it flowed to the sea within the event (Préfécture des Alpes-Maritimes 2016). A quick analysis was performed using the FEV (Flood-Excess-Volume) framework of Bokhove et al. (2019, 2020). It enabled estimating the cumulated retention volume that would be necessary to protect Biot and Antibes. For an equivalent event, the FEV would be 1.3–2.7 million m³ at the catchment scale (see detailed estimation for a subcatchment in Bokhove et al. 2019).

During stakeholder workshops, citizens asked that classical flood retention dams be studied. It would represent quite large structures considering the retention volume required. This option became strategy Grey #2 (see later) and was studied objectively to be compared to NBS strategies. Storing the FEV in smaller natural areas requires room: 1.3–2.7 million m³ is equivalent to a lake, 2-m deep, with a square shape of side length 800–1160 m. This virtual object helps assimilating the huge excess of water during extreme flash floods and is to be compared with the catchment topography. The conclusion is straightforward: if huge dams are not used, there is no room in the catchment to store such a great amount of water, as a whole or split in smaller volumes.

An inventory of areas eventually suitable to implement natural water retention measures (NWRM – Strosser et al. 2015) in the Brague catchment was reviewed (Fig. 13.4). Then, hypothesis regarding the average water depth that could be stored in each measure were performed (Gnonlonfin et al. 2019). It provides a first order estimate of the cumulated retention volume of all NBSs and existing retention dams. Overall, about 100 ha have a high potential to host NWRM and more than 200 ha should be further studied but could also be relevant sites. On these more than 300 ha, it seems reasonable to plan buffering at least 100,000 m³ of water and up to 3–5 times this volume providing that all potentially viable areas would be used.

In essence, the potential of retention by NBS varies in the range 0.1–0.5 million m³ while extreme flood events involve excess of several millions m³. These first order estimates merely demonstrate that upstream retention by NBS could be non-negligible but is certainly not sufficient to manage the flood and runoff risk: increasing channel capacity in the lowlands would also be necessary.

In addition, measures should be taken to prevent adverse consequences of NBS side effects, e.g., wilder rivers means more large wood in river channels during floods. Hybrid, i.e., green – grey strategies were consequently thought to be the most suitable for flash flood-prone rivers. Regarding grey measures, the most promising were:

- Replacement of existing bridges prone to trigger large wood jams or high backwater; and/or
- Racks to trap large wood during floods upstream of bridges likely to be clogged;
- Retention dams with large outlets enabling the reservoir to be dry in normal time and to fill only for severe floods, i.e., higher than 10 years return period events.

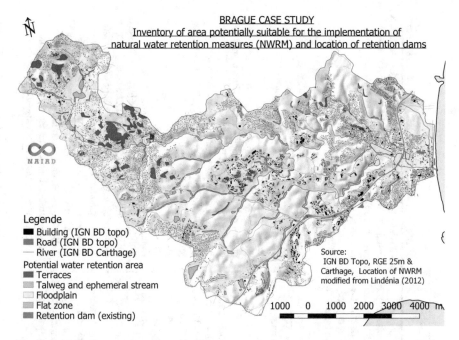

**Fig. 13.4** Map of areas suitable to implement retention measures (grey or NBSs) (Source: adapted from Lindénia 2012)

Regarding green measures, the most promising were:

- Small NWRM implemented in flat lands that would be transformed in wetlands and in talwegs were leaky dams could be built;
- An integrated floodplain management called "giving-room-to-the-river" gathering river channel widening, aquatic and riparian vegetation restoration, restoration of secondary arms and small tributaries connecting the main channel bed to restored wetlands in the floodplain.

During stakeholders workshops and a survey performed in the catchment in 2019, citizens were asked to choose between various ambitions of using green or grey measures (Arfaoui and Gnonlonfin 2020a). Four strategies were proposed (Table 13.1 and Fig. 13.5):

- Grey #1 was simply the building of large wood traps;
- Grey #2 was the wood traps plus very large retention dams on the two mains sub-catchments;
- NBS #1 merged NWRM and a scenario giving-room-to-the-river on its current axis;
- NBS #2 was based on NBS #1 but more ambitious by adding a large bridge on the highway to prevent the clogging of the current culverts and also moving a road located along the river further in the floodplain to reconnect several natural patches.

**Table 13.1** Summary of measures implemented in each strategy

| | Grey #1: Large wood traps | Grey #2: Large retention dams | NBS #1: Reopening a river corridor | NBS #2: Restoring an integrated floodplain to the river |
|---|---|---|---|---|
| Principle | Preventing bridges to be clogged by large wood transported during floods and aggravating flood levels | As Grey #1 plus building of two large flood control dams on the Brague River and Valmasque River. The dams are design to store the Oct. 2015 flood hydrograph | Implementing small natural water retention measures in the upper part of the catchment and giving-room-to-the-river in the lowlands by widening the bed, restoring the riparian forests and wetlands | As NBS #1 plus removing more lateral and longitudinal constraints (road, highway culverts), recreating a wide natural bed in the central natural area of the lowlands |
| Key measures | Building of 5–7 racks made of steel pipes with space 2–3 m between each other, 2–5 m high above ground level and founded deep in the sediment. Creating pedestrian paths out of private lands. No building expropriation required. | Flood control dam on the Brague River: retention capacity 880,000 m³. Outlet letting normal flows pass, up to the 10 years return peak discharge; Flood control dam on the Valmasque River: retention capacity 560,000 m³. Outlet letting normal flows pass, up to the 10 years return peak discharge; 200 m of dikes to build on the Brague River right bank, close from the train line. No building expropriation required. | Small natural retention areas, cumulated area ≈ 200 ha. 10–40 m of widening of the Brague river bed; Restoring riparian forests (13 ha); 3 bridges to demolish and rebuild wider and without central pile (bridge Brejnev, bridge Murator and golf course first pedestrian bridge); 5 Large wood trapping structures on the Brague River (3) and Valmasque River (2), height above ground level: 2–5 m; 11 ha of wetland restoration; 50–70 acquisitions and demolitions of houses; Continuous pedestrian and bicycle tracks along all banks. | Small natural retention areas, cumulated area ≈ 200 ha. 10–40 m of widening of the river bed; Restoring riparian forests (13 ha); 3 bridges to demolish and rebuild wider and without central pile (bridge Brejnev, bridge Murator and golf course first pedestrian bridge); 5 Large wood trapping structures on the Brague River (3) and Valmasque River (2), height above ground level: 2–5 m; 20 ha of wetland restoration; 55–75 demolitions of houses; Continuous bicycle tracks along all banks. Building a new high bridge for the highway #A8 and create a second channel in the natural central part of the floodplain with better connectivity with wetlands; Removing and rebuilding further from the river 1.4 km of the departemental road RD504 between roundabouts of route des Colles (access to Sophia-Antipolis) and of chemin de la Romaine (Pont Brejnev): rebuilding behind the golf course at the hillslope toe and through or close to the quarry. |

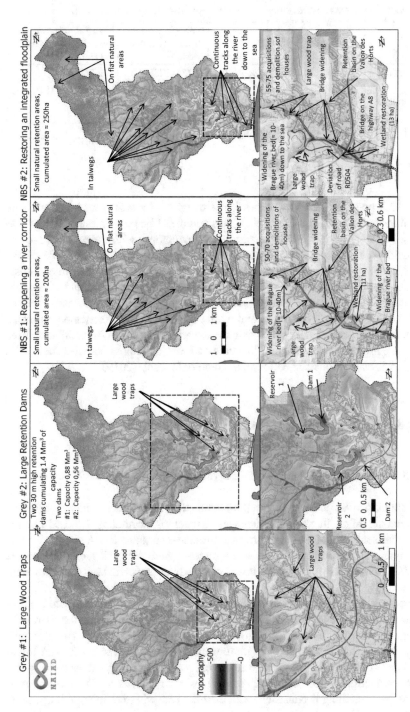

**Fig. 13.5** Maps of the four strategies. (Source: IGN BD Carto, BD Topo and BD Alti)

**Table 13.2** Discounted costs of strategies of a 50 years lifetime: mean (min; max)

| Strategy | Grey #1: Large wood traps | Grey #2: *Large retention dams* | NBS #1: *Reopening a river corridor* | NBS #2: *Restoring an integrated floodplain to the river* |
|---|---|---|---|---|
| Investment [M€] | 3.6 (2.5; 5.4) | 171 (89; 270) | 76 (56; 128) | 119 (90; 206) |
| Maintenance [M€] | 3.7 (2.8; 4.7) | 105 (63; 147) | 4 (3; 5) | 4 (3; 5) |
| Total [M€] | 7.3 (5.3; 10.1) | 66 (26; 123) | 80 (59; 133) | 12 (3; 211) |

### 13.3.2 Costs of Strategies

Table 13.2 presents the costs of the four strategies over a 50-year life cycle (see guidelines in Chap. 6 – Le Coent et al., this volume). The cost estimates were made from a literature review and local surveys. They comply with the national recommendations, including a 2.5% discount rate (Langumier et al. 2014; CGDD 2018). These costs include investment and maintenance costs. The investment costs include: (i) the costs of acquisition and compensation of the land and residences affected estimated from the market price of real estate (PERVAL database, prices of 2018) and (ii) the costs of studies and works. Maintenance costs include estimates of maintenance and repair costs. Due to the uncertainties in the cost estimates, three values are provided: an average followed by a lower and upper bounds. The details of the costs estimations can be found in Gnonlonfin et al. (2019).

### 13.3.3 Estimating Physical Efficacy for Hazard and Risk Reduction

#### 13.3.3.1 Protection Efficacy of Small Natural Water Retention Measures

A computation was performed with the catchment-scale model to study the effect of upstream retention in reducing damages. Modelling accurately NWRMs in the catchment-scale model with its 25 m size pixels was complicated. A simpler inverse procedure was thus followed: runoff coefficients were reduced, the model on the Oct. 2015 event was then run and it was checked how much overall damages were reduced (Fig. 13.6).

This straightforward analysis was performed to check how much should the runoff be reduced (whatever be the measures to do so) to reduce the runoff-related damage by a certain amount. It was actually performed through the runoff coefficient, i.e., a virtual way to say that more water is retained in the catchment. It was concluded that a reduction of 20% of runoff reduced insured damages by ≈ 7% while a reduction of 50% of runoff reduced insured damages by 45%. The conclusion of this modelling is consistent with the FEV analysis stressing that if upstream

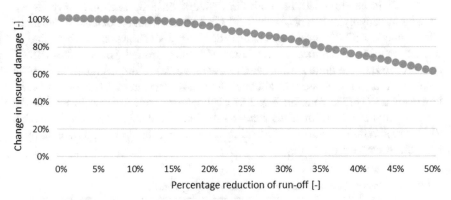

**Fig. 13.6** Effect of hazard reduction (reduced runoff coefficients) in case of Oct. 2015-type runoff events on insured losses for residential homeowners of the Brague River catchment

retention measures can be useful, they are not sufficient to achieve satisfying protection needs.

It is worth being stressed that the numbers provided in Fig. 13.6 are not mean annual values but values related to the Oct. 2015 event, which was rather extreme. The protection efficacy would be higher for events of smaller magnitude but this analysis was not performed.

### 13.3.3.2    Protection Efficacy of Large Dams and Giving-Room-to-the-River

Strategies Grey #2 (large dams) and NBS #1 emerged early in the project, during 2017, and were studied with more advanced models (Iber flood model and CCR runoff model). Strategy Grey #1 came out during the survey performed during autumn 2019, its protection efficacy was assessed by expert knowledge. Strategy NBS #2 also emerged during discussion with stakeholders in 2019. It could not be modelled as accurately as NBS #1 and its performance is also partially based on expert assessment.

The estimation of the avoided damages followed the standard protocol recommended by the French state (CGDD 2018 and Chap. 6, Le Coent et al., this volume):

(i) Modeling of the flood levels related to several flood events of known return period, this, in the current state and with protection strategies;

(ii) Assessment of damage to buildings using damage curves, i.e., relationship between damage and hazard intensity (e.g., flood level) for each asset;

(iii) Estimate of the mean annual avoided damage by difference between the mean damage in the current state and that assuming the strategy implemented.

Several damage curves provided by French guidelines and calibrated by CCR were used to provide low, medium and high damage estimates (Table 13.3).

One could be surprised that both Grey #2 and NBS #1 strategies only reduce damage by roughly one third. As the works to be carried out are extensive, the modelled reduction of 30% is less than expected. Figure 13.7 shows areas where significant influence on flood level are modelled with the proposed works in the NBS #1 strategy (results are quite similar with Grey #2 works). Cautious inspection of the model results demonstrate that although flow level are effectively reduced in the upstream area (green zone on Fig. 13.7), numerous assets are also located in the downstream part of the lowlands were the studied works have much less influence. There are several reasons to this situation, reasons that should be used later to optimize the strategies:

**Table 13.3** Damage per events and mean annual (avoided) damage

| Scenario | Current situation | | | Grey #2 | | | NBS #1 | | |
|---|---|---|---|---|---|---|---|---|---|
| Damage curve | Min | Mean | Max | Min | Mean | Max | Min | Mean | Max |
| Damage related to $Q_{20years}$ [M€] | 6.0 | 15.0 | 23.2 | 4.9 | 12.1 | 19.0 | 4.1 | 10.2 | 16.1 |
| Damage related to $Q_{100years}$ [M€] | 11.4 | 27.5 | 42.5 | 7.0 | 17.2 | 26.8 | 7.2 | 17.5 | 27.8 |
| Damage related to $Q_{500years}$ [M€] | 15.4 | 37.2 | 52.4 | 9.8 | 23.8 | 37.4 | 12.1 | 29.3 | 43.0 |
| Mean annual damage [M€] | 0.6 | 1.6 | 2.4 | 0.5 | 1.1 | 1.7 | 0.4 | 1.1 | 1.7 |
| Mean annual avoided damage [M€] | | | | 0.2 | 0.5 | 0.7 | 0.2 | 0.5 | 0.7 |
| Avoided/mean annual [−] | | | | 30% | 29% | 28% | 32% | 32% | 31% |

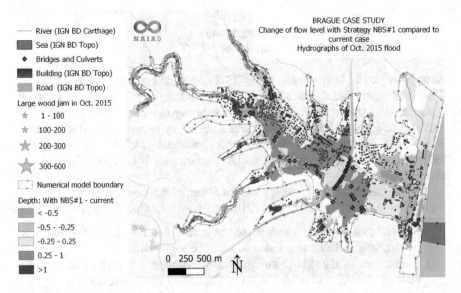

**Fig. 13.7** Changes in flood level with NBS#1 compared with current case. Large wood jam observed in Oct. 2015 are assumed trapped by large wood trap in NBS #1 strategy. Color code depends on decrease or increase of water depth

- Most measures being located on the main stems of the Brague River and its Valmasque tributary, flows coming from the numerous small tributaries are still present and generate their share of damage;
- The culverts at the highway A8 and the bridge constraining the river width at the sea mouth (and their associated backfill) are key bottleneck sections on the Brague main stem. They dam the flow and slow down the lowland drainage. Thus even though the upstream and eventual downstream sections of the river are widened, remaining bottleneck sections trigger extensive backwater and dramatically reduce the flood hazard protection efficacy.

Strategies Grey #1 and NBS #2 were not studied in such details. Based on the knowledge gained during the project on the river functioning, we consider that Grey #1 can only very locally reduce flooding. Consequently, its effect on damage reduction is considered to be negligible: the areas where flood level would be reduced are spatially limited and limited to the close vicinity of bridges and culverts.

NBS #2 is more ambitious than NBS #1 in terms of removing barriers formed by road networks. It should be optimized at a later stage to address the previously mentioned bottleneck section near the sea mouth (the other one at the highway if fixed by a new large bridge). The railway and roads bordering the coast actually dam the valley. This is the main reason of the aggravated flood level in the downstream area (orange zone in Fig. 13.7). These obstacles should be equipped with several discharge structures or culverts to prevent this side effect of a higher channel capacity upstream. The cost related to these structures were not included in estimates of Table 13.2. Although removing the barrier related to the highway culvert will reduce flood levels in the upstream part of the lowlands, it is difficult to predict how downstream flood levels will change. They will likely decrease if sufficient discharge capacity is found through the railway and road backfills but we cannot be sure without further detailed modelling. The avoided damage for NBS#2 are consequently not accurately estimated, just expected to be higher than for NBS#1.

It is worth noting that, a dual strategy combining the strategies Grey #2 and NBS #2 was proposed by citizens during the survey. Large civil engineering structures degrading nature in forested parks upstream of an ambitious project of river restoration seem a bit inconsistent. In France, it is now mandatory both to protect river environment and to mitigate flood risks in projects to be consistent with the EU Water Framework Directive and the EU Flood Directive (somehow merged in the "GEMAPI law", see Marchal et al. this volume; Vigier et al. 2019). This dual strategy cannot be considered an integrated strategy because the environmental alterations related to the large dams would only be partially offset by the environmental improvements associated to strategy NBS #2. Its protection efficacy would conversely be higher than each strategy taken alone. The issue related to the flows coming from the small tributaries would nonetheless remain but the main stems would be much less prone to flooding. Overall, the avoided damages would be higher than the 30% computed for each single strategy but would not reach 100% of the 1.6 (0.6; 2.4) M€/year (mean (min; max)).

### 13.3.4 Co-benefit Estimations

Two different methods were used and compared to estimate co-benefits: a top-down approach based on the transfer of values published in literature and a bottom-up approach based on a survey.

#### 13.3.4.1 Top-Down Approach

A benefit transfer method developed by Arfaoui and Gnonlonfin (2020b) was used to estimate the economic value of the co-benefits of environmental restoration. For this, a meta-regression of 49 studies of restoration of river ecosystems, conducted from 1996 to 2018 in Europe, East Asia and America, was carried out. A value transfer function from NBS strategies and their co-benefits was built upon an analysis of 187 values of willingness to pay (WTP) for ecological restoration measures. Input parameters are (i) the type of measures implemented (management of the riparian forest, restoration of the river bed, restoration of the floodplain, agricultural practices), (ii) the ecosystem services provided by the strategies and (iii) an indicator of the ambition of the project (normal or strong).

The estimation of the value of co-benefits in the different strategies at the scale of the Brague basin was estimated by WTP per year and per household. The methodological variables and the co-benefits were set at the average value of the database and the upper and lower bounds were estimated through the uncertainty of the statistical adjustment whose correlation coefficient was $R^2 = 0.38$. For strategy NBS #1 and NBS #2, the resulting WTPs were 28 (2; 353) and 75 (4; 608) €/household/year (mean (min; max)), respectively. Grey strategies are not restoration strategies and have consequently no co-benefits.

#### 13.3.4.2 Bottom-Up Approach

Unlike a top-down approach, the bottom-up approach incorporates additional knowledge from different stakeholders on a representative scale. In this perspective, our approach follows the three conditions recommended by Carolus et al. (2018): (i) definition of strategies by local stakeholders, (ii) participation of key players concerned by the environmental problem and (iii) relevant geographic scale.

Key stakeholders represent all the actors whose interests can be affected, positively or negatively, by the strategies. We distinguish public actors, representatives of civil society and the population. With semi-structured interviews and focus groups conducted between July 2017 and December 2018, stakeholders participated in the identification of co-benefits and in the preparation of the survey questionnaire. The survey was conducted face-to-face from September 6, 2019 to October 15, 2019 in a representative sample of 405 people (Gnonlonfin and Douai 2019; Arfaoui and Gnonlonfin 2020a). Respondents were recruited using a random sampling procedure in public places in order to independently respect three representativeness criteria: geographic location, gender and age.

The survey aimed at evaluating the WTP for strategies using the contingent valuation method in the perspective to measure the social preference for NBS and/or grey strategies and the willingness of the population to financially contribute to their implementation. For the sake of simplicity and because of the method constraints, both the risk reduction effect and co-benefits related to environmental and life quality improvement were merged in this WTP estimate.

Respondents were first asked to select their preferred level of ambition in the two strategy categories (grey and NBS). Respondents were informed about the level of socio-economic costs (investment and maintenance, expropriation and demolition of houses) and the ranges of potential benefits of all the measures (reduction of the risk of flooding and co-benefits). In a second step, the respondents had to indicate, on the one hand, whether they were willing to financially contribute for the preferred strategy and, on the other hand, to express their level of "bounded WTP" to take into account uncertainties related to the purchasing decision (Pondorfer and Rehdanz 2018). In addition, respondents who refused to contribute financially were asked to justify their choice in order to identify the protest responses.

The survey demonstrated the preferred strategies were (Table 13.4): NBS #1 for 44% of respondents, Grey #1 (28%), and NBS #2 (10%). No respondent prefers the Grey #2 strategy alone but 18% of respondents preferred that all measures be implemented to the highest possible ambition, i.e., that both Grey #2 and NBS #2 be done. We also observed that 69% of respondents refuse to contribute financially to their preferred strategy. However, the analysis of the reasons justifying this refusal showed that 31% were responses to protests linked to local governance and the methodology of the survey (Gnonlonfin and Douai 2019). This rate is similar to those reported in the literature (Meyerhoff and Liebe 2010). The Heckman (1976) model was used to correct the selection bias and predict the empirical mean of the WTP (Table 13.4).

### 13.3.5   Cost-Benefit Analysis (CBA) to Assess Strategy Efficiency

The WTP estimated by both approaches were aggregated at the scale of the watershed considering 28,874 households (INSEE database, inventory of 2014).

**Table 13.4** Preferences and willingness to pay per strategy in euros/household/year according to the bottom-up estimated by survey

| Strategy | Grey #1 | Grey #2 | Grey#2 + NBS#2 | NBS #1 | NBS #2 |
|---|---|---|---|---|---|
| Preference rate | 28% | 0% | 18% | 44% | 10% |
| Willingness to pay: **mean** (min; max) in €/household/year | **57** (31; 81) | 0 | **156** (63; 240) | **83** (31; 125) | **116** (2; 173) |

**Table 13.5** Cost benefit analysis on a 50 year time window: **mean estimate** [min; max]

| Strategy | Costs | Top down analysis | | | Bottom-up analysis | |
|---|---|---|---|---|---|---|
| | Total costs (M€) | Avoided damage (M€) | Co-benefits by transfer method (M€) | B/C (−) | Contingent analysis of total economic value (M€) | B/C (−) |
| GREY #1 | 7 | ≈ 0 | 0 | ≈ 0 | 47 | 6.7 |
| | [5; 10] | | | | [25; 66] | [2.5; 13.2] |
| GREY#2 | 170 | 13 | 0 | 0.1 | 0 | 0 |
| | [88; 271] | [5; 19] | [0; 0] | [0; 0.2] | | |
| GREY#2 + NBS#2 | 294 | <45 | ≈ 0 | <0.15 | 128 | 0.4 |
| | [182; 481] | | | | [57; 200] | [0.1; 1.1] |
| NBS #1 | 80 | 14 | 23 | 0.5 | 68 | 0.9 |
| | [59; 132] | [6; 21] | [6; 34] | [0.1; 0.9] | [25; 103] | [0.2; 1.7] |
| NBS #2 | 122 | 14+ | 40 | >0.4 | 95 | 0.8 |
| | [93; 211] | [6+; 21+] | [3; 498] | [>0.1; >5.6] | [51; 142] | [0.2; 1.5] |

Table 13.5 summarizes the cost-benefit estimates over a 50-year period and presents the benefit/cost ratios (B/C) in the two approaches. With the exception of the bottom-up estimate of the Grey #1, all strategies have an average values of B/C < 1. Strategy Grey #1 has a relatively low cost. Its supporters gave quite high WTP for it, thus its high B/C ratio. To the opinion of the authors, this result is related to an overestimated protection efficacy of large wood trap in the perception of local citizens. The trauma related to large wood obstructing bridges on October 2015 was often reported during interviews and stakeholder workshops.

Avoided damage are in any case much lower than total costs. Financing the works just based on the risk reduction potential appears to be economically inefficient. Co-benefits actually weight more in the cost-benefit balance. They consequently deserve effort to quantify them. It is worth stressing that estimation of the WTP varies greatly depending on the method used. Much higher WTP values were provided in the preliminary analysis of Gnonlonfin et al. (2019) resulting in ratio B/C > 1 for both NBS strategies. The key difference was that we used the results coming from stakeholder workshops, i.e., the list of ecosystem services that stakeholders considered relevant in the catchment. On the contrary, values in Table 13.5 were computed assuming a fully top-down analysis performed without stakeholder workshops. Using such an approach decreases dramatically the weight given to ecosystem services and consequently the WTP. In essence, the transfer method used in the top-down approach is highly sensitive to stakeholder feedback: if ecosystem services are reported to be important in the citizen perception, the computed WTP might be multiplied by 3–5.

Another parameter that could be discussed is the number of households considered. The rigorous way to determine it would be by estimating the critical distance from the site above which households no longer benefit from the project. We had neither time nor funds to gather sufficient data to do so. In such case an area has to be chosen somewhat arbitrarily. Administrative limits are usually considered in the literature (Logar et al. 2019). Computing the social benefit at the scale of the 11 municipalities intersecting the Brague basin would increase the number of households and thus the co-benefit estimates by a factor 3.4. This would result in a B/C > 1 balance for the NBS strategies in the two approaches. We used the conservative assumption that only households residing within the catchment geographical limit represents the population impacted by the strategies, thus ignoring the large population visiting the catchment to work, hike, play golf, camp or come in the parks. The level of aggregation of social benefit is therefore determinant for the output of a CBA.

## 13.4   Lessons Learnt and Replication/Scaling/ Re-scaling Issues

### 13.4.1   Should We Perform Top-Down or Bottom-Up CBA or Both?

One can wonder if it is worth performing all these modelling if a survey is sufficient to provide elements to perform a CBA. We think that both approaches worth being performed because elements on avoided damages and co-benefit in the light of a standard method or based on local perception are used to aid several different decisions taken nationally, regionally and locally.

The top-down CBA first covers the risk reduction by demonstrating the impact of strategies on risks without taking into account local perceptions. It is used by the French State, in addition to other criteria, to decide whether or not the strategy can be funded with the Barnier Funds (a national fund for natural hazard protection, see Chap 3 – Marchal et al., this volume; CGDD 2018). The top-down approach also allows to provide economic value to environmental impacts (highly uncertain but useful in some contexts, see Kallis et al. 2013) and to provide additional information for the decision-making of regional actors such as the Water Agency, whose financing decision relies on the environmental impact of strategies.

Our results show that, from a top-down perspective, no strategy is worth the investment from an economical point of view. Therefore, other criteria e.g., safety or indirect damage to the environment, will be decisive in the financing decision under the Barnier Funds (CGDD 2018). Economical valuation of co-benefits are not as standardized as for avoided damages. So far, the Water Agency relies more on other criteria related to environmental quality and restoration potential of strategies to support and finance strategies (e.g. Piton et al. 2018b). It can also be noted that other frameworks such as multicriteria decision-making methods have been developed to aid decisions on such situation involving multiple benefits and values. It allows for instance to consider global effectiveness of NBS combining e.g., physical, environmental, economic and social values. The implementation should be based on a close collaboration with stakeholders (Philippe et al. 2018; Tacnet et al. 2018, 2019). In this case-study application, the process was not advanced enough to implement it in practice but it will be possible to implement the methodology if needed in the future.

Conversely, the bottom-up approach provides information on local perception and the social acceptability of the strategies. The basin agency, in charge of the river management is responsible of both risk reduction and river restoration. The recent GEMAPI law enables basin agencies to raise a tax of a maximum of 40 €/person/year to finance their mission of managing watercourses. The survey provides information on the social acceptability of the tax. It gives clues on (i) the strategies for which the local population is ready to contribute financially and (ii) the amount of the socially acceptable contribution. Indeed, the median WTP is considered to be a good indicator of the acceptable financial contribution for the majority of the population (OECD 2018). For the Brague case study, the WTP for the sole Grey #2 strategy is null highlighting the social unacceptability of a financial contribution for this strategy despite its benefits as avoided damage. On the other hand, the median WTP (lower than the mean values provided in Table 13.4) are estimated at 27 (7; 47) €/household/year for the strategy NBS#1 (uncertainty range between brackets); 75 (38; 100) €/household/year for the strategy NBS#2 and 59 (8; 100) €/household/year for the dual strategy Grey#2 + NBS#2, i.e., 0.8–2.1 M€/year at the catchment scale. Basin agency have thus potentially significant funding opportunity, although the support of other actors as the French State and the Water Agency are still necessary.

Hence the three key partners with financial power, namely the regional Water Agency, the State, and the local basin agency need three assessments, namely avoided damages, environmental restoration ambition and local perception of both for each of them to take the decisions to support a given strategy.

### 13.4.2  Evidence of the Importance to Give Room to Rivers Prone to Flash Floods

A broad lesson learnt on this case study is that, in rivers hit by Mediterranean thunderstorms of high magnitude, even high ambition on retention measures in the upper and mid-catchment is insufficient to prevent flooding of downstream lowlands. Therefore, a sufficiently large corridor must be maintained for such rivers to convey flows down to the sea or to the downstream bigger river. Such corridor can be natural and/or comprise several flood resilient activities; however, built assets in those corridors create long lasting constraints. Stakeholder involvement and news in media demonstrate that people working and living there, initially not informed or unaware of the risks, regret that the authorities enabled them to settle. Then, protecting such areas is extremely expensive and regularly not feasible regarding high magnitude events. Buying such assets to remove them is another very expensive solution.

Maintaining on the contrary large corridors is certainly more resilient and provide numerous co-benefits. From a broader point of view, evidences of physical effectiveness and limits of NBS appeared as an essential criteria for their mainstreaming. NBS have co-benefits but their real physical effect on hazards reduction (storing a water volume, reducing runoff) had first to be demonstrated before moving to decision-aiding approaches such as economic approaches presented above. That is why some modelling is usually required to demonstrate the rational in the ambition of the strategy compared to the magnitude of the flooding. The FEV approach is a simple yet powerful tool to do so (Bokhove et al. 2019, 2020).

## 13.5   Conclusion

This chapter synthesizes some conclusions of 3 years of work involving five multidisciplinary research teams. Stakeholder participation through workshops and surveys helped us to define several protection strategies against flash floods in the Brague River catchment. A multi-scale modelling approach was implemented to compute flood hazards at the catchment scale and in the lowlands with a higher accuracy. The effects of NBS and grey measures involved in the various strategies were assessed in term of hazard reduction and avoided damages through model tuning. Total costs of each strategies were evaluated as well as co-benefits. Co-benefits were both locally assessed by survey (bottom-up approach) and valuated using transfer functions (top-down approach). Cost benefit analysis were performed using both the top-down and the bottom-up approaches. Both are useful because they provide different perspectives usable by the great variety of stakeholders involved in flood risk and river management, notably the local basin agency, the regional Water Agency and the French State.

**Acknowledgments**  This work was funded by the H2020 project NAIAD (Grant 730497) from the European Union's Horizon 2020 research and innovation program. The authors are thankful to the officers working at the CASA (Brague Basin Agency), at the French State (DDTM06) and at the Water Agency (AE-RMC) as well as to all participants to the survey, stakeholders meetings and workshops. We would to thanks R. YORDANOVA, C. CURT, M.-B. MUNIR, S. DUPIRE, P. ARNAUD, A. MAS and Z.X. WANG who helped to perform several preliminary steps of this final analysis. We are also grateful to Nora VAN CAUWENBERGH and to Philippe LE COENT who reviewed this chapter and provided many comments to help improving it.

# References

Arfaoui N, Gnonlonfin A (2020a) Supporting NBS restoration measures: a test of VBN theory in the Brague catchment. Econ Bull 40(2):1272–1280. [online] https://ideas.repec.org/a/ebl/ecbull/eb-20-00134.html. Accessed 19 Mar 2021

Arfaoui N, Gnonlonfin A (2020b) Testing Meta-Regression Analysis in the context of NBS restoration measures: the case of Brague River. Working paper ESDES n°2020-02. [online] Available from: https://www.esdes.fr/medias/fichier/wp-esdes-2020-01-arfaoui-gnonlonfin_1579084075949-pdf. Accessed 16 May 2020

Bladé E, Cea L, Corestein G, Escolano E, Puertas J, Vázquez-Cendón E, Dolz J, Coll A (2014) Iber: herramienta de simulación numérica del flujo en ríos. Revista Internacional de Métodos Numéricos para Cálculo y Diseño en Ingeniería 30:1–10. https://doi.org/10.1016/j.rimni.2012.07.004

Bokhove O, Kelmanson MA, Kent T, Piton G, Tacnet JM (2019) Communicating (nature-based) flood-mitigation schemes using flood-excess volume. River Res Appl 35:1402–1414. https://doi.org/10.1002/rra.3507

Bokhove O, Kelmanson MA, Kent T, Piton G, Tacnet JM (2020) A cost-effectiveness protocol for flood-mitigation plans based on Leeds' boxing day 2015 floods. Water 12:1–30. https://doi.org/10.3390/w12030652

Carolus JF, Hanley N, Olsen SB, Pedersen SM (2018) A bottom-up approach to environmental cost-benefit analysis. Ecol Econ 152:282–295. https://doi.org/10.1016/j.ecolecon.2018.06.009

CGDD (2018) Analyse multicritère des projets de prévention des inondations – Guide méthodologique 2018. Commissariat général au développement durable. [online] Available from: https://www.ecologique-solidaire.gouv.fr/sites/default/files/Th%C3%A9ma%20-%20Analyse%20multicrit%C3%A8re%20des%20projets%20de%20pr%C3%A9vention%20des%20inondations%20-%20Guide.pdf. Accessed 16 May 2020

Coent L et al (this volume) Economic assessment of nature-based solutions for water-related risks. In: López-Gunn E, van der Keur P, Van Cauwenbergh N, Le Coent P, Giordano R (eds) Greening water risks: natural assurance schemes. Springer, Cham

DREAL PACA (2013) Territoire à Risque Important d'inondation (TRI) de Nice – Cannes – Mandelieu- Cartographie des surfaces inondables et des risques. Direction Régionale de l'Environnement, de l'Aménagement et du Logement PACA, Service Prévention des Risques. [online] Available from: https://rhone-mediterranee.eaufrance.fr/cartographie-des-risques-dinondations-du-tri-de-nicecannesmandelieu. Accessed 16 May 2020

Gnonlonfin A, Douai A (2019) Quel système de valeur pour l'analyse coût-bénéfices de l'aménagement des cours d'eau? L'exemple de la Brague dans les Alpes Maritimes. Research Report. NAIAD H2020 project (Grant Agreement n°730497). [online] Available from: https://hal.archives-ouvertes.fr/hal-03030376. Accessed 19 March 2021

Gnonlonfin A, Piton G, Marchal R, Munir MB, Wang ZX, Moncoulon D, Mas A, Arnaud P, Tacnet JM, Douai A (2019) DELIVERABLE 6.3 DEMO insurance value assessment – Part 7: France: Brague. NAIAD H2020 project (Grant Agreement n°730497). [online] Available from: http://naiad2020.eu/wp-content/uploads/2020/10/D6.3.pdf. Accessed 19 March 2021

Heckman JJ (1976) The common structure of statistical models of truncation, sample selection and limited dependent variables and a simple estimator for such models. Ann Econ Soc Meas 5:475–492

Kallis G, Gómez-Baggethun E, Zografos C (2013) To value or not to value? That is not the question. Ecol Econ 94:97–105. https://doi.org/10.1016/j.ecolecon.2013.07.002

Langumier A, Jaffres G, Luczyszyn H, Gruffaz F (2014) Détermination des coûts de référence des travaux de restauration hydromorphologique des cours d'eau et conception d'une base de données de projets et d'un outil d'estimation du coût du volet hydromorphologie des programmes de mesures 2016–2021. AE-RMC Agence de l'Eau Rhône Méditerrannée Corse. [online] Available from: https://www.eaurmc.fr/upload/docs/application/pdf/2017-05/2014-etude-c-restau-hydromorpho.pdf. Accessed 16 May 2020

Lindénia. (2012) Plan De Gestion 2011–2018 Des Cours D'eau Du Bassin De La Brague Études Et Inventaires Complementaires – Etude Complementaire Des Zones D'expansion Des Crues Du Bassin Versant. SIAQUEBA

Logar I, Brouwer R, Paillex A (2019) Do the societal benefits of river restoration outweigh their costs? A cost-benefit analysis. J Environ Manag 232:1075–1085. https://doi.org/10.1016/j.jenvman.2018.11.098

Marchal et al (this volume) Insurance and the natural assurance value (of ecosystems) in risk prevention and reduction. In: López-Gunn E, van der Keur P, Van Cauwenbergh N, Le Coent P, Giordano R (eds) Greening water risks: natural assurance schemes. Springer, Cham

Meyerhoff J, Liebe U (2010) Determinants of protest responses in environmental valuation: a meta-study. Ecol Econ 70:366–374. https://doi.org/10.1016/j.ecolecon.2010.09.008

Moncoulon D, Labat D, Ardon J, Leblois E, Onfroy T, Poulard C, Aji S, Rémy A, Quantin A (2014) Analysis of the French insurance market exposure to floods: a stochastic model combining river overflow and surface runoff. Nat Hazards Earth Syst Sci 14:2469–2485. https://doi.org/10.5194/nhess-14-2469-2014

OECD (2018) Cost-benefit analysis and the environment – further developments and policy use. Organisation for Economic Cooperation and Development https://doi.org/10.1787/9789264085169-en

Pengal et al (2018) DELIVERABLE 6.1 catchment characterization report. NAIAD H2020 project (Grant Agreement n° 730497). [online] Available from: http://naiad2020.eu/wp-content/uploads/2018/07/D6_1.pdf. Accessed 16 May 2020

Philippe F, Piton G, Tacnet J-m, Gourhand A (2018) Focus – Aide à la décision par l'application de la méthode AHP (Analytic Hierarchy Process) à l'analyse multicritère des stratégies d'aménagement du Grand Büech à la Faurie. Science Eaux Territoires 26:54–57. https://doi.org/10.14758/set-revue.2018.26.10

Piton G, Dupire S, Arnaud P, Mas A, Marchal R, Moncoulon D, Curt T, Tacnet J (2018a) DELIVERABLE 6.2 from hazards to risk: models for the DEMOs – Part 3: France: Brague catchment DEMO. NAIAD H2020 project (Grant Agreement n° 730497). [online] Available from: http://naiad2020.eu/wp-content/uploads/2019/02/D6.2_REV_FINAL.pdf. Accessed 16 May 2020

Piton G, Philippe F, Tacnet J-m, Gourhand A (2018b) Focus – Caractérisation des altérations de la géomorphologie naturelle d'un cours d'eau Application du Morphological Quality Index (MQI) aux projets d'aménagement du Grand Büech à La Faurie. Science Eaux Territoires 26:58–61. https://doi.org/10.14758/set-revue.2018.26.11

Pondorfer A, Rehdanz K (2018) Eliciting preferences for public goods in nonmonetized communities: accounting for preference uncertainty. Land Econ 94:73–86. https://doi.org/10.3368/le.94.1.73

Préfécture des Alpes-Maritimes (2016) Inondations des 3 et 4 octobre 2015 dans les Alpes-Maritimes – retour d'experience. République Française [online] Available from: http://observatoire-regional-risques-paca.fr/evenement/rex-inondation-des-3-et-4-octobre-2015-alpes-maritimes. Accessed 16 May 2020

Strosser P, Delacámara G, Hanus H, Williams H, Jaritt N (2015) A guide to support the selection, design and implementation of Natural Water Retention Measures in Europe. Capturing the multiple benefits of nature-based solutions. Publications Office of the European Union [online] Available from: http://nwrm.eu/guide/files/assets/common/downloads/publication.pdf. Accessed 16 May 2020

Tacnet J-M, Piton G, Philippe F, Gourhand A, Vassas C (2018) Décider dans le contexte de la GEMAPI: exemple de méthodologie d'une approche intégrée d'aide à la décision et application aux projets d'aménagements. Science Eaux Territoires 26:48–53. https://doi.org/10.14758/set-revue.2018.26.09

Tacnet JM, Van Cauwenbergh N, Gomez E et al (2019) DELIVERABLE 5.4 integrative modelling framework and testing in the DEMOs: a global decision-aiding perspective (Main report: parts 1–5). EU horizon 2020 NAIAD project, Grant agreement N°730497

Vigier E, Curt C, Curt T, Arnaud A, Dubois J (2019) Joint analysis of environmental and risk policies: methodology and application to the French case. Environ Sci Pol 101:63–71. https://doi.org/10.1016/j.envsci.2019.07.017

# Chapter 14
# Can NBS Address the Challenges of an Urbanized Mediterranean Catchment? The Lez Case Study

**Philippe Le Coent, Roxane Marchal, Cécile Hérivaux, Jean-Christophe Maréchal, Bernard Ladouche, David Moncoulon, George Farina, Ingrid Forey, Wao Zi-Xiang, and Nina Graveline**

## Highlights

- We carry out an integrated evaluation of the impact of two types of NBS in the Lez watershed (South of France): (i) the conservation of agricultural and natural land through the control of urbanization and (ii) the development of green infrastructure.
- Using insurance data on damages, we establish that the most ambitious green infrastructure scenarios can reduce up to 20% of the mean annual damages due to urban runoff.
- Using a stated-preference survey with 400 inhabitants of the watershed, we estimate that co-benefits generated by NBS scenarios are very significant with 180€/household/year for the most ambitious strategy.
- The cost-benefit analysis of green infrastructure strategies reveals that benefits overweight the sum of the cost of construction and maintenance and land related costs.
- Urban communities are in the driving seat for the development of NBS. To achieve this objective, urban master plan need to be updated and urban communities should tap into diverse source of financing and work across services.

P. Le Coent (✉) · C. Hérivaux · J.-C. Maréchal · B. Ladouche · G. Farina · I. Forey
G-EAU, BRGM, Université de Montpellier, Montpellier, France
e-mail: p.lecoent@brgm.fr

R. Marchal · D. Moncoulon · W. Zi-Xiang
Caisse Centrale de Réassurance, Paris, France

N. Graveline
University of Montpellier, UMR Innovation, INRAE, Montpellier, France

© The Author(s) 2023
E. López Gunn et al. (eds.), *Greening Water Risks*, Water Security in a New World, https://doi.org/10.1007/978-3-031-25308-9_14

## 14.1 Introduction

The Lez river, which spring is the outlet of karst aquifer, is a small coastal Mediterranean river (29 km long – 746 km$^2$) that crosses the city of Montpellier (Fig. 14.1). The urban community of Montpellier (457,000 inhabitants) is character-ized by the largest population growth in France and a rapid urbanization with mas-sive soil-sealing (−2920 ha of agricultural and natural areas from 1990 to 2012). The Lez catchment is exposed to a typical Mediterranean weather marked by repeated droughts, heavy rainfalls and storms in very short time scale in autumn. It has faced major flood events in the history and in the last years notably in 2014, with three successive large events that led to 65 million euros of damages, only for pri-vate housing and businesses.[1] Large investments have been carried out to manage overflow risk but runoff risk, accentuated by the recent urbanization, remains a major challenge with 78% of damages in the recent large events of 2014.

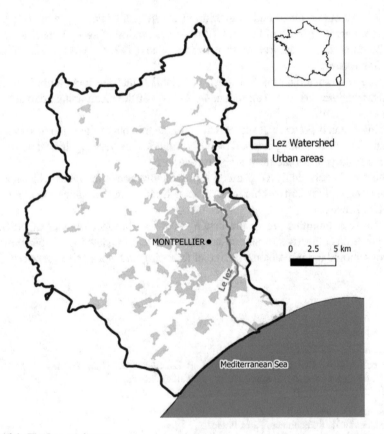

**Fig. 14.1** The Lez catchment

---

[1] CCR data.

Urban areas, especially the urban community of Montpellier, is facing several other challenges typical of large Mediterranean cities. Air pollution mainly due to the commuting of an increasing number of urban workers living in individual housing outside the main city centre and the vicinity of a major highway remain a large issue. Heat island effect is also a growing challenge with the increase of temperature peaks due to climate change, with a historical record of more than 45 °C reached in 2019. Finally, the Lez catchment is characterized by a very rich biodiversity due to its diversity of habitats and the inherent diversity of the Mediterranean biodiversity hotspots. This diversity is however also particularly threatened by the rapid urban sprawl observed in the last decades.

The issues and solutions studied in this case may be relevant to most urbanized catchments of the Mediterranean region, which are largely exposed to rapid urbanization, due to the attractiveness of the Mediterranean basin and the prevalence of generally dry climate with violent storms generating flash floods (Cramer et al. 2018).

NBS is considered as a potential means to address the flood risks of the territory and other urban challenges. The French Geological Survey (BRGM) and the Caisse Centrale de Réassurance (CCR, French reinsurance company), in close collaboration with local stakeholders, especially the urban community of Montpellier, decided to evaluate the interest of NBS to address these challenges. This chapter focuses on the early stages of NBS project cycle, i.e. the identification of NBS strategies and their integrated evaluation and the launching of initial pathways for their implementation.

## 14.2   Overall Methodology

The overall methodology developed in the Lez case study is presented in Fig. 14.2.

Participatory methods based on scenario planning were used to identify NBS strategies to be tested in the Lez case study (in orange). Spatial modelling was then used to translate NBS strategies into usable inputs for physical modelling (in orange).

Physical modelling (in blue) was used to evaluate the impact of NBS strategies and scenarios on flood hazard in terms runoff and river overflow hazard (Cf. Chap. 4 of this publication).

Economic methods (in green) were finally used to assess NBS strategies and scenarios (Cf. Chap. 6 of this publication)

- Damage assessment was carried out mainly based on insured damages data and flood modelling.
- The implementation costs and opportunity costs of NBS were assessed using value transfer from other reference studies.

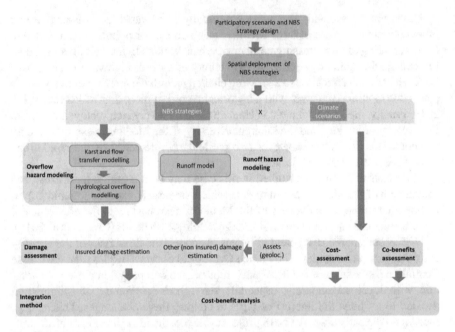

**Fig. 14.2** Modelling framework of NBS Strategies in the Lez Basin

- Co-benefits valuation was undertaken through a choice experiment[2] carried out with 400 citizens of the Lez Basin. This survey evaluates the socio-cultural and monetary value of co-benefits associated with two types of NBS strategies: (i) conservation of natural and agricultural land by limiting urban sprawl and (ii) development of green infrastructure (GI) in the city.
- A cost-benefit analysis (CBA) was finally used to compile the monetary valuation results.

Stakeholders were extensively involved throughout the process, especially for: (i) the identification of NBS strategies (ii) co-benefits valuation (iii) the evaluation of NBS assessment results and (iv) the identification of opportunities and barriers for the development of NBS including funding strategy. More details on stakeholders' involvement is provided throughout the chapter.

This chapter is structured as follows. Section 14.2 presents the identification of NBS strategies. Section 14.3 described how the impact of NBS strategies on flood risk reduction was evaluated. Section 14.4 is dedicated to the economic methods developed to assess NBS strategies and their results. Finally, Sect. 14.5 concludes with key steps towards the implementation of NBS strategies.

---

[2] Method described in Sect. 14.3.

## 14.2.1   Identification of NBS Strategies

The identification of NBS strategies mainly relied on a stakeholder consultation process. NBS strategies are combinations of NBS individual measures. Two workshops gathered different departments of the urban community of Montpellier "Montpellier Mediterranée Metropole", a citizen's association, public and private developers, region and State representative as well the Lez watershed Authority. The first workshop aimed to (i) identify the main challenges of the territory in terms of water risks, (ii) highlight the drivers of scenarios for the evolution of the territory at the 2040 time horizon and, (iii) identify the advantages and disadvantages of individual NBS measures for risk reduction and (iv) their respective co-benefits. The second workshop was then mainly focused on the elaboration of NBS strategies (combination of NBS) and their validation as well as a discussion of the integrated assessment methodology.

Three types of NBS strategies with the main objective to reduce flood risks and address territorial challenges were identified:

1. The active management of the Lez karst aquifer, i.e. an increased pumping of the karst aquifer during summer time, to reduce overflow risk at the peak period of storm events.
2. The conservation of agriculture and natural lands through the implementation of urbanization strategies aiming at limiting urban sprawl. These strategies will limit soil sealing and therefore avoid the increase of urban runoff and the destruction of ecosystem services linked to agricultural and natural land.
3. The development of green infrastructure in the city to improve stormwater management and to reduce runoff-flooding risk. These green infrastructures are detailed below and represent combinations of small scale measures spread throughout the territory (Sect. 14.2.4). We use the term green infrastructure to differentiate this NBS from strategy 2.

## 14.2.2   Active Management of the Karst

The karst, when not saturated, plays a buffer role and limits Lez flow and subsequent overflow. The level of saturation can be influenced by active water pumping used for drinking water. Currently, 33 $mm^3$ are abstracted each year (reference strategy). It corresponds to a pumping rate able to supply the Montpellier population with drinking water most of the time while maintaining a karst water level above the authorized threshold (35 m above sea level). An alternative strategy was identified to increase the capacity of the karst to reduce overflow risk. This strategy considers an increase in pumping (45 $Mm^3/y$) which is compatible with the pump elevation and natural recharge of the aquifer. Both strategies (reference and alternative) were compared to a theoretic situation without any pumping (0 $Mm^3/y$) in order to estimate the impact of abstraction strategies on flood hazards and damages.

### 14.2.3  Urbanization Strategies

Another type of NBS strategy relates to different urbanization strategies resulting in different levels of conservation of natural and natural land. The conservation of these areas indeed directly falls under the definition of NBS established by the IUCN (Cohen-Shacham et al. 2016). With the help of the stakeholders three different strategies characterized by different targets in terms of population growth and the level of urban densification.

The resulting effect on land artificialization, i.e. the transformation of natural or agricultural land into urban land, was estimated based on a simple model that estimates land requirement depending on additional population to accommodate, the target density of new neighbourhoods and the rate of new housing to be built in existing neighbourhoods:

- Level 1: a "laissez-faire" strategy in which the population growth remains high and new urbanization mainly relies, as at present time; on individual housing that leads to the artificialization of 4000 ha.
- Level 2: a "central" strategy with a lower rate of population growth and efforts of densification that lead to the artificialization of 1600 ha. This scenario takes on the hypotheses of the current urban strategy of the urban communities of the catchment.
- Level 3: A "green" strategy with a lower rate of population growth and an objective of almost no additional artificialization. This scenario is considered as a highly virtuous scenario. Although very ambitious, it reflects the 0 net artificialization policy ambitioned by the government.

In order to evaluate the impact of the different urbanization strategies on flood hazard, it was necessary to identify the spatial impact of these urbanization strategies on land use. This simulation was done through the application of an urban planning model (Calvet et al. 2020).

The results of the urban planning model provided land use maps for the three different urbanization strategies. We present in Fig. 14.3 a focus on one zone of the Lez Basin that shows differences of urbanization in the three strategies in the west of Montpellier. The figure especially shows the development of large patches of discontinuous urban housing (dense) around the peri-urban municipalities: Lavérune, Pignan, Saussan, Fabrègues, Saint Jean de Védas in the Laissez-faire strategy and to a lesser extent in the central strategy. In the green strategy, most new housing is rather made through the development of discontinuous collective housing.

### 14.2.4  Green Infrastructure Strategies

**Green infrastructure (GI) strategies** were developed mainly to address runoff-flooding risks, considered prevalent in the watershed. It was collectively decided during the participatory process to evaluate the effect of GI which main benefit is

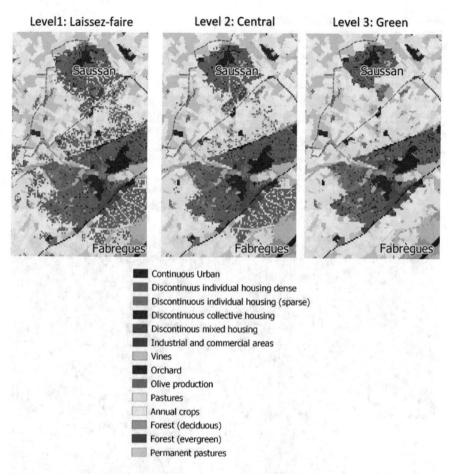

**Fig. 14.3** Detail of the Lez catchment urbanization in the laissez faire, the central and the green strategies based on the SimUrba model (Calvet et al. 2020). (Data: IGN: BD Ortho BD Topo, Corine Land Cover, Montpellier 3)

the retention of water. The combination of GI evaluated in this document are presented in Table 14.1.

The potential extent of implementation of these solutions was evaluated based on photo interpretation of four sample neighbourhoods. The available space identified in the sample neighbourhoods was subsequently extrapolated to the whole urban areas of the watershed, considering the extent of the different types of neighbourhoods throughout the urban areas of the catchment (Fig. 14.4).

For the measures that present a retention capacity, it is directly estimated based on the technical characteristics of the individual measures (Depth, porosity etc.). Table 14.2 presents the estimation of the extent (in $10^3$ m$^2$) of the measures that only reduce soil-proofing and directly the retention capacity (in $10^3$ m$^3$) for measures that generate a retention capacity.

**Table 14.1** Description of green infrastructure strategies

| NBS measure | Description | Level 1 | Level 2 |
|---|---|---|---|
| City deproofing + greening | Deproofing of large areas of urban concrete soils. | X | X |
| Green parking spaces | Waterproof concrete parking pavements are replaced by porous "green" pavements. | X | X |
| Bioswale small | 0.5 m large bioswales are to be constructed along roads. | X | |
| Bioswale large | 2 m large vegetalized bioswales are to be constructed alongside roads except in continuous habitat | | X |
| Vegetated retention basin | 25% of parking areas are transformed into vegetated multi-purpose retention basins. | | X |
| Green roofs | 50% of flat roofs are transformed in vegetated green roofs | | X |

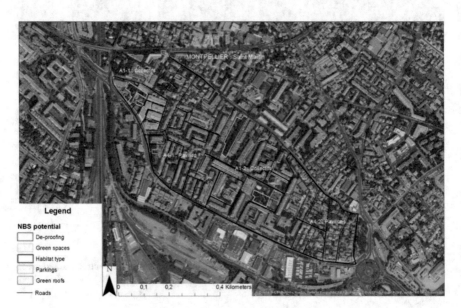

**Fig. 14.4** NBS potential in one sample neighbourhood: example of Saint Martin neighbourhood– Montpellier

**Table 14.2** Potential extent and retention capacity of the NBS strategies

| GI Strategy | NBS measure | Unit | |
|---|---|---|---|
| Level 1 | Deproofing | $10^3$ m$^2$ | 98.5 |
| | Green parking spaces | $10^3$ m$^2$ | 352.1 |
| | Bioswale small | $10^3$ m$^3$ of retention | 24.5 |
| Level 2 | Deproofing | $10^3$ m$^2$ | 98.5 |
| | Green parking spaces | $10^3$ m$^2$ | 352.1 |
| | Bioswale large | $10^3$ m$^3$ of retention | 190.3 |
| | Vegetated retention basin | m$^3$ of retention | 44.0 |
| | Green roofs | $10^3$ m$^3$ of retention | 5.9 |

We finally estimate the retention capacity of NBS strategies in $L/m^2$ of water-proof area for the different types of habitats of the watershed. This estimation is necessary in the next step for the evaluation of the impact of NBS on the reduction of damages. On average, the GI level 2 strategy brings additional 30.3 L of water retention/$m^2$ of waterproofed area whereas the GI level 1 strategy mainly reduces water proofing and yields only 3.6 $L/m^2$.

## 14.3   Risk Modelling and the Impact of NBS Strategies

The main aim of NBS strategies identified in the Lez catchment is the reduction of flood risk. The estimation of the impact of NBS on the reduction of this risk was therefore an important focus of the research carried out in this case study. Two approaches were pursued for the assessment of NBS: the evaluation of the impact of the active management of the karst on overflow risk based on BRGM modeling and the evaluation of urban flood risk and the impact of urbanization and green infra-structure strategies based on CCR modeling. Considering the approaches developed by the CCR, we include the assessment of damage cost avoided thanks to NBS although this also belongs to the economic assessment (Sect. 14.4).

### 14.3.1   The Impact of the Active Management of the Karst on Overflow Risk

The high infiltration capacity of karst aquifers usually contributes to increasing the retention capacity of karst catchment areas during heavy rains. This is linked to the absence of soil and the presence of surface karst phenomena that facilitate water infiltration. During heavy rains, floods are generally of lesser importance in karst basins as long as their aquifer are not fully saturated. Indeed, during flooding, the karst aquifer is recharged quickly until it is fully saturated: its infiltration capacity decreases and in some cases, rapid underground circulation within karst conduits contributes to worsening the surface flooding. Depending on the initial state of satu-ration of the karst aquifer, its ability to reduce flooding varies. This has been dem-onstrated in many karst areas of southern France in particular, where frequent heavy rains are present (Maréchal et al. 2008; Fleury et al. 2009).

It is the case of the Lez river where the karst aquifer, located upstream of the Montpellier city, is used to supply drinking water to the city. The active manage-ment practiced on this aquifer consists in pumping a flow higher than the natural flow of the spring in summer in order to reduce water levels at the spring and in the karst conduits and thus mobilize the water reserves located in the less permeable compartments of the aquifer. At the end of summer, the water level is lowered by about 30 m, creating an unsaturated zone capable of absorbing the first autumn rains.

In this study, a historical analysis of water level data from the karst aquifer, rainfall, observed flows in the river and damage caused by several flood episodes was conducted. This was complemented by a modeling approach to test the impact of the pumping strategy on floods and induced damage. A cascade of hydrogeological and hydrological models, coupled with an estimate of the damage generated, has been developed and applied to test various hydrological and aquifer management scenarios.

The results show that pumping the aquifer contributes to reducing the impact on flows and damage of the first rainfall event. However, once the karst is full, it no longer helps to reduce the flood. Therefore, from a statistical point of view, over a full hydrological year, active aquifer management has very little impact on floods and damage in the city of Montpellier (Fig. 14.5). Although differences can be observed between the three pumping strategies they remain very limited, especially to very high intensity rainfall (>50 mm/h) which are very rare.

Although the karst aquifer can play a significant role in flooding the Lez catchment, in specific cases, our results show that the alternative pumping strategy in the karst aquifer does not have a significant impact on average. In addition, the increase in water pumping, may generate side effects, such a reduction of water levels in connected aquifers which may have adverse effects on other water users or the environment. Based on the limited impact of this strategy on flood risks, we decided not to assess further the impact of intensified pumping strategies in the karst aquifer because its impact is limited to very rare specific events.

# Lavalette

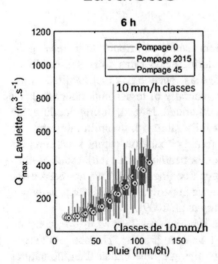

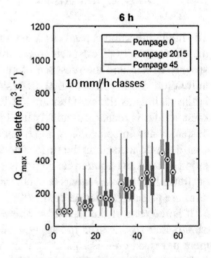

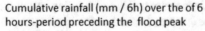

Cumulative rainfall (mm / 6h) over the of 6 hours-period preceding the flood peak

Maximum rainfall (mm / h) over the 6 hours-period preceding the peak of flood

**Fig. 14.5** Maximum inflow of the Lez River (Qmax) at Lavalette station (Entry of Montpellier) according to the cumulative rainfall over a 6 hours period preceding the flow peak and the maximum rainfall over the 6-hours preceding the flow peak for the no pumping (pompage 0) the reference pumping (Pompage 2015) and the 45 mm$^3$ strategy (Pompage 45)

## 14.3.2    The CCR Risk Modelling Approach and Its Use to Evaluate the Impact of NBS Strategies

### 14.3.2.1    General Overview

The CCR catastrophe risk model has been the basis of the evaluation of the runoff flooding risk and the evaluation of the impact of urbanization and GI strategies as described in Chap. 6 of this publication.

The catastrophe loss risk model is composed of: the hazard unit, the vulnerability and damage units (Fig. 14.6).

The hazard unit is based on a runoff model a model developed by CCR, which is a 2D rainfall/runoff spatialized production and transfer model based on hourly-spatialized rainfall data. It uses a 30 sec time step and a 25 m altitude grid, GIS data related to large watercourses, Météo-France rainfall data and Corine Land Cover data. Each land cover class is associated with a runoff coefficient that reflects the capacity of soil to infiltrate water (natural cover has the highest infiltration rate while continuous urban habitat has the lowest) (Moncoulon et al. 2014).

The vulnerability unit of the model gathers information based on the historical flood claims database (insured damage data) collected by CCR. It is called an insurance portfolio which contains address-based insured claims data such as the amount of the claims, insured value, risk location, portfolio exposure (number of policy contracts).

In the damage unit, the link between hazard and vulnerability is made to estimate damages with damage functions. There are no damage curves allowing the estimation of runoff damages in the Lez watershed. Indeed, the French national guidelines for flood damage (CGDD 2018) focus only on overflow hazards and do not consider the runoff hazard in the calibration of the curves. Specific damage curves were therefore developed for the Lez watershed. In the NAIAD project, the damage functions

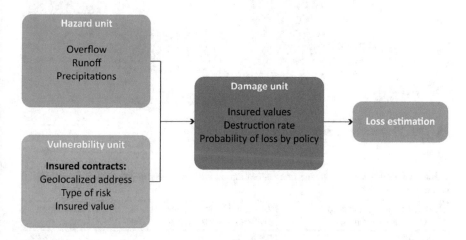

**Fig. 14.6** CCR catastrophe loss risk structure

are calibrated on the 2014-events on claims related only to residential homeowners. They are expressed in terms of damage rate (probability of damage) and destruction rate (amount of claims divided by the insured value) according to runoff flow.

Once calibrated, the catastrophe risk was used to estimate the impact of various modifications of the system (climate change, urbanization strategies, and green infrastructure strategies).

### 14.3.2.2 Calibration of Damage Curves in the Lez Watershed

The statistical analysis of predicted urban runoff with insured loss of residential homeowners provides the correlation between the runoff (expressed in m³/s) and the damage rate and between runoff and the destruction rate. These correlations are fitted in damage curves as illustrated in Fig. 14.7. Damages associated to a runoff below 0.07 m³/s are considered null.

These damage curves were used for an assessment of flood damages on the 2014-events. The validation of the damage rate curve has been done by comparing the real costs for the residential homeowners to the simulated costs (Table 14.3).

As the calibrated damage rate provides relevant and close results of the real 2014 flood losses, the damage could be used for subsequent estimations.

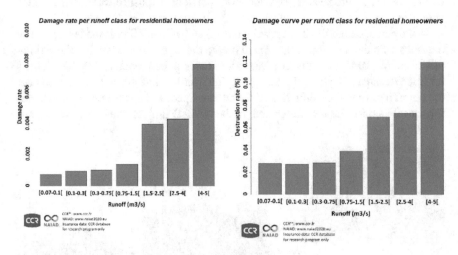

**Fig. 14.7** Flow damage function for the Lez. (Source: CCR)

**Table 14.3** Validation of damage rate calibration on the Lez case study for runoff on residential home owner

| Real 2014 damage costs for residential homeowners | Simulated 2014 damage costs for residential homeowners | Simulation error |
|---|---|---|
| 3,353,146 € | 3,298,343 € | −1.63% |

Source: CCR

### 14.3.2.3   Impact of Urbanization Strategies on Runoff Hazard

The hazard model was used to evaluate the impact of the three urbanization strategies defined in Sect. 14.2. This impact is evaluated with predicted land use maps from the GIS model and associated infiltration coefficient obtained in the three strategies and the estimation of runoff hazard on the 2014 flood events.

Figure 14.8 represents an example of hazard modeling for the laissez faire and the green strategies in the municipality of Cournonterral (periurban).

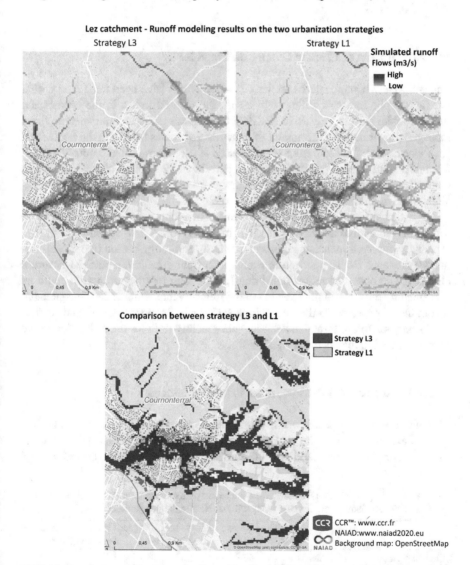

**Fig. 14.8** Comparison of flood maps between two urbanization strategies in Cournonterral simulated on the 2014-flood events

The evaluation does reveal an impact in the urban areas that are expected to be built by 2040. The yellow areas reveal areas that would be impacted by urban runoff with the "laissez-faire" (L1) strategy and not in the (L3) green strategy.

However, the impact of strategies in the other highly urbanized area, which are largely dominant in the areas at stake of the catchment, did not reveal significant effects. At the catchment scale, the urbanization strategies therefore did not reveal a significant modification of runoff flood hazard. It was therefore decided not to evaluate further the impact of urbanization on the reduction of flood risks in terms of avoided damages.

### 14.3.2.4 Impact of Climate Change on Flood Risk

The calibrated damage curves on the 2014-Lez events have been integrated within the catastrophe loss risk structure to assess the insured losses in the current and future climate for the year 2050 (RCP8.5 climate scenario) without specific flood management strategies. The annual average insured losses (AAL) in the Lez watershed was assessed based on the simulation of 400 years of climatic hourly rainfall from ARPEGE-Climat (Meteo-France) at current and 2050 conditions. Within that simulation, we detected and simulated extreme events, estimated damages and classified them in terms of return periods (see Table 14.4).

We estimated that, in the future, the number of events per year will rise from 43 to 57 and the annual average losses will increase from 7.2 to 9.2 €M (30%). We especially observed increasing damage for short-term return period (the model estimate 0 damages for 10 year return events in current climate but 53.5€M in future climate). The observed reduction of damage for long-term return period could be explained by the uncertainties related to the future events. Thus, it can be concluded that in the Lez case study the future flood events will be more frequent and costly.

This estimation of total damages are subsequently used for the estimation of the damages avoided thanks to GI strategies.

### 14.3.2.5 Impact of GI Strategies on Urban Flood Risk

The initial aim of the study was to stimulate the impact of GI strategies on urban runoff hazard. However, the research on the integration of NBS within the CCR runoff model is complex and still on going. To avoid this difficulty and make a

**Table 14.4** Comparison of total damage costs per return period of extreme events between current and future climate damage on Lez case study without NBS strategies (source: CCR)

|  | AAL | 10-year cost | 20-year cost | 50-year cost | 100-year cost | Number of simulated extreme events |
|---|---|---|---|---|---|---|
| Current climate | **7.2 €M** | 0 | 67.7 €M | 89.3 €M | 98.9 €M | **43** |
| Future climate (2050) | **9.2 €M** | 53.5 €M | 73.1 €M | 87.3 €M | 98.6 €M | **57** |

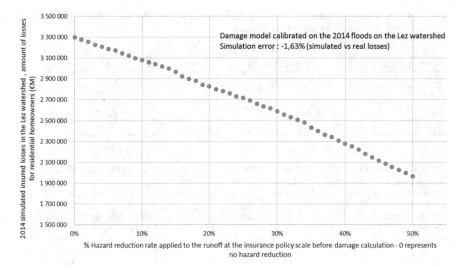

**Fig. 14.9** Effect of hazard reduction on 2014-flood insured losses for the Lez case study. (Source: CCR)

coarse estimation of damages avoided thanks to NBS strategies, the following procedure was used.

1. A relation between the percentage of reduction of runoff hazard and the related effects on avoided damage costs was estimated (Fig. 14.9)
2. A simple link between water retention resulting from GI strategies and the reduction of runoff was established by BRGM and used to estimate the impact of NBS strategies on the reduction of flood damages.

Using the results of the damage model, we estimated the effect of hazard reduction on flood damage (avoided damage) at 25 m resolution. For a reduction of 50% of hazard, the damage will be reduced by 1.9 €M (or −40.45%), a reduction of 20% of hazard reduces the damage to 2.8 €M (or −14.2%). These elements provide an overview of the necessary effect of NBS on the reduction of runoff to be effective to reduce damages.

As a simplification, we subsequently used this relationship for the estimation of damages in all events.

In order to estimate the avoided damages generated by the Green Infrastructure strategies, our assumption is that the retention they generate (30.3 L/m² for level 2 and 3.6 L/m² for level 1 (cf. Sect. 14.2.3)) stores all the rain that falls on a unit of land until its capacity is filled and that the % of rain taken out of the system is directly equivalent to the % of reduction of the resulting runoff. Concretely, considering that the average retention of water generated by the GI level 2 strategy is 30.3 L/m², we consider that if for example 100 mm of rain falls, 30.3 mm of rain is retained in the GI (for the level 2 strategy) which represents a 30.3% reduction of runoff. The argument that can justify the use of this method is that the GI strategies are spread out in a relatively homogeneous manner on the watershed.

**Table 14.5** Estimation of return period of rainfall events at the Montpellier Frejorgues weather station for the 1958–2008 period (Meteo France)

|                      | Montpellier Frejorgues (1958–2008) | | |
| -------------------- | -------------- | -------------- | -------------- |
| Return period (years) | 1H event (mm) | 2H event (mm) | 6H event (mm) |
| 10                   | 56             | 71             | 96             |
| 20                   | 65             | 84             | 117            |
| 50                   | 76             | 103            | 150            |
| 100                  | 84             | 120            | 179            |

**Table 14.6** Estimation of the damages with and without Green Infrastructure for current climate (Infinite event is considered to be 1.5 damages of the centenal event)

|                      | Damages in M€ | | | | |
| -------------------- | ------------------- | ------------------- | ------------------- | ------------------- | ------------------- |
| Return period (years) | Current climate No GI | GI L1 strategy (2H) | GI L2 strategy (2H) | GI L1 strategy (6H) | GI L2 strategy (6H) |
| 10                   | 0     | 0     | 0     | 0     | 0     |
| 20                   | 67.7  | 65.5  | 49.4  | 66.2  | 54.6  |
| 50                   | 89.3  | 87.0  | 69.6  | 87.7  | 75.7  |
| 100                  | 98.9  | 96.7  | 80.2  | 97.4  | 86.3  |
| Infinite             | 148.4 | 145.0 | 120.3 | 146.1 | 129.5 |

The relationship between runoff reduction and damage reduction is estimated in Fig. 14.9. The damages without NBS for different return period are also defined in Table 14.4. We identified the return-period of rainfall events based on the data of the Montpellier Frejorgues station. We obtain the following return periods in Table 14.5.

Based on these different elements, we estimate in Table 14.6 the damages with no GI, GI level 1 and GI level 2 strategy using the 2H and 6H event rainfall information.

Based on this estimation, we can infer a mean avoided damages of 1.02 to 1.45 M€/year for the GI level 2 strategy (see Chap. 6), i.e. a reduction from 14 to 20% of annual damages, and 0.12 to 0.17 M€ for the GI Level 1 strategy for insured damages. If we include an estimation of 28% of additional damages (public and agriculture) obtained from data collection on the 2014 events, we obtain a mean avoided damage of 1.30 to 1.86 M€/year for the GI level 2 strategy and 0.15 to 0.22 M€/year for the GI Level 1 strategy.

This monetary assessment of the damages avoided thanks to NBS strategies is subsequently used in the overall economic assessment of NBS strategies.

## 14.4 Economic Valuation of NBS Strategies

The economic valuation of NBS strategies is undertaken according to the methodology described in Chap. 6 of this publication. We especially present here: the assessment of implementation and opportunity costs, the economic valuation of co-benefits

**Table 14.7** Economic assessment methodologies used for the different NBS strategies

|  | Avoided damage costs | Implementation and opportunity costs | Co-benefits assessment | Cost- Benefit Analysis |
|---|---|---|---|---|
| Urbanization strategies |  |  | X |  |
| Green infrastructure strategies | X | X | X | X |

and the integration of the economic assessment through a cost-benefit analysis. As mentioned earlier, for the sake of clarity, the assessment of damage costs avoided thanks to NBS, which is the primary benefit of NBS strategies, is already described in Sect. 14.3.

We present the elements of the economic assessment that were implemented for the urbanization and the green infrastructure strategies. As mentioned earlier the active management of the karst was not evaluated in the economic assessment due to its lack of significant impact on flood hazard (Table 14.7).

Considering the lack of significant impact of urbanization strategies on hazard, the avoided damage cost can be considered negligible. Implementation and opportunity costs could not be estimated for urbanization strategy, as this would have required sophisticated research beyond the scope of this work. The complete economic assessment was therefore only carried out for GI strategies.

## 14.4.1  Assessment of Implementation and Opportunity Costs of NBS Strategies

We present in this section the method and results of the estimation of costs for the GI strategies.

### 14.4.1.1  Method

The estimation of implementation costs included the estimation of capital expenditures and operation maintenance (O&M) costs over a 20-years lifetime. These costs were estimated through a literature review and value transfer from other studies on GI costs, or grey literature from practitioners (Appendix A). As precise costs cannot be established based on a literature review, given the variability of land costs, the precise characteristics of GI and economies of scale, costs were set as ranges. These costs were estimated both for situations in which GI are implemented in existing urban areas, through urban requalification, and for situations in which GI are implemented in entirely new urban areas. This has a great impact on cost estimations, as requalifying already urbanized areas with GI often requires removing concrete pavement, thus implying extra investment costs. For this reason, requalification costs are often higher than new area development costs. In the case of green roofs, requalifying existing roofs requires changes in the load-bearing structure of the buildings, implying as much as 40% extra costs. Therefore, unit cost ranges are especially large for those categories.

Opportunity costs represent the costs associated with the foregone alternative, which can be measured by the net benefit foregone because the resources that provide the services cannot be used in their next beneficial use (Tietenberg and Lewis 2016). Considering that NBS generally require large amount of land for their implementation, compared to grey solutions, it is of utmost importance to consider them. We estimated the cost implied by choosing to deploy NBS instead of other land uses by using the average land market price, as a proxy of the sacrifice costs of not having this land usable for alternative profitable investments. This could be considered an upper bound estimation as it is not obvious whether these areas may have an alternative profitable use. These opportunity costs were added only to some NBS: city deproofing, bioswales and vegetated retention basins. It was indeed considered that roofs do not have alternative profitable uses.

### 14.4.1.2 Results

The cost estimates are presented in Table 14.8. They are expressed as much as possible in terms of $€/m^3$ of water retention, which is a good proxy of the cost-effectiveness of individual NBS measures to reduce flood risks.

Costs ranges are very wide as economies of scale can greatly reduce marginal costs for surface infrastructures. A vast range of technology is available for many GI. The level of cost varies greatly depending on the type of cover included in green

**Table 14.8** Investment and annual Operation and Maintenance costs. Units depend on the type of GI

| NBS | Unit | O&M | | Investment requalification | | Investment new areas | |
|---|---|---|---|---|---|---|---|
| | | Low | High | Low | High | Low | High |
| City deproofing + greening | $€/m^2$ | 1.05 | 1.05 | 69.9 | 93.6 | 53.0 | 75.0 |
| Green parking spaces | $€/m^2$ | 0.65 | 1.00 | 66.9 | 128.6 | 50.0 | 110.0 |
| Bioswale small | $€/m^3$ | 1.20 | 1.80 | 95.5 | 103.5 | 28.0 | 36.0 |
| Bioswale large | $€/m^3$ | 8.2 | 11.9 | 102.0 | 131.4 | 63.5 | 93.0 |
| Vegetated retention basin | $€/m^3$ | 7.2 | 10.4 | 45.0 | 143.0 | 12.0 | 120.0 |
| Extensive green roofs | $€/m^3$ | 167 (2 years) | 301 (2 years) | 484 | 1322.6 | 417.5 | 1002.0 |
| Intensive green roofs | $€/m^3$ | 83 | 150 | 1282.8 | 1982.5 | 916.3 | 1416.1 |

infrastructure, from a basic herbaceous cover to systems that include trees (high level cost for large bioswales and vegetated retention basins).

We also see a large heterogeneity of cost-effectiveness among the different individual NBS evaluated in the project (Fig. 14.10). This heterogeneity raises questions especially on the opportunity of integrating green roofs in future strategies considering their limited effect on water retention and their large costs.

In order to evaluate the cost over the lifetime of the project, the net present value of costs (see Chap. 6) was calculated and aggregated for the two GI strategies and gave the following estimates for the two strategies (Table 14.9).

The figures show that the GI strategies represent very large investments for the Lez catchment reaching 73–148 €M for the level 2 of GI. When opportunity costs are included, the amounts considered are largely superior. This underlines the fact that GI have a strong spatial extent, which can represent a challenge for their generalization. This also goes in favour of implementing NBS in places that are not suitable for other uses either because of the space or of spatial characteristics. However, although it is recommended to include opportunity costs in CBAs, it is questionable whether these areas all have alternative profitable use and therefore represent an opportunity cost. In the final CBA, we will therefore present results with and without opportunity costs. These costs need to be confronted to an estimation of the benefits brought by GI strategies.

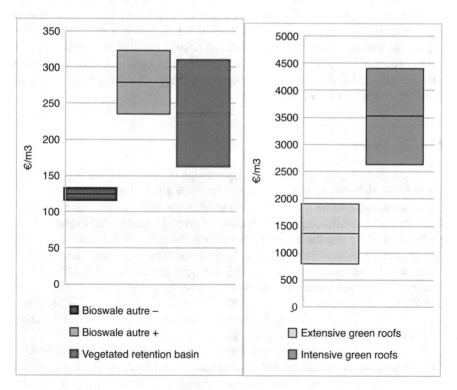

**Fig. 14.10** Cost-effectiveness of individual measures in the Lez Case study. Cost includes the investment and maintenance costs discounted for the next 20 years

**Table 14.9** Overall actualized costs of GI strategies over a 20 years lifespan (in M€)

| Strategy | GI type | Without opportunity costs | | With opportunity costs | |
|---|---|---|---|---|---|
| | | Low | High | Low | High |
| GI level 1 | Deproofing | 8.6 | 10.9 | 32.8 | 50.9 |
| | Permeable parking pavement | 27.3 | 50.1 | 113.9 | 194.0 |
| | Bioswale small | 2.8 | 3.2 | 26.9 | 105.4 |
| | **Total** | **38.6** | **65.1** | **173.6** | **350.4** |
| GI level 2 | Deproofing | 8.6 | 10.9 | 32.8 | 50.9 |
| | Permeable parking pavement | 27.2 | 50.9 | 113.8 | 194.1 |
| | Bioswale large | 44.6 | 61.3 | 151.6 | 238.3 |
| | Vegetated retention basin | 7.1 | 13.7 | 28.8 | 49.5 |
| | Green roofs | 4.8 | 11.3 | 4.8 | 11.3 |
| | **Total** | **73.4** | **148.2** | **331.9** | **544.1** |

## 14.4.2   Economic Valuation of Co-benefits

The economic valuation of co-benefits is fundamental in the evaluation process of NBS. The multifunctionality of NBS is one of their key advantages as compared to grey solutions for flood control. The NAIAD project generally adopted an integrated valuation approach of co-benefits, that considers that co-benefits also have a physical and a socio-cultural value (Jacobs et al. 2016). Nevertheless, we will focus in this report on the research developed for the monetary valuation of co-benefits. In the Lez basin, we used a stated-preference approach through the implementation of a choice experiment (CE) to valuate co-benefits. Details of this work are presented in Hérivaux and Le Coent (2021).

### 14.4.2.1   Method

Stated-preference approaches rely on representative surveys of the population to estimate people's willingness to pay (how much they would contribute in terms of fee or tax increment) for a hypothetical modification of the environment (here the implementation of NBS strategies). The survey gives the opportunity to evaluate the preferences of the population for different NBS strategies, their flood risk perception and the importance they grant to ecosystem services. It provides socio-cultural and monetary indicators for different NBS strategies and associated bundles of ecosystem services, without seeking to evaluate ecosystem services one by one. In the Lez catchment, we used a choice experiment to evaluate two types of NBS strategies: (1) the conservation of agriculture and natural areas through urbanization strategies and (2) the development of green infrastructure.

The elaboration of the survey was first based on a participatory process, involving two preliminary workshops with local stakeholders, in order to identify expected co-benefits, NBS implementation levels and potential barriers, and to introduce the

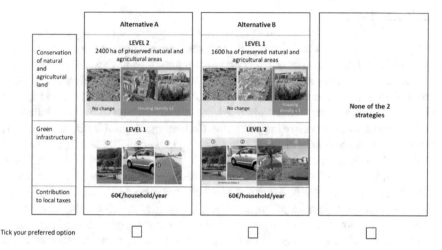

**Fig. 14.11** Example of a choice set in the Lez survey

CE method to local stakeholders (See Sect. 14.2 for a description of stakeholders' involvement). A questionnaire was then elaborated and tested with 29 respondents (face-to-face interviews with residents of the Lez catchment). The survey was subsequently administered on-line and yielded 400 valid responses from residents of the Lez case study.

In the main section of the questionnaire, the CE itself, respondent make choices between hypothetical flood management strategies for the Lez catchment presented in the form of choice cards (Fig. 14.11). In each choice card, respondents choose between two flood management strategies that achieve the same level of flood risk management but differ in the levels of implementation of NBS and in the level of contribution, in terms of tax increase. If neither of the two alternative is suitable for respondents, they can choose "Neither of the two strategies" (*status quo* situation). In this case, in which no payment is included, we emphasize that the level of flood control is not guaranteed. In the survey, respondents have to respond to six choice cards.

Each flood management strategies are characterized by three attributes. Attribute 1 is a simplification of the urbanization strategies, mentioned here as the conservation of agricultural and natural land, with a fixed population growth rate. Attribute 2 represents the GI strategies. For simplification of the questionnaire, green roofs were excluded from the strategies and therefore are slightly different from the strategies identifies in the other components. Attribute 3 is the financial contribution that respondents are willing to pay for financing the flood management strategy. The payment vehicle was identified as a 10-year yearly increase in local taxes. It is either 20, 40, 60, 80, 100 or 120€/household/year. These amounts were adjusted according to the test survey.

The questionnaire also included questions that allowed the identification of the main advantages (co-benefits) and disadvantages perceived by urban residents.

### 14.4.2.2   Results

The main co-benefits and constraints that respondents perceive for the two NBS strategies and their level of implementation are presented in Figs. 14.12 and 14.13.

On average, three co-benefits are associated with the level 1 of conservation of natural and agricultural land (similar to our urbanization strategies), and 2.5 for level 2. The three most cited co-benefits are climate change mitigation, landscape conservation and air quality improvement. On average, 1 and 1.7 disadvantages are respectively associated with level 1 and level 2. Lower quality of life, traffic problems and landscape deterioration are quoted by more than 20% of the respondents.

Co-benefits associated with **green infrastructure** are quite similar between level 1 and level 2 (respectively 3.2 and 3.1 benefits on average). More than half of the respondents quote landscape conservation, air quality improvement, biodiversity conservation, local urban temperature regulation and climate change mitigation. The number of disadvantages is quite low (0.4 and 0.7 on average respectively for level 1 and 2). Traffic and car parking problems is the most frequently quoted disadvantage for level 2 (18% of the respondents).

The results of the econometric analysis (mixlogit model) of respondents' choice in the CE allows us to estimate the preference for the different levels of NBS strategies.

This analysis reveals that respondents prefer the level 2 of implementation of the two NBS types to the level 1 (and the level 1 over no implementation of the NBS).

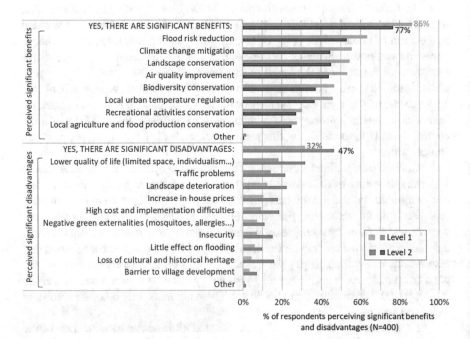

**Fig. 14.12** Perception of significant benefits and disadvantages associated to conservation of natural and agricultural land

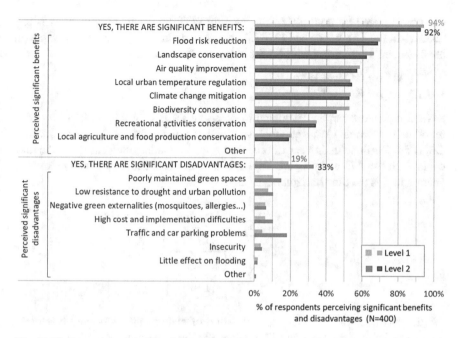

**Fig. 14.13** Perception of significant benefits and disadvantages associated to green infrastructure

- On average, respondents are willing to pay 141€ and 179€/household/year respectively if the level 1 and 2 of conservation of agricultural and natural land is implemented instead of the level 0.
- On average, respondents are willing to pay 143€ and 180€ /household/year respectively if the level 1 and 2 of green infrastructure is implemented instead of no green infrastructure.

Results also show an important heterogeneity of preferences among respondents, influenced by socio-demographic and housing environment characteristics. The analysis of this heterogeneity of preferences is beyond the scope of this chapter, but readers interested in more details can refer to Hérivaux and Le Coënt (2020).

### 14.4.2.3 Integration of the Economic Assessment

As mentioned before, the cost benefit assessment could only be carried out for the GI strategies. The various assessment described in Sects. 14.3 and 14.4 provide the building blocks for the economic assessment of NBS strategies as per the method described in Chap. 6 of this publication. Some elements are however missing to perform this cost benefit analysis.

First, in order to carry out the assessment, we need to extrapolate the co-benefits value estimated with the choice experiment to the whole watershed. A first provisional estimation of the value granted to co-benefits associated with NBS strategies can be estimated by multiplying the average WTP by the number of households residing in the Lez Watershed (230,000 households in 2019). The annual value of the co-benefits associated with NBS can therefore be estimated at:

- 32.9 M€ for GI level 1;
- 41.5 M€ for GI level 2.

Several indicators can be calculated to carry out a CBA. In this study, we mainly report on the Benefit Cost Ratio that is estimated with the following formula, where $CB_t$ is the Co-Benefits in year t, $AD_t$ is the Avoided Damage in year t, $r$ is the discounting factor,[3] $C_t$ and $OC_t$ are implementations Costs and Opportunity Costs in year t and T is the time horizon of the assessment.

$$BCR = \frac{\sum_{t=0}^{T} \left[ \frac{AD_t + CB_t}{(1+r)^t} \right]}{\sum_{t=0}^{T} \left[ \frac{C_t + OC_t}{(1+r)^t} \right]} \tag{14.1}$$

We therefore obtain the following estimation of benefits and costs and economic indicators for the GI level 1 and 2 strategies (Table 14.10 and Fig. 14.14).

A first key conclusion of the Lez GI economic valuation is that the cost-benefit analysis reveals close or slightly superior to 1. Investing in GI is therefore economically efficient and should be part of the priority of investments of the urban communities of the Lez catchment. This picture is even clearer, when opportunity costs are excluded from the analysis. Finally, an overwhelmingly large share of the value granted by residents of the Lez watershed to NBS is due to their co-benefits.

Table 14.10 Overall actualized costs of GI strategies over a 20 years lifespan (in M€)

| Strategy | GI level 1 | GI level 2 |
|---|---|---|
| Implementation costs (M€) (Sect. 14.4.1.2) | 52 39–65 | 120 92–148 |
| Opportunity costs (M€) (Sect. 14.4.1.2) | 210 135–285 | 318 239–396 |
| Avoided damages (M€) (Sect. 14.3.2.5) | 3.4 | 29 |
| Co-benefits (M€) (Sect. 14.4.1.2) | 287 | 363 |
| Avoided damages/Costs (rate) | 0.07 0.05–0.09 | 0.24 0.2–0.3 |
| BCR | 1.3 0.8–1.7 | 1.0 0.7–1.2 |
| BCR without opportunity costs | 6.0 4.5–7.5 | 3.5 2.7–4.3 |

---

[3] We use the standard rate recommended in the Quinet report of 2.5%.

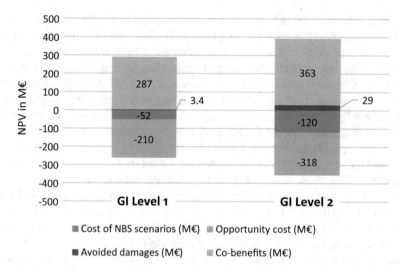

**Fig. 14.14**    Cost-Benefit analysis of GI strategies in the Lez catchment

Another conclusion of our study is that ambitious GI strategies (level 2), involving the implementation of green infrastructures at a large scale could mitigate 14–20% of damages related to floods. This is quite significant and should support the interest for green infrastructure in the future. The benefits arising from the reduction of flood risks are nevertheless largely inferior to the implementation costs and even more so if opportunity cost of land are included. The inclusion of co-benefits is therefore fundamental for NBS to be perceived beneficial.

## 14.5    Towards Implementation of NBS Strategies in the Lez Catchment

In the Lez case study, the main objective was to co-design and evaluate NBS strategies aiming at mitigating flood risk and addressing other urban challenges. The strategies we have designed are hypothetical macro strategies, at the city scale. Our study has therefore desmonstrated the potential interest of these measures and call for the development of a practical implementation program that could be the framework for the implementation of neighbourhood scale projects, using the same approach developed in the Copenhagen's cloudburst programme. At present, the NBS strategies are currently not included in local development plans or investment plans for the city. The consultation process with local stakeholders has nevertheless allowed us to identify some key information on the pathway towards implementation. We especially present here some of the key opportunities and constraints for the development of NBS as well as potential policy instruments that could be mobilized for their development.

One of the constraint of NBS development is directly related to the specificity of NBS: the multiplicity of benefits they generate. Dealing with NBS therefore requires a multiplicity of skills and responsibilities that are usually fragmented in the local administration: green space management, flood management, biodiversity, climate change etc. Considering the diversity of benefits of NBS and the limited space available in cities, the development of NBS should be planned based on an optimization of the diversity of benefits and not on one single benefit such as flood protection. Our cost-benefit analysis shows that NBS may be economically efficient when all benefits are considered, but not necessarily for sectorial challenges such as flood management, which may also complicate their acceptance by an administration still characterized by silos. A transition is nevertheless currently happening with the transfer of the responsibility for the management of aquatic ecosystems and flood management (GEMAPI) to urban communities (Montpellier Méditerranée Metropole (3 M) and Communauté de Commune du Grand Pic Saint Loup (CCGPSL) in the Lez catchment) and the possibility to perceive a local tax to finance projects in line with both objectives. This transition forces urban communities to address these challenges in an integrated manner, which should favour the development of NBS. Unfortunately, stormwater management responsibility, currently being transferred to urban communities, is still treated separately from other flood risks which currently limits the opportunities for funding the NBS we have studied.

This diversity of benefits may however be an opportunity to facilitate political support for these measures. In the Lez case study, the conservation of biodiversity remains low in the agenda of local decision makers but risk management, the reduction of heat island effects and air pollution may be good entry points to ensure political buy-in for NBS development.

Several current policy instrument may be mobilized to facilitate the development of NBS. The Water Development and Management Plan (SDAGE) of the Rhone-Mediterranean and Corsica basin includes the measure 5A-04 *"Avoid, reduce and compensate the impact of new soil sealing"* that sets ambitious objective for the limitation of soil sealing in the basin. Considering that all documents developed at the territorial level should be in conformity with the SDAGE, this document provides an excellent opportunity for the development of NBS (limitation of urban sprawl and development of green infrastructure). In addition, the water basin agency can provide 50% of funding of infrastructure investment aiming at reducing soil sealing, the green infrastructure we have studied are eligible.

For the practical implementation at the territorial level, urban communities (3 M and CCGPSL) and cities are in the driving seat for the implementation of the urban NBS strategies we have evaluated. This is especially true since the recent modifications initiated by the territorial reform law of 2015, includes the transfer of water and wastewater management to urban communities. Urban communities are also in charge of the development of urban master plans (SCOT, PLUi) that could be the main instrument for the development of NBS by setting rules for the construction of new neighborhood, which should promote the limitation of soil sealing and the use of green infrastructure, in agreement with SDAGE recommendations. Finally, cities intervene directly as public developer for the creation of new neighborhoods and rehabilitation programs. Including ambitious NBS in these programs would also create an example to be followed by private developers.

## Appendix A: Sources for the Estimation of GI Costs

| NBS | Source |
|---|---|
| City deproofing | https://Construction.info/renovation/VRD_et_amenagements_exterieurs/Revetement_de_sols_exterieurs/Case studylition/ASD020_Case studylition_d_un_revetement_de_sol_e.html |
| Green Parking spaces | Guide technique Ecovegetal (2017)<br>KURAS, Maßnahmensteckbriefe der Regenwasserbewirtschaftung – Ergebnisse des Projektes KURAS, Berlin, 2017 |
| Bioswale large | Grand Lyon, fiche technique n°2, Fossés et noues, 2016<br>Daniel Johnson, Sylvie Geisendorf, Are Neighborhood-level SUDS Worth it? An Assessment of the Economic Value of Sustainable Urban Drainage System Scenarios Using Cost-Benefit Analyses, Ecological Economics, Volume 158, 2019, Pages 194–205, ISSN 0921–8009, https://doi.org/10.1016/j.ecolecon.2018.12.024<br>ARB, Etude comparative des coûts des infrsatructures grises hybrides et vertes |
| Bioswale small<br>Vegetated retention basin | Royal Haskoning DHV, Costs and Benefits of Sustainable Drainage Systems, Committee on Climate Change, July 2012, Project number 9X1055.<br>KURAS, Maßnahmensteckbriefe der Regenwasserbewirtschaftung, Ergebnisse des Projektes KURAS, Berlin, 2017<br>ARB, Etude comparative des coûts des infrsatructures grises hybrides et vertes<br>Grand Lyon: Guide pratiques de gestion des eaux pluviales Fiche 5: Bassins de rétention et/ou infiltration |
| Extensive green roofs | Mairie de Paris, Végétalisation des murs et des toits, 2016<br>IBGE, Formation Bâtiment Durable: Toitures vertes: du concept à l'entretien, 2012<br>Direction de l'Environnement et de l'Energie Nice Côte d'Azur, Etude pour la définition d'une démarche de développement des toitures végétalisées, 2009 |

## References

Calvet C, Delbar V, Chapron P, Brasebin M, Perret J, Moulherat S (2020) Modélisation prospective des dynamiques urbaines et de leurs conséquences sur les dynamiques écologiques. Application à la région Occitanie. Sci Eaux Territ

CGDD (2018) Analyse multicritère des projets de prévention des inondations – Guide méthodologique 2018

Cohen-Shacham E, Walters G, Janzen C, Maginni S (2016) Nature-based solutions to address global societal challenges. IUCN

Cramer W, Guiot J, Fader M, Garrabou J, Gattuso JP, Iglesias A, Lange MA, Lionello P, Llasat MC, Paz S, Peñuelas J, Snoussi M, Toreti A, Tsimplis MN, Xoplaki E (2018) Climate change and interconnected risks to sustainable development in the Mediterranean. Nat Clim Chang 8:972–980. https://doi.org/10.1038/s41558-018-0299-2

Fleury P, Ladouche B, Conroux Y, Jourde H, Dörfliger N (2009) Modelling the hydrologic functions of a karst aquifer under active water management – the Lez spring. J Hydrol 365:235–243. https://doi.org/10.1016/j.jhydrol.2008.11.037

Hérivaux C, Le Coent P (2021) Introducing nature into cities or preserving existing Peri-urban ecosystems? Analysis of preferences in a rapidly urbanizing catchment. Sustainability 13:1–34

Jacobs S, Dendoncker N, Martín-lópez B, Nicholas D, Gomez-baggethun E, Boeraeve F, Mcgrath FL, Vierikko K, Geneletti D, Sevecke KJ, Pipart N, Primmer E, Mederly P, Schmidt S, Aragão A, Baral H, Bark RH, Briceno T, Brogna D, Cabral P, De Vreese R, Liquete C, Mueller H, Peh S, Phelan A, Rincón AR, Rogers SH, Turkelboom F, Van Reeth W, Van Zanten BT, Karine H, Washbourne C (2016) A new valuation school: integrating diverse values of nature in resource and land use decisions. Ecosyst Serv 22:213–220. https://doi.org/10.1016/j.ecoser.2016.11.007

Maréchal JC, Ladouche B, Dörfliger N (2008) Karst flash flooding in a Mediterranean karst, the example of Fontaine de Nîmes. Eng Geol 99:138–146. https://doi.org/10.1016/j.enggeo.2007.11.013

Moncoulon D, Labat D, Ardon J, Leblois E, Onfroy T, Poulard C, Aji S, Rémy A, Quantin A (2014) Analysis of the French insurance market exposure to floods: a stochastic model combining river overflow and surface runoff. Nat Hazards Earth Syst Sci 14:2469–2485. https://doi.org/10.5194/nhess-14-2469-2014

Tietenberg TH, Lewis L (2016) Environmental and natural resource economics. Routledge

# Chapter 15
# Glinščica for All: Exploring the Potential of NBS in Slovenia: Barriers and Opportunities

Polona Pengal, Alessandro Pagano, Guillaume Piton, Zdravko Kozinc, Blaž Cokan, Zarja Šinkovec, and Raffaele Giordano

## Highlights

- An overarching and comprehensive participative process can result in a risk-management NBS scheme accepted by the stakeholders
- Citizen science can support risk management
- Spatial planning in Slovenia is not yet aligned with the European NBS agenda
- Decision makers rather than the general public fail to accept NBS as an alternative to grey solutions
- Methods for assessing economic value of NBS co-benefits need further development in the field of ecological and cultural benefits

## 15.1 The Glinščica Catchment Characterization

Glinščica catchment is situated within the borders of the City Municipality of Ljubljana (MOL) that spans roughly 275 km² and has a population of about 284,000 inhabitants (Fig. 15.1). The case study site covers 7% of Ljubljana's surface area, it

P. Pengal · B. Cokan
Institute for Ichthyological and Ecological Research (REVIVO), Dob, Slovenia
e-mail: polona.pengal@ozivimo.si

A. Pagano · R. Giordano
Istituto di Ricerca Sulle Acque del Consiglio Nazionale delle Ricerche (IRSA-CNR), Bari, Italy

G. Piton (✉)
Université Grenoble Alpes, INRAE, CNRS, IRD, Grenoble INP, IGE, Grenoble, France
e-mail: guillaume.piton@inrae.fr

Z. Kozinc · Z. Šinkovec
Iskriva, Ljubljana, Slovenia

© The Author(s) 2023
E. López Gunn et al. (eds.), *Greening Water Risks*, Water Security in a New World, https://doi.org/10.1007/978-3-031-25308-9_15

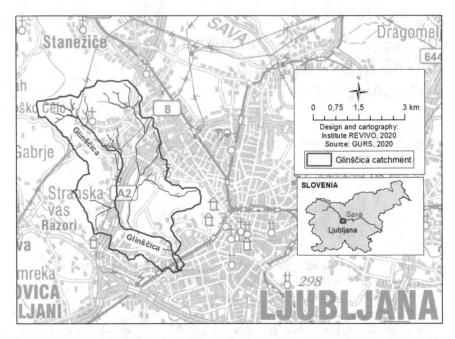

**Fig. 15.1** Glinščica catchment location and extent

includes five of its districts (Dravlje, Šiška, Rožnik, Vič, Šentvid) and accounts for 8.17% (23,200) of its population. Ljubljana has spread extensively over the flood-plains of rivers like Glinščica, Gradaščica and Ljubljanica during the past decades (Komac et al. 2008). Furthermore, the spatial planning process allowed properties to be built right on the banks of the watercourses, leaving no space for flood waters. Consequently, both hazard and vulnerability increased significantly, multiplying the flood risk in the catchment. Nevertheless, natural areas still cover approximately 50% of the catchment, agricultural land about 20% and urban areas about 30%, which allows for the planning and implementation of NBS. Over the last decade, the City Municipality of Ljubljana has implemented numerous urban green measures and was designated the European Green Capital in 2016.

Urban watercourses in Ljubljana are an important component of the urban green system, primarily as a network of natural areas which stretch into the urban fabric and introduce natural landscape elements in the urban area. However, the lower reaches of Glinščica have been lined with concrete for several decades and other forms of regulations extend upstream to the mountainous headwaters. Inappropriate regulations coupled with urbanization of flood plains are reflected mainly in the high frequency of flood events and low species diversity. Fast drainage through the straightened stream channel conveys flood waves directly into the city center. The most recent devastating floods of 2010 cost an estimated 14.74 million € in dam-ages, mainly to watercourses, houses and buildings (Benedičič 2011).

Straightened channels and the continuous removal of riparian vegetation have also reduced the aesthetic and educational value, as well as the status of the water environment and thus the experiential value of the stream. Therefore, Glinščica ceased to provide many of its functions as an urban green corridor: hydrological, ecological, spatio-structural, aesthetic, sports and recreational and social.

Several local initiatives have been put forward during the last 10 years to restore parts of the Glinščica stream to a more natural status. Together with the EU requirements arising from the Water Framework Directive (WFD) and Floods Directive (FD), these resulted mostly in changes on strategic level. Sustainable development is one of the main targets of the 2014–2020 Strategy of the City Municipality of Ljubljana (Trajnostna urbana strategija ... 2015). Furthermore, restoration of natural features has been integrated as one of the priorities in water management in the strategic part of the Municipal Spatial Plan (Odlok o občinskem ... – strateški del 2018) and in the implementation part of the Plan, through the conservation of the ecological status of water bodies (Odlok o občinskem... – izvedbeni del 2018). Glinščica River is also one of the priority streams for river restoration in the Program of Fish Management in Freshwaters of Slovenia for the period until 2021 (Program upravljanja... 2015). However, based on monitoring data and actual conditions in the field these changes have thus far had little or no effect on the practical planning and implementation.

Glinščica was selected as the target catchment by the NAIAD project in Slovenia due to several reasons. First of all, it is a small catchment, located entirely within the borders of one municipality, which is exceptional in Slovenia, having 212 municipalities on a land surface area of 20,3 km². Second, the series of floods since 2010 illustrates the insufficient capacity of the current risk management measures. Third, the Glinščica catchment is defined as the green wedge of the city of Ljubljana, both locals and tourists using it for recreation and relaxation, but the stream itself is completely channelized and void of riparian vegetation. Last, but not least, the number of previous local initiatives and the extent of agricultural land (Fig. 15.2) indicate the desire and potential for restoration.

The NAIAD process was therefore applied to offer an alternative strategy, based in the concepts of NBS and catchment management approach, to Glinščica stakeholders, identify the potential of NBS for risk management in the Glinščica catchment and to demonstrate participative planning process. However, the long-term willingness to implement the NBS strategy developed by the stakeholders, remains to be achieved.

## 15.2 Risk Assessment and Perception

Floods are quite frequent and severe in the Glinščica catchment, with increasing damages to communities and the built environment, as a consequence of historical regulations. Continuously, grey measures are being implemented to reduce flood risk with limited effectiveness and a detrimental impact to the environment (Griessler

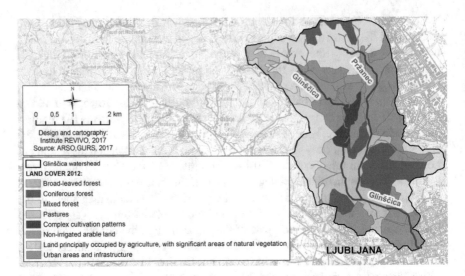

**Fig. 15.2** Land use in the Glinščica catchment shows great potential for using NBS for flood risk management

Bulc et al. 2017; Žaberl et al. 2011). Meanwhile, NBS and hybrid solutions that could simultaneously contribute to flood risk reduction and achieve good environmental status have not been considered in the current risk management planning.

Moreover, local knowledge and initiatives are not considered in flood protection planning and public participation in flood risk management in Slovenia is limited to submission of suggestions after the course of action and/or design has already been decided. However, experience shows that flood management measures need to be considered as a collective decision-making process characterized by multiple-actors with different, and often conflicting, risk perceptions (Santoro et al. 2019).

In an institutional decision environment, where the presence of ambiguity is unavoidable, the different roles played by the decision-actors affect the lens through which these actors give a meaning to a certain situation. Evidence demonstrates that making the decision actors aware of the existence of ambiguous problem framing is the key to enable creative and collaborative decision-making processes (e.g. Giordano et al. 2017). Addressing the existence of different and equally valid problem framings (unilateral decision-making process) in the initial stage of the participatory process increases the time required for making the decision. However, the diversity in frames also offers opportunities for innovation and the development of creative solutions (Brugnach and Ingram 2012), thus facilitating the implementation phase and the measure's effectiveness. The process and the underlying scientific methods are explained in detail in Chap. 5 (Giordano et al., this volume), while this chapter focuses on the implementation of these methods in the Glinščica case study. This chapter is therefore intended mainly for the decision makers and managers wishing to transition to a modern water management approach, but scientists will also find information on how theoretical knowledge fairs in practice and hopefully work on further developing the methods accordingly.

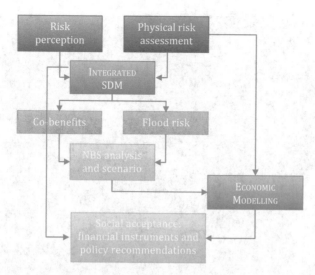

**Fig. 15.3** The full participative process as implemented in the Glinščica catchment. (please refer to Chap. 5 (Giordano et al., this volume) for methodological explanation)

As already stated in other parts of this book (Chap. 5, Giordano et al., this volume), stakeholder engagement in defining risks and designing NBS was a key step in the process. The main aim of the Glinščica case study was to enhance the future NBS implementation by investigating the potential impacts from NBS measures, facilitating a dialogue between stakeholders, aligning divergences and promoting the social acceptance of NBS measures for risk reduction and co-benefit generation (i.e. a natural assurance scheme) at different levels (local, regional, national) and sectors (e.g., municipality, civil protection). To this end, the process was implemented as a fully participative (Fig. 15.3; please consult Chap. 5 (Giordano et al., this volume) for methodological details of the process). This also allowed for the most important impact of the activities, a raised awareness about NBS and demonstration of how participative planning can ensure the most acceptable solution is developed and accepted by all.

### 15.2.1   Physical Flood Risk Assessment

First of all, a physical flood risk assessment was performed to determine the hazard and vulnerability levels. It was based on a full hydrological study (see Chap. 4 (Mullingan et al., this volume) for the overall approach), which considered a statistical regional analysis of rainfall and discharge data of a river station, with the identification of peak flood values for different return periods. The model was calibrated to the official flood maps available for the larger Gradaščica catchment (Fig. 15.4).

Expectedly, flood risk is most extensive in the lower reaches, but several areas at risk were identified far upstream. These are generally areas where historical

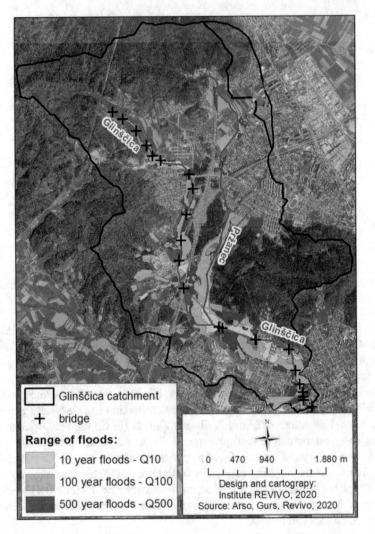

**Fig. 15.4** Flood extent in the Glinščica catchment, for 10-, 100- and 500-year return periods. Bridges play an important role in flood waters distribution

regulations have impacted the river channel the most and where buildings and infra-structure have spread to its banks. Comparison with the land use distribution was performed to reveal the highest vulnerability areas and areas with highest NBS potential.

Additionally, it was found that the bridges on the Glinščica stream and the relevant roads play an important role in distributing flood waters during high discharge. They act as bottlenecks, stopping and redirecting flood waters to the flood plains and thus controlling the downstream discharge.

## 15.2.2  Risk Perception

Watercourses are considered public goods in Slovenia and the Ministry for Environment and Spatial Planning (MOP) is directly responsible for their management and maintenance. Therefore, MOP was identified as the most important stakeholder in NBS solution planning in the Glinščica catchment. MOP is also responsible for the transfer of WFD, FD, Birds, Habitats and all other water related Directives into Slovenian legislation, as well as their implementation throughout Slovenia. To understand and map the many number of different functions and/or roles of MOP in the NBS solution planning in Glinščica, 12 different departments were identified within the Ministry, one regional department and three of its agencies and institutes directly involved in water management. However, it was difficult to map and determine the exact responsibilities of each of these actors within the water management system. In addition, at least 16 other stakeholder groups were identified and targeted following the snowball approach (see Chap. 5, Giordano et al., this volume, for details on the process), including governmental institutions, research institutes, chambers, recreational associations, city quarters, civil initiatives.

The initial stakeholder participation was performed through a series of individual interviews, through which different risk perceptions, existing cooperation, responsibilities and level of NBS awareness were collected and used as part of the overall catchment characterization. A large number of individual stakeholders (over 50) were initially contacted and invited for an interview, of which only a handful accepted participation. Of those that agreed to participate, only a couple of stakeholders understood the principles and were able to indicate examples of NBS, but all of them considered flooding as the main risk for the Glinščica catchment, which was in agreement with the physical risk assessment performed simultaneously.

## 15.3  The Participative Search for Solution

Research shows that solutions developed through public participation are more likely to be trusted and accepted by individuals within the network, prompting individual and collective action. The initial interactions described above revealed a considerable lack of awareness of modern water management concepts such as NBS, adaptive management, catchment approach among the participating stakeholders. The first workshop was consequently structured to provide an extensive explanation of these concepts, including best practice examples from abroad. Furthermore, the interviews resulted in a list of issues that need to be addressed in order to support the flood risk management in the Glinščica catchment. The workshop thus included a ranking of these issues and defining the main goals for flood risk management in the Glinščica catchment (Table 15.1).

These results defined and directed the following work and development of the solutions in the Glinščica catchment case study.

**Table 15.1** The five most important issues and their transition to management goals for the Glinščica catchment as defined by the stakeholders through participative process

| Issue | Goal |
|---|---|
| Flood plain occupation | Maintain flood plain occupation (do not increase) |
| State of the ecosystem | Improve the state of ecosystem |
| Lack of public funding | Increase public funding |
| Community safety | Increase community safety |
| Watercourse speed | Decrease watercourse speed |

## 15.3.1  Identifying Potential Solutions

The process started with experts developing a comprehensive list of available measures, including grey, hybrid and green (Chap. 5, Giordano et al., this volume). These were presented during the first workshop, when the stakeholders were encouraged to propose, discuss and finally rank the different potential solutions (NBS, hybrid and grey) in relation to their contribution to achieving the five set goals for the Glinščica catchment. The stakeholders mostly selected hybrid solutions with a general notice that the correct design and location of these measures is of utmost importance for achieving their effectiveness. Moreover, they felt that the measures should be designed in harmony with each other and help to achieve multiple goals simultaneously (co-benefits). The following five measures were selected as the most promising:

1. Dry retention areas
2. Re-meandering of the stream (including revegetation)
3. Opening of the flood plains
4. Widening of the stream channel
5. Small multi-functional wet retention areas

It was suggested by the stakeholders that the dry retention areas and opening of the floodplains should be implemented in the spaces upstream of the built-up areas. The stakeholders explained that flood risk management measures have been planned for the Glinščica catchment since the 2010 floods and that one of the dry retention areas had already been built. Re-meandering has somewhat contradictory expected impacts on the five main goals according to the stakeholders. Re-meandering will greatly improve the state of ecosystem and slow the water flow, but should be implemented within the opened-up flood plain or within a dry retention area, because it might increase the risk of flooding by slowing the flow and hence will not attribute to community safety. Widening of the stream channel was suggested for the stretch of the Glinščica within the urbanized areas, where buildings and other infrastructure prevent other restoration measures. The concrete lining should be removed and the more natural two-level channel restored to maintain the ecological flow in the lower, smaller channel during low flows, but to allow the larger volumes during flood events to be discharged efficiently. As the last suggested measure, wet retention areas were seen as the least effective in flood risk management, but as an important factor for improving the state of ecosystem and an important addition to the green areas of the city.

### 15.3.2  Identifying and Modelling Co-benefits

Following the first workshop, the stakeholders were again approached individually to identify and rank the different co-benefits and to identify and rank the soft measures[1], intended to enable the implementation and enhance the efficiency of physical measures (Chap. 4, Mullingan et al., this volume, for a description of method to assess NBS effects). The objective of this step was not only to understand which of the selected measures provide the most benefits, but also to be later used in the valuation of the different solutions, both in monetary as well as non-monetary terms (see Chap. 6, Le Coent et al., this volume, for a description of the valuation methods and Chap. 5, Giordano et al., this volume, for a description of the stakeholder engagement methods).

During the second round of semi-structured interviews, individual stakeholders were first requested to rank the level of individual co-benefit production for each of the five measures, selected during the first WS. This step was highly controversial for the stakeholders in that they felt the co-benefit production depends heavily on the exact location and design of the selected measure, an issue already raised during the first workshop. Moreover, the stakeholders also identified overlap or counteracting impacts of the measures or they were not aware of the functioning and hence, they couldn't predict the co-benefits. In some cases, they refused to perform the ranking and so the scores were not taken into account. Finally, although the stakeholders were encouraged to expand the list of co-benefits, no new suggestions of co-benefits were given. The results were later grouped and averaged to obtain the common score of five highest-ranking co-benefits, which would be included in the next steps of the process. The five most important co-benefits identified by the stakeholders were reduction of flood extent and damages to the built environment, enhancement of biodiversity and the state of ecosystems, improvement of community safety and increase of the social value of ecosystems, which were well aligned with the main goals for the Glinščica catchment management. A similar, but simplified process was applied to define the five most important soft measures, which were intended to support and enhance the effectiveness of the physical measures. These were: enforce land protection planning strategies, enforce urban planning strategies, territory control (illegal activities), implementing projects that target the

---

[1] A socio-institutional measure to support or enhance the functioning or impact of its opposite, a hard measure, either NBS, hybrid, grey. Examples include policy and legislation, enforcement and financing, but also behavioural change, capacity building, education.

involvement of local communities, defining innovative protocol of interaction among different institutions.

All the information collected was eventually integrated by experts into the System Dynamic Model (SDM) for the Glinščica catchment, developed in order to allow a comparative analysis of the different strategies. The SDM is based on the integration of different stakeholders' risk understandings and problem perception and the physical assessment of the water-related risk (see Chap. 5, Giordano et al., this volume, and Pagano et al. 2019 for a description of the SDM approach). On the one hand, the model was used to support the development of an integrated community-based evaluation method, drawing both on scientific evidence (e.g. deriving from physical risk assessment activities and economic analyses) and on the local/expert knowledge. On the other hand, the SDM enabled a participative (semi-) quantitative simulation of the impacts of specific strategies to deal with water-related risks, supporting a comprehensive analysis of trade-offs among different stakeholders and analysing costs and benefits (including co-benefits) at different scales and on different issues.

### 15.3.3 Identifying and Selecting Indicators

The identification of the most useful set of indicators for evaluating NBS effectiveness (Table 15.2) was not required for the continuation of the process as such. However, the task implemented at the start of the 2nd workshop encouraged the stakeholders to discuss again the desired (co-)benefits and agree on the future evaluation and monitoring after the implementation of the NBS measures. This served to further consolidate their acceptance and ownership of the developing solutions as well as to re-confirm the most important benefits sought.

Additionally, during the discussion on indicators and monitoring, the lack of water level monitoring station was identified as a barrier for improved flood risk management on the Glinščica stream. Although a discharge gauge existed on the Glinščica stream in the past, it was dismissed in the 1990s and so no current reference discharge data are available for either up to date hydrological/hydraulic modelling, as an early-warning support system or for tracking the impact of climate change now and in the future.

The Department for civil protection of the City Municipality of Ljubljana is, among others, responsible to monitor, report and issue early warnings in cooperation with the national civil protection agency (URSZR) in case of natural disasters. Since there had been no gauging station on the Glinščica stream (neither on other streams in the municipality) before this project, the department officers were required to regularly monitor the water levels during high rainfall events in-situ, or by deploying officers of other available civil protection departments or services (e.g. local volunteer firefighter community personnel). In either case, the personal field observation in time of emergency is a waste of valuable and limited time that

**Table 15.2** The three most important indicators according to the stakeholders' opinion were determined for each of the five most important (co-)benefits

| | |
|---|---|
| Reduced flood extent | Runoff coefficient in relation to precipitation quantities (mm/%) |
| | Flood peak reduction (e.g. Qmax,0/Qmax,1) and increase in time to peak [s] |
| | Volume of increased storm water retention |
| Reduced damages to built environment | Reduction in human casualties n° or ratio |
| | Value of damages on public infrastructure |
| | Value of damages to buildings |
| Enhanced biodiversity and ecosystems state | Species richness and composition |
| | Biodiversity index |
| | Water flow speed (in relation to natural) |
| Improved community safety | Flood peak reduction (e.g. Qmax,0/Qmax,1) and increase in time to peak |
| | Extent of urbanized floodplain areas |
| | Number and extent (number of people affected) of intervention events |
| Increased social value of ecosystems | Distribution of public green space – total surface or per capita |
| | Improved human health and wellbeing |
| | Urban green: index of biodiversity, provision and demand of ecosystem services |

the personnel could be using to organize and implement mitigation and/or rescue activities.

Therefore, the utilization of the FreeStation approach in the Glinščica catchment was initially suggested to collect the reference data for improving the hydrological/hydraulic models used in assessment of the NBS solutions. However, the stakeholder participation process revealed that it can also be used to monitor the impact in case of implementation of the selected NBS and to establish a remote sensing location to support civil protection service of the City Municipality of Ljubljana in monitoring water levels and issuing flood warnings.

### 15.3.4  Freestation as a Multi-functional Monitoring Tool

The FreeStation open-source initiative enables different stakeholders and communities to build and deploy reliable environmental data loggers with the lowest cost and easiest DIY build possible (please consult Chap. 4, Mullingan et al., this volume, for a full description of the Freestation initiative). Its modular design allows the user to select and install the assortment of sensors most suitable for their specific purpose.

**Fig. 15.5** The Freestation was installed on the driftwood barrier at the lower end of the newly built Brdnikova reservoir

The first FreeStation monitoring station was installed on a suitable bridge over the Glinščica stream upstream from the vulnerable urban area to test its efficiency and usefulness for the above-mentioned purposes (Fig. 15.5). The data collected from the Glinščica monitoring station will be incorporated into the existing flood monitoring system of the MOL, available online to the general public and used in various analyses and forecasts. Besides water level, the station is also collecting data on air temperature, humidity and pressure, and it can be upgraded with modules for rainfall and wind speed monitoring if requested by the stakeholders. The solar panel powered FreeStation is completely independent and requires minimal maintenance. It is thus not surprising that the City Municipality of Ljubljana has deployed 7 additional FreeStations to the different observation locations throughout the municipality as of 2022.

## 15.4 Developing, Testing and Selecting the Most Suitable Strategy

Eventually, the 2nd workshop aimed to co-design the most effective combination of NBS, hybrid and soft measures for achieving the selected benefits (strategies). The participants were required to create three different boxes, each representing a strategy and each of which should contain five different actions selected among the potential NBS, hybrid and soft measures identified in previous steps (Table 15.3).

**Table 15.3**  The measures included in the three developed strategies

| Renaturation | Bureaucratic | Bottom-up |
|---|---|---|
| Retention areas effectiveness River renaturation with re-meandering Wetlands restoration Physical risk management infrastructure maintenance Funding opportunities for infrastructure | Opening floodplains Territory control Community involvement Monitoring and warning system effectiveness Insurance policy effectiveness | Retention areas effectiveness Community involvement Institutional cooperation Training Funding opportunities for infrastructure |

Once the measures had been selected, they named the three strategies the "Renaturation", the "Bureaucratic" and the "Bottom-up", their names indicating the main measures considered.

The results were integrated in the SDM model described above to support interactive comparison of these strategies with a real-time visualization of their impacts on the selected parameters of co-benefit production (Fig. 15.6). In other words, the simulation of strategies in the SDM model over a 50-year period provided relative information about the impacts of applying the different strategies on the flood risk as the primary goal and on the environmental, social and economic co-benefits, selected by the stakeholders. The key result of this exercise was that the stakeholders recognized the relevant importance of combining the physical and soft measures. In the specific case of Glinščica catchment, stakeholders understood the important role that institutional measures play in either preventing or increasing the implementation and effectiveness of urban and regional management plans and measures.

The results were discussed among the stakeholders, who were asked to rank the strategies and choose the best one according to their opinion. An agreement emerged that, while the "Renaturation strategy" was the most promising one, none of the strategies adequately addressed the challenges of the Glinščica catchment management. The stakeholders felt restrained by the number of measures allowed per strategy, so an additional strategy was proposed by them with no constraints on the number of measures. To develop this strategy, the "Renaturation strategy" was complemented with seven additional soft measures, since the simulation results indicated that their simultaneous implementation with the proposed NBS and hybrid solutions could help achieve the target benefits in the long term. This strategy was named "Glinščica for all" and was approved by all the participating stakeholders (Table 15.4, Fig. 15.6).

As the last step of the 2nd workshop the stakeholders were provided with a map of the Glinščica catchment and required to indicate the locations where the selected three physical measures should be implemented. This map was used to develop the final Glinščica catchment management plan (Fig. 15.7).

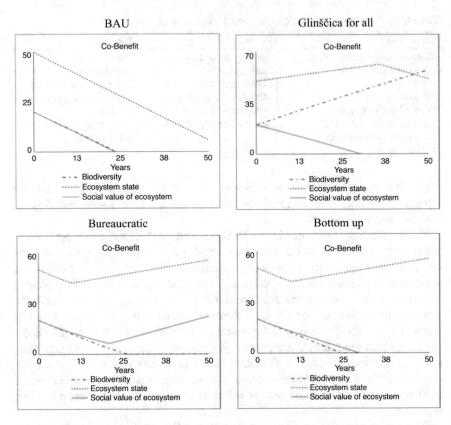

**Fig. 15.6** The output of the SDM model predicting the changes in the 3 ecological co-benefits production, depending on the Glinščica catchment management strategy applied

**Table 15.4** The list of NBS/hybrid and soft measures defining the Glinščica for all strategy developed by the stakeholders

| NBS measure | Soft measure |
|---|---|
| Regular maintenance of retention areas | Territory control |
| Wetlands restoration | Funding opportunities for IRR |
| River renaturation with re-meandering | Launch of an effective monitoring and warning system |
| | Community involvement |
| | Insurance policy effectiveness |
| | Training |
| | Institutional cooperation |

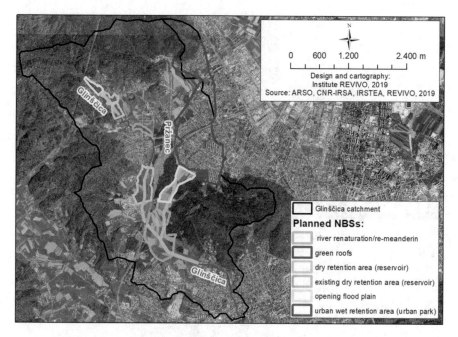

**Fig. 15.7** Locations of proposed NBS and hybrid measures as predicted in the Glinščica for all strategy developed by the stakeholders

## 15.5 The Road to Implementation (E/Valuation)

The physical solutions of the Glinščica for all strategy selected during the 2nd stakeholder workshop were detailed by experts and integrated into the hydrological/hydraulic model built in HEC-HMS and HEC-RAS and combined with the results of the Flood Excess Volume (FEV) methodology (Fig. 15.8).

More specifically, the hydrological/hydraulic models (HEC-HMS and HEC-RAS) were used to assess the impact of individual NBS and to produce maps of the flood extent, whereas the FEV methodology was applied to provide a simplified assessment of the synergistic effects of all the measures. The results of the integrated modelling (HEC-HMS, HEC-RAS, FEV) were used to analyse and evaluate the impacts of strategies co-designed with the stakeholders on flood risk. The most important aspect and advantage was the possibility of modelling a complex (and variable) set of measures keeping on the one hand the flexibility and modularity given by FEV, while relying on the other hand on a solid rainfall/runoff model to build maps of the flooded areas. In the end, this was highly relevant to support a strong and reliable analysis of economic benefits of NBS in terms of flood risk reduction.

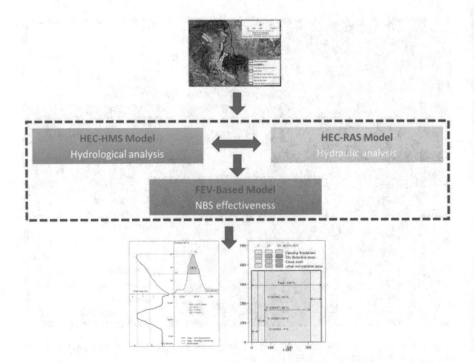

**Fig. 15.8** The impact of the "Glinščica for all" strategy on flooding was evaluated using HEC-HMS, HEC-RAS and FEV-based model. Please consult Chap. 4 (Mullingan et al. this volume) and Bokhove et al. (2019, 2020) for description of the FEV and hydrological modelling applied to the Glinščica case study

### 15.5.1 Economics of Glinščica for All Strategy

Relying on the previous research, it was expected that the added value of NBS in terms of providing co-benefits would outweigh the known limitations of the BAU strategy. The BAU strategy was defined as a continuation of current management practices and urbanization trends, with no or only grey measures applied. On the other hand, future conditions, under which the stakeholder defined NBS strategy was developed, is one of increasing flood risks, combined with the pressures of further urbanisation and population expansion in the area. Therefore, the selected strategy, named by the stakeholders "Glinščica for all", was recognised as the most suitable, especially in the light of future pressures of green taxation or even penalties, deriving from EU policy.

The economic valuation of the chosen strategy was performed as an ex-ante analysis with the goal to compare the chosen NBS and BAU strategies as alternative strategies for the development and management of the Glinščica catchment over a 30-year time scale.

Main objective of performing the economic analysis was to show that in the long term, proposed NBS strategy will be a more rational choice for the following reasons:

1. The future area development, under BAU, foresees continuation of housing construction and grey measures as flood protection measures, with maximum profit in mind. This in long term is not favourable for a capital city, aiming for green and sustainable development practices.
2. The preservation of fully functional ecosystems, supported by NBS measures and alternative development of the case study site will result in preservation of the area's potential for a combination of quality living area, recreational area as well as support local green economy in terms of urban farming.

Calculations were performed following the:

1. Proposed methodology and guidelines, on LCA and CBA assessment (please consult Chap. 6 – Le Coent et al., this volume). Given the drawbacks of both LCA and CBA approaches, a combination of top-down and bottom-up approaches was used, where all proposed elements of calculation were tested with the stakeholders. This especially refers to the co-benefits identification and assessing the viability of their development potential in terms of there being a realistic chance of being accepted as credible alternative to BAU. This also reflects in the nature and cost assessment of soft measures supporting or at least promoting the implementation of selected NBS strategy.
2. National legislation and recommendation under the Decree on the uniform methodology for the preparation and treatment of investment documentation in the field of public finance (Official Gazette No:. 60/06, 54/10 in 27/16), proposing a discount rate of 3%.

Under this methodology the investment costs, taken into calculations include:

1. Costs of land works for implementation of proposed measures
   This includes: (a) costs of investment documentation and studies, (b) preparatory work, (c) building of the infrastructure and (d) maintenance of the built infrastructure
2. Costs of soft measures in support of NBS strategy implementation
   These costs were estimated as costs of (a) awareness raising and competence building for public sector stakeholders to increase their understanding of NBS as fully credible alternative to BAU in flood protection. This would be implemented in series of workshops (five foreseen in calculations); (b) cost of developing suitable sustainable plans for spatial planning and land use, possible due to the restored ecosystem potential following implementation of selected NBS; (c) costs of preparation of individual projects (three annually, using Local Action Group funding opportunity, financing scheme being part of National Cohesion Funds) with relevant stakeholder for future development of an area.

### 15.5.2 Co-benefits

In order to fully exploit and present the alternative NBS scenario as opposed to BAU, a special attention was given to the identification and (e)valuation of co-benefits. A supporting template was used, allowing to systematically elaborate every co-benefit for its individual impact on the ecosystem status, integrated further on into economic impact as well.

The co-benefits identified by the stakeholders were then translated into ecosystem services and used as the foundation for monetary valuation (Table 15.5). Combining the two approaches and wide literature review on monetary value of use or non-use value of the ES that provide these co-benefits, led to the calculation of net present value (NPV) of ecosystem services production potential in the 30-year period, which amounts to cca. 0,5 MIO EUR annually.

In addition, results of the SDM semi-quantitative model, aligned to the Glinščica case study, showed the following correlation (Table 15.6), which is based upon stakeholder responses and aggregated input from literature and was used as an input for economic valuation of co-benefits.

### 15.5.3 Costs of Strategies

It is important to highlight that assessed values of co-benefits (Table 15.7) are calculated as the developed potential of ecosystems in good status, which would be enabled by the implementation of all measures in the selected strategy of Glinščica

**Table 15.5** A generalized list of services (translated from benefits) adjusted from Wright (2007) and used for economic valuation in the Glinščica catchment

| Physical benefits | Ecological benefits | Societal benefits |
|---|---|---|
| Store and convey floodwaters thus, reducing flood velocities and food peaks | Enable biological productivity | Sources of wild and cultivated plants |
| Control water temperature | Enable biodiversity | Fertilize agricultural lands for higher productivity |
| Manage sediment flows | Sustain critical habitats for aquatic organisms (from birds and fish to dragonflies and frogs), including rare, threatened, and endangered species | Provide sites for aquaculture |
| Filter nutrients and pollutants thereby enhancing the quality of surface waters | | Provide forest lands |
| Enhance infiltration that ensures groundwater and aquifer recharge, thus mitigating low surface flows | | Provide recreational opportunities |
| | | Provide aesthetic resources |
| | | Provide areas for scientific study and outdoor education |
| | | Contain cultural resources (historic and archaeological sites) |

**Table 15.6** Expected change in co-benefits production for Glinščica for all and BAU strategies. All variables related to co-benefits production are expressed in dimensionless form (ranging from 0 to 100, where 0 stands for 'extremely poor' and 100 for 'extremely high'). The 'expected change' compares the NBS scenario with the BAU condition both in absolute and % form. (Short – 10 years; Medium – 20 years; Long – 30 years)

| Co-benefits | BAU | | | NBS | | | Expected change | | | Expected change | | |
|---|---|---|---|---|---|---|---|---|---|---|---|---|
| | Short | Medium | Long | Short | Medium | Long | Short | Medium | Long | Short (%) | Medium (%) | Long (%) |
| Infrastructural damages | 45 | 72 | 72 | 45 | 28 | 10 | 0 | −44 | −62 | 0.0 | −61.11 | −86.11 |
| Biodiversity | 12 | 10 | 8 | 29 | 38 | 67 | 17 | 28 | 59 | 141.67 | 280.00 | 737.50 |
| Ecosystem state | 41 | 32 | 14 | 54 | 60 | 76 | 13 | 28 | 62 | 31.71 | 87.50 | 442.86 |
| Community safety | 35 | 10 | 10 | 49 | 51 | 52 | 14 | 41 | 42 | 40.00 | 410.00 | 420.00 |
| Social value of ecosystems | 12 | 8 | 8 | 23 | 30 | 53 | 11 | 22 | 45 | 91.67 | 275.00 | 562.50 |

**Table 15.7** Discounted costs of strategies over 30 years lifetime (in MIO EUR)

| Strategy | BAU | Glinščica for all |
|---|---|---|
| Investment | 3,9 | 2,6 |
| Maintenance | 0,6 | 0,2 |
| Total | 4,5 | 2,8 |
| Comment: | Estimation on proposed solution, prepared by Municipality of Ljubljana in 2013/2014<br>Includes the cost of buying land and properties needed for construction<br>Does not include any soft measures<br>Does not include any economic activities to be developed as support for maintenance of the solution or area as such<br>No co-benefits included | Co-benefits in estimated value of 0,5 MIO EUR / annually in ecosystem services potential preserved |

for all. This means that the implemented measures will preserve the ecosystems' potential, while the BAU strategy would continue their degradation at a current rate. This difference allows for the development of activities that lead to an economic benefit for the area and the stakeholders, like the increase in the attractiveness of the area which is translated into higher market prices of exiting real estate (non-use value) or the area developed into sustainable farming land, using suitable crops, which represent further revenue flows.

The economic comparison of the BAU and the Glinščica for all strategy revealed that the NBS approach to flood risk management in Glinščica catchment is:

I.  The more rational policy choice in light of predicted climate change by taking into account and responding to future environmental challenges, calling for adaptive planning and ensuring climate resilience[2]. By this we imply that the implementation of NBS now, will result in positive effects in opportunity costs in comparison to the BAU or other alternatives that do not build on ecosystem approach and climate change trends in the next 30 years.

II. The more rational investment choice in terms of financial investment and financial burden for the municipality and the state, since it preserves the use value of the area for implementation of different development strategies/land use practices. Investing in proposed NBS measures and land use practices opens the potential for utilisation of co-benefits, either as avoided costs (or damages) or added value in terms of land use.

III. The more holistic approach to flood risk management by engaging the relevant stakeholders to jointly set the goals, develop and assess the impacts of proposed

---

[2] See NAIAD Deliverable 6.2., Chap. 1 on EU policies and regulation that foresee principles of climate change adaptive planning. Alongside that, Ljubljana has committed to development vision, building on green, sustainable and preserved natural and cultural heritage, which call for ecosystem preservation and policies that take into account the need for climate resilient, adaptive planning principles.

solutions, allowing them to adjust and accept the reality of the changing climatic and social conditions and embrace the adaptive water and land management principles. Financial means already exist that can complement the implementation of NBS at the stakeholder level, thus reducing the operating cost of the proposed NBS.

IV. Encouraging the long-term holistic spatial planning approach, which would support preservation of the natural capital, added value and attractiveness of an area to live and work in.

## 15.6 Conclusions and Lessons Learnt

The Slovenian case study is specific in the size and the governing structure of the country. Slovenia's socio-cultural background that shapes its political and governing processes developed a rigid and overt (as opposed to public) governing structure, ill-suited to solve the inter-connected and often contradictory challenges of the twenty-first century which require a high level of adaptive capacity and cooperation across all sectors and actors. In addition, it was found that the fear of flooding is used in political risk discourses to downplay other challenges, such as biodiversity loss and/or ecological degradation, and to eventually support urbanization of the floodplains. Therefore, it was not surprising that the last workshop, designated to fine tuning the economic assessment and identifying potential barriers to implementation, revealed several previously raised issues about the national water management system in Slovenia. This chapter thus provides a short overview of the main issues raised by the stakeholders that are believed to be effectively preventing the implementation of the chosen Glinščica for all strategy.

### 15.6.1 Barriers to Implementation

All freshwaters are centrally managed by the Ministry of environment and spatial planning (MOP), but the risk management, watercourse maintenance, water quality monitoring and disaster recovery are managed independently by three different ministerial agencies and twelve different departments within the ministry. The lack of communication among these bodies has been recognized as the main barrier to implementing nature-based risk management in Glinščica by the stakeholders participating in the NAIAD process. Moreover, the institutional knowledge gap has been put forward both as the reason and the consequence of the lack of institutional cooperation. At the onset of this process of development of a natural assurance scheme, only individual stakeholders had heard about the nature-based solutions, natural water retention or adaptive water management, with only two experts having a thorough understanding of these concepts and were able to give us examples.

Therefore, despite focusing on NBS solutions to flood risk in the Glinščica catchment, our research revealed a broader issue of failed transition to an adaptive integrated water management in Slovenia. As other researchers have found before, the technologies (including NBS) already exist, but the barriers to their implementation are socio-institutional rather than technical and include uncoordinated institutional frameworks, ineffective regulatory frameworks, limited community engagement, empowerment and participation, unclear, fragmented roles and responsibilities, technocratic path dependencies and little or no monitoring and evaluation (Godden et al. 2011). Unfortunately, the Glinščica catchment is a perfect example of all these barriers. However, the applied process was successful in demonstrating the potential effectiveness for flood reduction and other co-benefits as well as several aspects and/or principles that derive from an integrated adaptive planning approach (see Chap. 6 Basco et al., this volume for more details).

Since scientific research and knowledge is undervalued in Slovenia, participation by decision makers and/or public institutions in large collaborative EU research projects is very limited. Our experience shows that at the decision-maker level, these stakeholders perceive research projects as a theoretical exercise that does not produce any viable and/or applicable solutions and fail to see the benefits from their participation. These stakeholders participated in the first individual interview, but later never joined any other part of the participative process. In addition, it was highly surprising that the Civil initiative for the flood safety of Ljubljana (CIPVLJ) did not wish to participate in the process. On the other hand, the public officers and experts working in risk management, nature conservation, spatial planning and other disciplines, as well as individual residents and their representatives (including farmers) were more willing to participate, contribute and learn throughout the process. Overall, the stakeholder participative process revealed that stakeholders of all levels lack the awareness and understanding of a fully participative process. They have never experienced this kind of participation themselves nor were they ever involved in capacity building in participative planning and/or management. Moreover, when they did participate in a typical public participation process led by Slovenian governmental institutions, they most often had the feeling that their participation was not appreciated, nor their suggestions taken into account. The stakeholders that did experience the whole participative process in the Glinščica case study found it very engaging and connective, they were happy with the results and have committed themselves to spread this knowledge further and use the principles in their future work.

The third most important barrier to efficient flood risk management in Slovenia that was revealed through the participative process was the lack of enforcement. The stakeholders note that the policy/legislation already in force has advanced much in recent years, but the implementation is far from meeting the standards. This only confirms the common knowledge, evident from the several pilots and infringement processes being held by the EU against Slovenia due to failed implementation of EU Water Framework, nature conservation and other directives. During the discussion about the reasons for this situation, the stakeholders were reluctant to discuss the failure of the relevant institutional frameworks, especially with the representatives of those institutions present. However, several examples of lack of knowledge from the Slovenian enforcement authorities were shared among the group, explaining the necessity for reaching out to the EU.

Finally, once the stakeholders developed and agreed on the best future strategy for the Glinščica catchment, we detected a mismatch between the perceptions of the different stakeholders regarding land ownership. While this is a broader issue, also connected to the struggles of the EU Common Agricultural Policy, the implementing institutions expressed the opposition from land owners (mostly farmers) as the main barrier, specifically preventing the implementation of NBS measures that usually require more space than grey solutions. Although the Ministry for agriculture, forestry and food was invited to participate in the process, they only took part in the first interview and expressed their opposition to using agricultural land for water risk management. On the other hand, while farmers require a specific participation approach, their opposition was more declarative than real and reflected their past bad experience with governmental institutions that usually fail to understand and accommodate their requirements. Surprisingly, one of the farmers expressed his anger with the governmental institutions that prohibited him from building small retention reservoirs on several of his fields, a practice that he learned from German colleagues was supported by the German government. We believe that this misunderstanding results from an inefficient institutional harmonization on the national level as well as the poorly executed public participation processes in Slovenia.

Unfortunately, the three main barriers identified, first, the failure of decision makers to participate and consequently learn from research results, second, the institutional knowledge gap and lack of cooperation and third, the inadequate public participation process, form a broader water management approach loop that is practically impossible to influence and/or adapt by a bottom-up approach. Therefore, as was also expressed by several of the participating stakeholders, the top-down guidance and pressure from the EU through the changes in its own policy, regulations and enforcement are a key leverage opportunity to support the required transition to integrated adaptive water management in circumstances such as the one in Slovenia.

---

**Box 15.1 Key Lessons Learnt**

- Awareness of NBS/green infrastructure, adaptive management and participative planning concepts as well as appreciation of co-benefits improved through the process.
- Fear of flooding is used in political risk discourse to downplay other challenges, such as biodiversity loss and/or ecological degradation.
- Including ecosystem state as the main benefit parallel to flooding helped keep the focus on NBS.
- Lack of participation opportunities for local people to engage in city planning/ management/development.
- Cooperation among governmental institutions hampered by political discourse.
- No R & I within the insurance sector in Slovenia.
- Decision makers not involved/interested.
- Freestation replacing work intensive personal observation for civil protection department of the municipality.

**Acknowledgements** The Authors would like to thank the local stakeholders of all NAIAD Demos for their support in the activities performed: without their invaluable contributions this work would have been impossible.

# Supplementary Material (Tables 15.8, 15.9 and Fig. 15.9)

A promotional video was produced to present the results and impact of the NAIAD project in the Glinščica case study, which can be accessed at https://www.youtube.com/watch?v=dT_zMHge-eM. In addition, the specific details of the NAIAD process for the Glinščica case study can be referenced according to the table below (Table 15.9).

**Table 15.8** Process flow of participatory approach employed in the Glinščica catchment

| STEP | Method | Information collected |
| --- | --- | --- |
| STEP 1 Case study characterization (problem framing) | 1st round of interviews | Study stakeholder risk perception. |
| | Literature review | Describe the climatic, meteorological, ecological, societal, political and cultural characteristics. |
| | Data analysis | Hydrological, agricultural, spatial characteristics. |
| | 1st SH workshop | Identify main goals in Glinščica catchment management. |
| STEP 2 Develop alternatives | | Identify potential solutions. |
| | 2nd round of interviews | Identify the co-benefits. |
| | 2nd SH workshop | Identify and select the most suitable indicators for measuring the efficiency of the solutions. |
| | | Develop, test and select the most suitable strategy. |
| | | Identify the most suitable locations for implementation of physical measures. |
| STEP 3 Value the chosen strategy(s) | 3rd SH workshop | Economic assessment and comparison of different strategies. |

**Table 15.9** Additional publications where specific detailed information about the Glinščica case study can be found

| Title | Form | Main topic | Access |
| --- | --- | --- | --- |
| Chapter 4.5 of the Catchment Characterization Report | NAIAD deliverable 6.1 | A full description of the catchment. | http://naiad2020.eu/ media-center/ project-public- deliverables/ |
| PART 5 of the From hazards to risk: models for the DEMOs | NAIAD deliverable 6.2 | A full description of the hydrological modelling applied. | |

**Fig. 15.9** A fact sheet on natural hazard insurance system in Slovenia as a result of public participation process and insurance analysis

# References

Benedičič U (2011) Poplavna varnost Ljubljane in poplave septembra leta 2010. Dipl. nal. – UNI. Ljubljana, UL, FGG, Študij vodarstva in komunalnega inženirstva, p 68

Bokhove O, Kelmanson MA, Kent T, Piton G, Tacnet JM (2019) Communicating (nature-based) flood-mitigation schemes using flood-excess volume. River Res Appl 35:1402–1414

Bokhove O, Kelmanson MA, Kent T, Piton G, Tacnet JM (2020) A cost-effectiveness protocol for flood-mitigation plans based on Leeds' boxing day 2015 floods. Water 12:1–30. https://doi.org/10.3390/w12030652

Brugnach M, Ingram H (2012) Ambiguity: the challenge of knowing and deciding together. Environ Sci Pol 15(1):60–71

Giordano R, Brugnach M, Pluchinotta I (2017) Ambiguity in problem framing as a barrier to collective actions: some hints from groundwater protection policy in the Apulia Region. Group Decis Negot 26(5). https://doi.org/10.1007/s10726-016-9519-1

Godden L, Ison RL, Wallis PJ (2011) Water governance in a climate change world: appraising systemic and adaptive effectiveness. Water Resour Manag 25(15):3971–3976. https://doi.org/10.1007/s11269-011-9902-2

Griessler Bulc T, Uršič M, Vahtar M, Krivograd Klemenčič A (2017) Ekosistemske tehnologije za trajnostno upravljanje z vodami. In: Globevnik L, Širca A (eds) Zbornik. Drugi slovenski kongres o vodah 2017. SLOCOLD – Slovenski nacionalni komite za velike pregrade/DVS – Društvo vodarjev Slovenije, Ljubljana, pp 391–396

Komac B, Natek K, Zorn M (2008) Geografski vidik poplav v Sloveniji. (Geographical view on floods in Slovenia). Geografski inštitut Antona Melika ZRC SAZU, Ljubljana, p 180

Odlok o občinskem prostorskem načrtu Mestne občine Ljubljana – izvedbeni del (2018) (Uradni list RS, št. 78/10, 10/11 – DPN, 22/11 – popr., 43/11 – ZKZ-C, 53/12 – obv. razl., 9/13, 23/13 – popr., 72/13 – DPN, 71/14 – popr., 92/14 – DPN, 17/15 – DPN, 50/15 – DPN, 88/15 – DPN, 95/15, 38/16 – avtentična razlaga, 63/16, 12/17 – popr., 12/18 – DPN, 42/18 in 78/19 – DPN). Municipality of Ljubljana, Ljubljana, 2018

Odlok o občinskem prostorskem načrtu Mestne občine Ljubljana – strateški del (2018) (Uradni list RS, št. 78/10, 10/11 – DPN, 72/13 – DPN, 92/14 – DPN, 17/15 – DPN, 50/15 – DPN, 88/15 – DPN, 12/18 – DPN in 42/18). Municipality of Ljubljana, Ljubljana, 2018

Pagano A, Pluchinotta I, Pengal P, Cokan B, Giordano R (2019) Engaging stakeholders in the assessment of NBS effectiveness in flood risk reduction: a participatory system dynamics model for benefits and co-benefits evaluation. STOTEN. https://doi.org/10.1016/j.scitotenv.2019.07.059

Program upravljanja rib v celinskih vodah Republike Slovenije za obdobje do leta 2021 (2015) Government of the Republic of Slovenia, 2015

Santoro S, Pluchinotta I, Pagano A, Pengal P, Cokan B, Giordano R (2019) Assessing stakeholders' risk perception to promote nature based solutions as flood protection strategies: the case of the Glinščica river (Slovenia). Sci Total Environ 655:188–201

Trajnostna urbana strategija Mestne občine Ljubljana 2014-2020 (SUS MOL) (2015) Municipality of Ljubljana, Ljubljana, 2015

Žaberl M, Pavšič Mikuž P, Perc A, Hodalič J et al. (2011) Okoljsko poročilo za Občinski podrobni prostorski načrt za območje zadrževalnika Brdnikova. EIA Report. E-NET OKOLJE d.o.o., p141

# Chapter 16
# The Opportunities and Challenges for Urban NBS: Lessons from Implementing the Urban Waterbuffer in Rotterdam

Kieran Wilhelmus Jacobus Dartée, Thomas Biffin, and Karina Peña

**Highlights**

- The UWB Spangen has reduced pluvial flooding, whilst also storing and supplying 15 million litres of water for irrigation, reducing the waste of drinking water for non-consumable purposes.
- The benefits of a realised pilot project include valuable lessons around the various motivations and disincentives for NBS uptake in the Dutch context, and the need for diverse stakeholders to work together.
- The broader neighbourhood assessment indicates that in certain contexts, NBS do appear able to deliver a comparable level of service for an equivalent or lower capital outlay than grey solutions, whilst also delivering a multiplicity of additional benefits, albeit often requiring more space.

## 16.1   Introduction

### 16.1.1   Background

In the coming years, The Netherlands will face a range of water challenges, not least of which concern rising sea levels, as well as risks from the large rivers Rhine, Meuse and Scheldt that flow through the country. However, within cities other obstacles are emerging. The onset of climate change is bringing increasingly frequent extreme precipitation, as well as higher temperatures and longer periods of drought (Aerts et al. 2012). In this context, Dutch cities are facing a need for renewed investment in water infrastructure and climate adaptation, if they are to

K. W. J. Dartée (✉) · T. Biffin · K. Peña
FieldFactors, Van Der Burghweg 1, Delft, The Netherlands
e-mail: kieran@fieldfactors.com

© The Author(s) 2023
E. López Gunn et al. (eds.), *Greening Water Risks*, Water Security in a New World, https://doi.org/10.1007/978-3-031-25308-9_16

maintain their renowned liveability and high quality of life. As such, cities are increasingly turning to emerging technologies and innovations with keen interest, particularly those that can address multiple issues integrally.

The city of Rotterdam lies in the province of South Holland, within the Rhine-Meuse-Scheldt River delta, near the North Sea. The city has a population of around 645,000, with an average density of 2963 inhabitants per km² (CBS 2019). Like the nation at large, much of Rotterdam's past and future successes are strongly linked to water; it is home to Europe's largest port, approximately 36% of the municipal area consists of water and 80% of the city lies below sea level. As such, the city is both highly dependent on, and yet vulnerable to threats from water (Gemeente Rotterdam 2013).

As with many cities, the proportion of impervious surfaces in Rotterdam has increased over time, due to urbanisation and changing land use priorities. Over a century ago, several canals, which previously took care of household water management, were filled up to make space for new infrastructure. Drainage became increasingly reliant on a sewage system, which now appears insufficient to cope with heavy rainfall events (De Greef 2005). With the aforementioned emerging climate change impacts, Rotterdam is now becoming susceptible to not only pluvial flooding, but also higher levels of heat stress, and degradation of building foundations (in case of wooden pilings) due to dropping ground water tables in many parts of the city (Fig. 16.1).

Acknowledging that this combination of urbanisation and climate change poses a significant threat to the city's future prosperity, the city of Rotterdam set the objective of becoming 100% climate-proof by 2025, through the Rotterdam Adaptation Strategy (Dircke and Molenaar 2015). This aims for a robust water system that protects the city and prevents climate-related nuisances, achieved through forming connections across disciplines, programmes and multiple stakeholders. Consequently, the water policy is shifting from a catch-store-dispose approach, towards local infiltration, retention, storage and reuse. As will be elucidated throughout this chapter, this broader policy objective provided an entry point for innovative approaches that aim to break away from the traditional water management paradigm.

### 16.1.2 The Case Study Area: Spangen Neighbourhood

The Rotterdam case study focuses on Spangen, a low-lying neighbourhood in the west of the city, spanning roughly 65 ha with around 10,000 inhabitants. In Spangen, as with much of the city, the effects of urbanisation are clear: hard surfaces dominate and there are few green areas (Fig. 16.2).

In recent years, Spangen has been suffering frequent nuisance during heavy rain events. Hydrological analysis confirmed the need for additional retention capacity in the neighbourhood, and the upcoming investments were taken as an opportunity to simultaneously tackle some of the underlying issues, such as the lack of green space. The Rotterdam case study in NAIAD has assessed the implementation of an

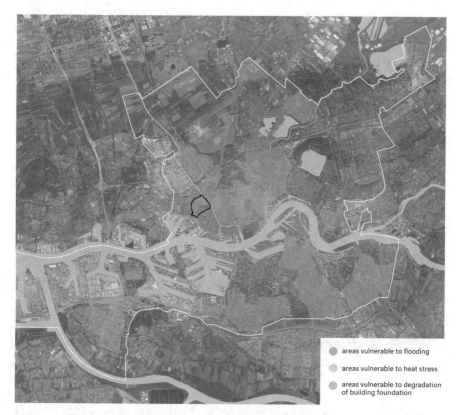

areas vulnerable to flooding

areas vulnerable to heat stress

areas vulnerable to degradation
of building foundation

**Fig. 16.1** A map of Rotterdam, showing areas vulnerable to flooding, heat stress and subsidence. (Adopted from Gemeente Rotterdam 2013). A detailed characterization of the case study area can be found in Pengal et al. (2017), Section 5.10

innovative NBS for rainwater retention and reuse, the so-called Urban Waterbuffer (UWB).

The UWB Spangen presents a solution for the localised pluvial flooding, whilst also providing additional green space, and large-scale seasonal water storage (see Fig. 16.3). In 2018, the first pilot application was built in Spangen. Additionally, the UWB has been assessed as one of the measures in the three neighbourhood wide strategies that complemented the Rotterdam Case Study.

The objectives in the Rotterdam case study were to perform:

1. An empirical assessment of the recently implemented UWB in Spangen. Spanning a 4 ha catchment area, this hybrid solution reduces pluvial flooding using an underground buffer tank for temporary retention, together with biofiltration and aquifer storage and recovery (ASR) techniques to provide long term storage and reuse of the captured rainwater.

**Fig. 16.2** A representative view of the Neighbourhood of Spangen, showing the high percentage of impermeable surfaces surrounding the Sparta Stadium. Photo by FieldFactors, 2017

2. A broader assessment of the impacts of the UWB, when nested within a hybrid water management strategy at the neighbourhood scale, and how this compares with fully green solutions and grey alternative strategies (see Fig. 16.4).

The assessments of the UWB Spangen and the three neighbourhood strategies have been executed in line with the methodologies developed within NAIAD. In Fig. 16.5, an overview of the NAIAD framework applied in the Rotterdam case study is presented. The scale at which the methods were applied (UWB project scope versus neighbourhood strategy) varied to accommodate the methodological approach and to deliver the relevant insights.

### 16.1.3 Chapter Outline

Drawing upon an examination of the factors leading to the implementation of the UWB and an assessment of long-term potential future costs and benefits of NBS, important insights have been derived from the Rotterdam case study. This chapter aims to share these insights through seven key lessons that have emerged from the research. These lessons result from utilising the methodologies introduced in this

# Urban Waterbuffer Spangen

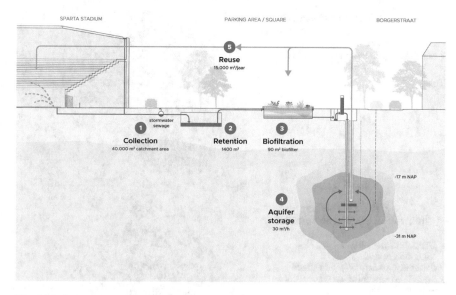

**Fig. 16.3** Schematic representation of the Urban Waterbuffer in Spangen, Rottterdam

volume. The aim is to support stakeholders in future assessment and decision-making on NBS implementation, by shedding new light on the impact of NBS in urban areas and their potential for wider uptake.

## 16.2 Lessons from the Rotterdam Case Study

### 16.2.1 The NBS Implementation Was Driven by Its Ability to Address Multiple Challenges Integrally

The UWB in Spangen benefited from a window of opportunity that arose when multiple initiatives coincided in 2016. The city had just launched the Water Sensitive Rotterdam program, aiming to improve the city's climate resilience; and the Spangen neighbourhood was already earmarked as in need of additional water retention measures. Simultaneously, residents of the neighbourhood had been requesting additional green space to counteract the dominance of paved surfaces. Additionally, a recently formed consortium of knowledge, market and public partners was searching for locations to pilot urban ASR under the umbrella of a research project (Urban Waterbuffer), subsidised through a national innovation fund. Finally, the local football club emerged as a potential end user for the large volumes of rainwater that could be treated and stored by the hybrid measure. The stored water is used to

# Demo Rotterdam Strategies

### 1 Grey Strategy:
### Traditional and centralised

- Separated sewer
- Permeable pavement in road edges

### 2 Hybrid Strategy
### Infiltration via public squares

- 2 Urban Waterbuffers
- Retention ponds
- Expansion of central green space
- Separated sewer for water collection and distribution

### 3 Green Strategy
### Local infiltration through green infrastructure

- Green roofs
- Façade gardens
- Green strips and swales
- Retention ponds
- Expansion of central green space

**Fig. 16.4** Overview of the three neighbourhood-wide strategies that were assessed in the Rotterdam Case Study

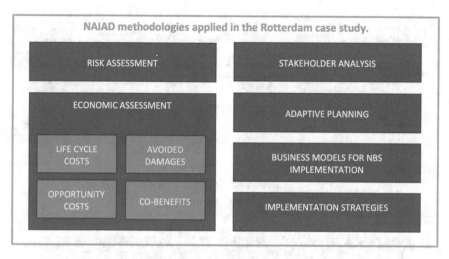

**Fig. 16.5** An overview of the methodologies applied in the Rotterdam case study

substitute the use of tap water for non-consumable purposes, like irrigation of the sports field, allowing the football club to reduce their drinking water footprint. The UWB offered an integral approach to these various issues, leading to its successful implementation in the summer of 2018.

### 16.2.1.1 Drivers of the Implementation Process

Based on interviews with key stakeholders, a timeline was developed to visualise critical milestones in the decision-making process and identify key success factors (Fig. 16.6 and Chap. 9 of this volume for further background on the Adaptive Planning framework). One of the notable insights from this process was the role of various policy programmes within the City of Rotterdam and Water Authority Delfland (e.g. Water Sensitive Rotterdam,[1] Resilient Rotterdam[2]; Waterbeheerplan 2016–2021, Delfland 2015) in fostering the adoption of measures beyond traditional sewer replacement. The authorities aimed to exploit multi-functional approaches to improve the city's climate resilience, and to increase the amount of green infrastructure in the city.

Besides the primary benefit of the UWB solution to reduce the risk of sewer overflow, the ability to use the collected rainwater locally, was considered as a critical co-benefit. This even led to a new business case involving the local football club as end-user. The creation of this new local fresh water source contributes to the city's resilience to future droughts.

---

[1] https://www.watersensitiverotterdam.nl

[2] https://www.resilientrotterdam.nl

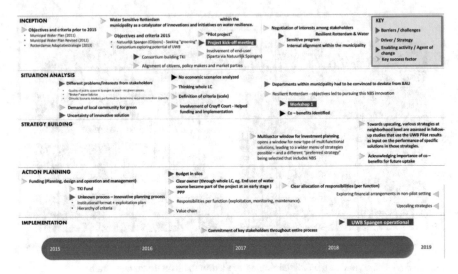

**Fig. 16.6** Analysis of adaptive planning process around the implementation of the UWB Spangen: Timeline of the decision-making processes. (Source: Dourojeanni Schlotfeldt 2019)

In addition, the spatial improvement, such as added green space, water playing features, and seating elements that the UWB brought to the neighbourhood, were an important motivation to invest in the UWB. Aside from their main objective to resolve the water nuisance in the area, the Water Authority in particular also had a policy objective to stimulate water awareness among inhabitants. This objective was fulfilled through the incorporation of an open and visible water treatment element and the ability for children to play with the harvested water through a water feature.

### 16.2.1.2 Stakeholder Perception Analysis

The multiple functions of NBS naturally leads to the involvement of many different parties with different interests and responsibilities. If these parties are well aligned, this can create opportunities to benefit from cost-savings through one integral solution. In contrast, the involvement of many stakeholders can also be a burden on the decision-making process, with transaction costs, risk of protracted disputes and missed opportunities. A reflection on the successful alignment of the stakeholders in the UWB project led to the identification of two key drivers. First of all, key individuals within the involved organisations were highly committed to making the project work from the outset and took a leading role in aligning stakeholders and their interests. Secondly, there was a common understanding between stakeholders that action was required. Various interviews were analysed using Fuzzy Cognitive Mapping to identify the different stakeholder perceptions regarding the main problems that needed addressing. The results indicated a large common understanding of the main issues at hand (sewer overflow), the factors that cause sewer overflow, and the potential for the UWB to reduce the risk of pluvial flooding (see Fig. 16.7).

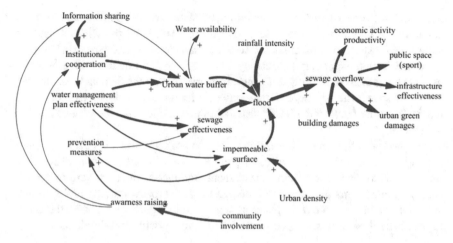

**Fig. 16.7** A Fuzzy Cognitive Map (FCM) showing the relationships between factors related to pluvial flood risk as perceived by stakeholders in Spangen (developed for the Rotterdam case study by Giordano, R. and Pagano, A. in 2018). The thickness of the arrows indicates the relative number of respondents who acknowledged the causal relations between the factors (thicker = mentioned by more stakeholders)

For the successful implementation of NBS, it is important that stakeholders share a common understanding that achieving climate resilience in urban areas will require new, cross sectoral approaches. Having to deal with more extreme rain events, longer lasting dry periods and increasing temperatures in urban areas requires measures that go beyond traditional sewer replacement and address these multiple issues integrally.

## 16.2.2   Current Planning Regulations and Policy Making Do Not Facilitate the Uptake of NBS

Despite the political willingness to integrate NBS in the climate adaptation agenda, the current planning regulations and processes in the Netherlands do not facilitate the uptake of NBS. Throughout the implementation process, various challenges arose that nearly blocked the realisation of the UWB in Spangen. For example, the business-as-usual approach was initially preferred by certain departments of the implementing authorities. During the stakeholder workshops, a special emphasis was put on the upscaling perspectives of NBS for climate adaptation in the Dutch urban context. What became evident was that the current frameworks in Dutch planning practices do not facilitate the uptake of NBS at the operational level. One of the barriers that was identified in the Rotterdam case study is the sectoral silos and split responsibilities among and within various public organisations. Typically, departments within organisations have a dedicated budget and responsibility for a "single" task. As NBS typically address multiple elements of urban planning and water management, they require the integration of tasks, responsibilities and budgets across the various silos. Without clear tactical and operational guidelines to assess a measure's impact across other silos, operational decisions typically remain based on a measure's mono-functional efficiency. Therefore, in addition to political willingness at strategic level, specific multi-purpose objectives at the tactical level are necessary to transform willingness into action, without having to be dependent on passionate individuals or dedicated cross-silo programmes, like the Water Sensitive Rotterdam.

For the UWB Spangen project, this cross-silo program was a key factor in overcoming barriers in the decision-making process. Additionally, a national research fund helped to lower the required investment and uncertainties for various project commissioners. This pilot approach helped to work cross-sectorally and step outside of some of the more rigid planning regimes and regulations. Lastly, the UWB project benefitted from strong commitment of some specific stakeholders.

### 16.2.3  NBS Can Compete on Life Cycle Cost with Grey Solutions, Though Strategies Must Be Carefully Designed and Assessed per Location

Although NBS are often claimed to be cost effective, there currently remains limited evidence in support of this (Le Coent et al. 2020). In the Rotterdam case study, uncertainty about the costs of NBS did indeed emerge as a concern for a wide range of stakeholders. These concerns included the potential impact of high upfront costs, uncertainties of long-term maintenance requirements, and even difficulties in internalising many of the acknowledged co-benefits. There was a clear desire amongst stakeholders for more information on how the costs of these new approaches compared to business as usual, not only initially, but also over the long term.

As part of the economic assessment in the Rotterdam case study (approach outlined in Chap. 6), the costs of three neighbourhood-scale water management strategies – grey, hybrid and green – were compared through the application of life cycle costing (LCC). The strategies were specified with similar retention capacity in order to provide equivalent flood risk reduction, which allowed for a comparison between the strategies on their overall costs for a certain level of service (see Dartée et al. 2019 for a more detailed overview of this approach).

The Rotterdam case study was unique as compared to other case studies presented in this volume, in that it was possible to draw upon real implementation costs for the UWB, which was incorporated as a central measure into the hybrid strategy. In order to best align these empirical cost figures with the values sourced from literature for other measures, local Dutch cost data were chosen wherever available. This aimed to improve the compatibility between the UWB and other cost data, but also increased the potential influence of outliers when compared to expressing results as a range. As with any such assessment, the results are therefore to be considered indicative rather than definitive, given the many context specific factors that can influence the costs of implementing public infrastructure. That is also why it is critical to design strategies carefully, as context specific factors can lead to significantly higher or lower costs for certain measures in relation to their hydrological effects on reducing flood risk (e.g. Low permeability of soil would require a bigger area of green infiltration strips to offer a similar retention/infiltration capacity and hence would lead to higher opportunity costs).

The results of the Rotterdam case study showed hybrid and green strategies to be marginally cheaper over a 50-year time period, than the grey strategy for the same level of service, both before and after the application of a 3% discount rate (Fig. 16.8), supporting the notion that NBS can compete with grey solutions on cost.

The results also highlight the importance of taking a long-term view to cost assessments, as maintenance expenditures represent a significant portion of the overall LCC. Given the long timeframe of the LCC, the reference values for maintenance costs are potentially strongly influential, yet there is a relative scarcity of reliable long-term maintenance cost data, particularly for green infrastructure. It is worth noting that, once discounted, these costs represent a smaller share of the LCC, yet the limited availability of relevant data serves to highlight the importance of further implementation of, and research on, NBS to address knowledge gaps and improve understanding and confidence in their costs over time.

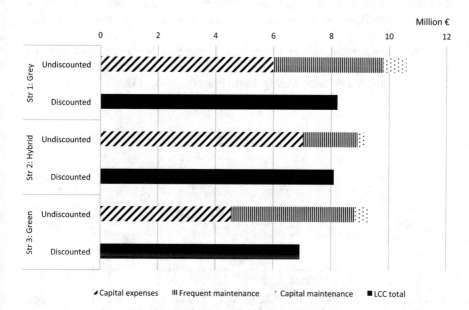

**Fig. 16.8** The Life Cycle Costs (LCC) of the three strategies in the Rotterdam case study over 50 years in Million EUR, showing a breakdown of costs drivers for each strategy. This overview presents summaries of both discounted and undiscounted assessments. Discount rate applied was 3%

### 16.2.4 Monitoring Co-benefits Is Critical to Support the Wider Uptake of NBS

Much of the debate around NBS centres on their multifunctionality, and indeed it was one of the key motivators for the implementation of the UWB in Spangen. In Rotterdam, the identification of co-benefits for the UWB began prior to implementation through stakeholder consultation, which identified the following potential co-benefits:

- Local temperature regulation
- Water quality regulation
- Cost reduction of water treatment
- Storm water re-use
- Spatial quality betterment
- Increased green
- Increased water awareness
- Increased social cohesion

The economic assessment of co-benefits was undertaken at the neighbourhood scale, so the initial list of co-benefits was expanded to account for the variety of different measures under consideration and the multiple impacts they can provide. Sixteen (16) co-benefits were identified and considered for assessment (Fig. 16.9), drawing predominantly from the EKLIPSE framework. Direct valuation was the dominant method used, drawing values from peer reviewed literature, except in the case of the UWB, where some data was available directly from the implemented project UWB Spangen.

Whilst our assessment aimed to monetise as many of the co-benefits as possible so as to capture the full value provided by NBS, some of the co-benefits were not able to be valued monetarily, mainly due to one of two reasons:

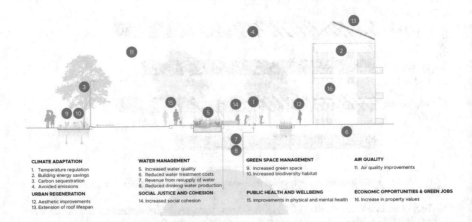

**Fig. 16.9** Overview of the co-benefits assessed in the Rotterdam Demo

- Insufficient data/reference literature
- Risk of double counting

The monetised results from the co-benefit assessment (Fig. 16.10) showed, firstly, that even at a small scale the value of co-benefits can be substantial, and secondly, that the kinds of co-benefits that deliver value in dense urban areas can be very different from those provided by large scale ecosystems. By far, the largest benefits were derived from avoided health care costs and labour loss, as well as property value increases. Other high value co-benefits included revenue from the supply of water from the UWB, and heating savings from improved insulation due to green roofs. Many other co-benefits provided marginal value at this small scale.

Given the relatively close LCC results between the strategies, co-benefits are a clear differentiator for NBS compared to grey solutions. However, as some of the NBS measures considered can provide co-benefits that were not monetised, the above results do not represent the full value of additional co-benefits provided. Examples from the case study include mitigation of the urban heat island effect through local temperature regulation, which in the wake of ongoing urbanisation and record temperatures is becoming increasingly central to the discussion on NBS in The Netherlands. Similarly, increased water awareness was not monetised, but was an important potential co-benefit to both the municipality and the Water Authority. This interest is likely linked to the fact that 60% of the land within the municipality is privately owned (De Doelder 2019), meaning that the authorities are partially dependent on private actions to increase the city's climate resilience.

As can be seen, in the current climate of enthusiasm for NBS, certain potential co-benefits, tied to specific contextual objectives, can prove decisive for project

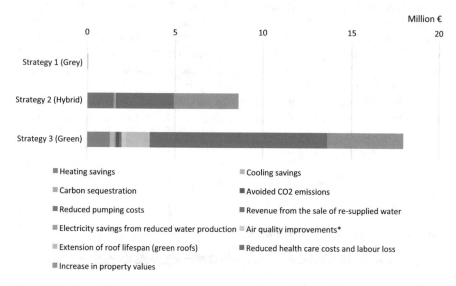

**Fig. 16.10** The undiscounted value of co-benefits delivered by each strategy in Spangen over 50 years in Million EUR

implementation even in the absence of clear quantification of their monetary value. However, evidence that NBS can actually deliver these potential co-benefits is likely to become increasingly important as the initial enthusiasm around NBS begins to fade. As such, it is important that monitoring and assessment of co-benefits will be prioritised, particularly for social co-benefits, to build a better understanding on performance in some of the areas where data is lacking. However, the evidence in support of the delivery of co-benefits need not be exclusively monetary. The multi-functionality of NBS is their key advantage, and the case for them is most compelling when impacts are considered holistically. This means that consideration of a range of environmental indicators alongside economic valuation may be most suited to capturing the full scope of NBS benefits and functions.

### 16.2.5 If There Is Space, Full NBS Is Ace; if Space Is Tight, Hybrid Might Be Right

The Netherlands is one of the most densely populated countries on earth, and as a consequence, the value of land in urban areas is high. This means that there is always competition for space among potential users, and that water management interventions that require additional space compared to business as usual will involve trade-offs. To fairly compare NBS with grey infrastructure, it is thus important to consider the opportunity costs associated with their implementation, which in the Rotterdam case study was achieved through using the value of the land required for each strategy as a proxy, taken from Levkovich et al. (2018). Whilst this is a widely used approach, it should be applied with caution, as NBS measures located in areas such as sidewalks would be unlikely to limit development potential. Thus, the use of a single land value risks overstating the opportunity cost. In the Rotterdam case study, given the use of a single high land price, the opportunity cost assessment should be considered as an upper bound estimate (Fig. 16.11), imposing significant negative impacts on the NBS strategies.

Consequently, when bringing together the various results of the economic assessment, it became clear that neither the equivalent avoided damages, nor the relatively

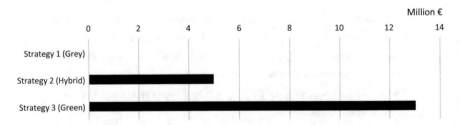

**Fig. 16.11** Opportunity costs for the three strategies, using value of the land required for implementation in Million EUR

small differences in implementation costs between the strategies would prove decisive. As a result, it was opportunity costs and co-benefits that emerged as the key differentiators (see Fig. 16.12).

Whilst the assessment was subject to substantial variability, particularly given the long timeframe and large number of measures, the order of magnitude difference between the various pillars does allow for the following lesson to be drawn with a relatively high level of confidence. When deciding between NBS and grey infrastructure to mitigate flood risk in dense urban areas, the question can be boiled down to whether the opportunity costs imposed by the space requirements of NBS can be offset by the co-benefits that they can provide. Table 16.1 shows the Benefit Cost Ratios (BCR) of the integrated economic results - the sum of all of the benefits divided by all of the costs for the three strategies, expressed in economic terms.

What can be seen is that when considering the economic results alone, no strategies resulted in a BCR above 1 – the point at which an intervention is generally considered cost effective. However, considering the difficulty in including all direct and indirect damages in the assessment and the strong societal relevance, responsible authorities in The Netherlands do at times invest in risk reduction measures with a BCR < 1 (Jonkman et al. 2004). This means that in this case the BCR is more useful in comparing the strategies to one another than in assessing cost effectiveness in absolute terms. When viewed in this light, the results show that the hybrid and green strategies bring significantly more benefits per Euro spent than the grey approach.

The results of the economic assessment support the notion that, at least in the Dutch context, if there is space available, NBS are likely a better choice than standard grey approaches, given that they can potentially provide similar core functionality for an equivalent or slightly lower cost, whilst also delivering an array of co-benefits. However, if the opportunity costs of implementation are high and space is scarce, hybrid solutions may prove a better choice, as they can provide many of

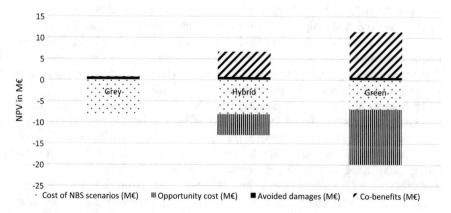

**Fig. 16.12** The integrated economic results for each strategy, showing the large influence of co-benefits and opportunity costs as presented in Le Coent et al. (2020)

**Table 16.1** The total costs, benefits, Net Present Value (NPV) and Benefit Cost Ratio (BCR) of the three strategies over 50 years

| Category | Strategy 1 – Grey | | Strategy 2 – Hybrid | | Strategy 3 – Green | |
|---|---|---|---|---|---|---|
| | Nominal | Discounted | Nominal | Discounted | Nominal | Discounted |
| LCC | – € 10,590,000 | – € 8,210,000 | – € 9,160,000 | – €8,100,000 | – € 9,280,000 | – € 6,920,000 |
| Opportunity costs | | – | | – € 4,980,000 | | – € 13,020,000 |
| Co-benefits | € 40,000 | € 20,000 | €8,570,000 | €6,014,000 | €17,700,000 | €10,730,000 |
| NPV | – € 8,190,000 | | – € 7,054,000 | | – € 9,070,000 | |
| BCR | 0.002 | | 0.46 | | 0.54 | |

Source: Le Coent et al. (2019)

the benefits with less space requirements. The single strongest case for grey infrastructure remains familiarity and institutional embeddedness. Nevertheless, in the Dutch context there is widespread and growing understanding that the multi-faceted challenges of the twenty-first century are likely to be best served by multifunctional approaches.

### 16.2.6   Multi-functionality of NBS Allows for the Development of New Business Cases

As an example of an effective Public-Private-Partnership, the implemented UWB Spangen offered valuable insights on potential business cases for NBS. The partners involved included two public authorities (municipality and Water Authority), a semi-public utility provider, and private companies (see Fig. 16.13).

The business case in this project could inspire future implementations of similar NBS, even though it was in the context of a pilot project. Through the pilot setup, the technical risk and investment threshold for the various stakeholders to collaborate was reduced, and typical concerns related to innovations were allayed by allowing for intense monitoring and research. The UWB in Rotterdam is the first urban implementation of this solution in the Netherlands, and its pilot status allowed for slightly different legal arrangements to be applied. These advantages might no longer be available in future implementations of the solution.

In order to reflect on the upscaling potential of the UWB, as well as what is needed to build sustainable business cases for NBS, a workshop was held with key stakeholders. One identified driver for the business case of the UWB specifically is the exploitation of the new water source, as this can generate direct cash flows through the water supply, though it must be noted that tap water rates in The Netherlands are currently extremely low. However, since projections show increasing pressure on freshwater availability and security of supply, the drinking water prices are expected to rise, generating a bigger saving to be achieved through the use of treated storm water. The societal value placed on measures that increase water

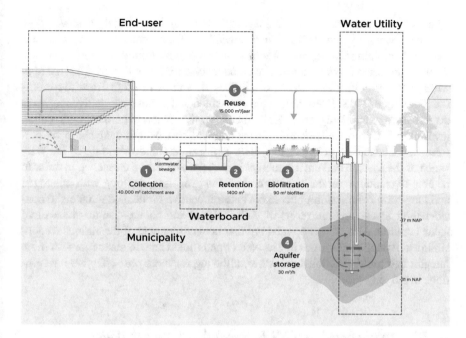

**Fig. 16.13**  Distribution of responsibilities between stakeholders in relation to UWB Spangen

availability is also likely to grow, considering the potential damages droughts might incur on existing green space, infrastructure and human health.

In the case of the implemented solution in the Rotterdam case study, the UWB, as it generates a saleable commodity in the form of the harvested water, there are opportunities for developing business models with private sector partners, creating a return on investment over time. As such, Public-Private-Partnerships are expected to remain the main implementation arrangement for the UWB.

In order to facilitate the cooperation of various stakeholders, it is important that NBS provide a distinct level of service that the relevant parties are willing to invest in. In current practice, the impact of co-benefits is seldom addressed in terms of an increased level of service. As a result, building financial and legal arrangements that are based on the exchange of these benefits is challenging, particularly since often the implementing party is not the same as who will receive the benefits. Another challenge in making the business case for NBS is the fact that there can be a vast difference between the monetary value of co-benefits in an impact assessment, and the tangible contributions to a budget bottom line that these co-benefits can actually provide. This means that for NBS in general, it appears that municipalities and other public bodies are well suited for implementing them, as the broader societal benefits could contribute to achieving other objectives within their responsibilities. Within NAIAD, special emphasis was put on the role of the insurance industry in future business cases for implementing NBS. Following the interaction with the Dutch insurance industry during this research, it is expected their role within funding and financing of Dutch urban water management will remain minimal. However, they do have means to nudge individuals towards more climate-adaptive behaviour through facilitating the development of water awareness among citizens.

What did come forward during stakeholder consultation in Rotterdam is the high willingness to collaborate across disciplines in order to benefit from the multi-functionality of NBS. Even though the assessments of various alternative measures do not always explicitly consider co-benefits, the majority of the larger public investments in The Netherlands require a social-cost benefit analysis to be performed. Stakeholders acknowledge there is a growing understanding of the relevance of the societal and environmental impacts of alternative measures. In order to be able to capitalise on the various co-benefits over time, working with integrated and long-term contracts might create a better incentive for decision-makers to consider their investments in terms of wider impact and potential multi-stakeholder value creation. Also, it was pointed out that better alignment of the various responsibilities of public authorities in The Netherlands could help to make room for more integral and multi-functional measures, allowing for more cost-effective interventions in the long-term.

### 16.2.7    Implementing NBS Is Needed to Catalyse Wider Acceptance, Interest and Future Uptake

One of the challenges with the implementation of the UWB in Spangen was the management of uncertainties related to the newness of the solution: actual effects on the water system, the water quality achieved, and the need for new operation and maintenance regimes. The best means to overcome these challenges was to implement the solution and consequently intensively monitor system performance. The availability of an innovation fund proved highly valuable as it allowed for some of the aforementioned uncertainties to be addressed in a pilot setting (Fig. 16.14), and the performance of research. Reflecting on the entire process, it was apparent how much impact the implementation of the UWB had on the wider acceptance of and interest in NBS. Being able to showcase the solution and prove the concept works was a critical milestone in order to build capacity for the larger uptake of other similar solutions.

Particularly in the Dutch water management sector, standardization, robustness and economies of scale have long dominated business as usual practices. Given that NBS offer completely different characteristics, the evidence of NBS effectiveness is often still limited. So long as this remains the case, it will be difficult to convince stakeholders of the efficacy of the investment in NBS. With the emergence of co-benefits as a crucial component of the case for NBS, how these are best to be assessed remains a key question. As discussed throughout this chapter, on several occasions over the course of the implementation of the UWB Spangen,

**Fig. 16.14** The Biofilter component of the Urban Waterbuffer in Spangen

stakeholders identified the difficulties in assessing co-benefits, and concerns about long term costs and reliability as key challenges for the upscaling of NBS. Much of this difficulty stems from a lack of comparable existing cases to draw from, meaning that reasonable estimates of the level of service that can be expected for key co-benefits, or the long-term costs for a given measure over time are not well established.

In the current climate in the Netherlands, there is a willingness to subsidise innovation and assumptions on co-benefits that may result from an implemented NBS appear sufficient, but this can be expected to change as novelty fades and budgetary realities set in. Therefore, building more NBS, monitoring their impact and using that to overcome uncertainties regarding effectiveness, costs and co-benefits is needed in order to provide the necessary evidence base to facilitate their wider uptake (Fig. 16.15). Being able to engage with stakeholders around the actual implementation of an NBS in Spangen and being able to communicate and discuss the NAIAD results with these stakeholders, might result in the beginning of a paradigm shift in which NBS are considered a worthy equivalent of and maybe even a better alternative to hard engineering structures.

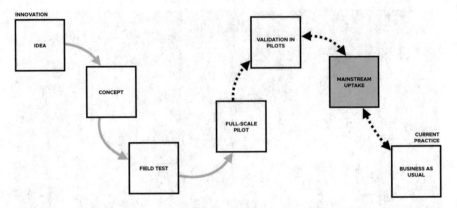

**Fig. 16.15** Pathway to mainstreaming NBS. Having validated the effectiveness in multiple pilots is needed to be able to take on business as usual practices

## 16.3 Conclusions

In the 18 months of monitoring, since its implementation, the UWB Spangen has performed strongly, having reduced pluvial flooding, whilst also storing and supplying 15 million litres of water for irrigation, reducing the waste of drinking water for non-consumable purposes. It has gained widespread interest from a range of stakeholders, both in The Netherlands and abroad.

The benefits of a realised pilot project extend beyond raising awareness and proving efficacy, as pilots also offer important research opportunities. With the Rotterdam case study built around this actual piece of hybrid infrastructure, it was possible to gain important insights that may not have been derived from modelling and forecasting alone. These include valuable lessons around the various motivations and disincentives for NBS uptake in the Dutch context, and the need for diverse stakeholders to work together. The smaller scale of the Rotterdam case study compared to other case studies in the NAIAD project, though posing some challenges, also made possible to highlight the similarities and differences in NBS effectiveness and viability across a wide range of spatial scales.

The broader neighbourhood assessment indicates that in certain contexts, NBS do appear able to deliver a comparable level of service for an equivalent or lower capital outlay than grey solutions, whilst also delivering a multiplicity of additional benefits. However, they often require more space, and in order to justify this use of space, it is important that the many co-benefits NBS can provide are valued and properly accounted for, despite the complexity of achieving this. Finally, the relatively high levels of uncertainty still involved with research related to the UWB type of NBS considered in this chapter (and others) can best be improved through a greater number of implemented projects. Which would be just one of many co-benefits of implementing more NBS to resolve pressing urban water challenges and design for climate resilient cities.

# References

Aerts J, Botzen B, Bowman J, Ward P, Dircke P (eds) (2012) 2012 climate adaptation and flood risk in coastal cities. EARTHSCAN Ltd., London

CBS (2019) Kerncijfers wijken en buurten 2019. www.cbs.nl/nl-nl/maatwerk/2019/31/kerncijfers-wijken-en-buurten-2019

Dartée KWJ, Biffin T, Peña K (2019) DELIVERABLE 6.3 DEMO insurance value assessment report - part 8: the Netherlands – Rotterdam DEMO. EU horizon 2020 NAIAD project, Grant agreement N°730497

De Doelder B (2019) UWB brengt de stad in waterbalans. Presentation on 3-12-2019 in Rotterdam

De Greef P (2005) Rotterdam Water City 2035

Delfland, Hoogheemraadschap (2015) Waterbeheerplan 2016–2012: Strategie richting een toekomstbestendig en samenwerkingsgericht waterschap

Dircke P, Molenaar A (2015) Climate change adaptation; innovative tools and strategies in Delta city Rotterdam. Water Pract Technol 10(4):674–680. https://doi.org/10.2166/wpt.2015.080

Dourojeanni Schlotfeldt PA (2019) Understanding the role of uncertainty in the mainstreaming of nature-based solutions for adaptation to climate change. Thesis, IHE Delft Institute for Water Education, Delft

Gemeente Rotterdam (2013) Rotterdamse adaptatiestrategie: themarapport stedelijk watersysteem

Jonkman SN, Brinkhuis-Jak M, Kok M (2004) Cost benefit analysis and flood damage mitigation in the Netherlands. Heron 49(1):95–111

Le Coent P, Hérivaux C, Farina G, Forey I, Wang Z-X, Graveline N, Calatrava J, Martinez-Granados D, Marchal R, Moncoulon D, Scrieciu A, Mayor B, Burke S, Mulligan M, Douglas A, Soesbergen A, Giordano R, Kieran D, Biffin T, Peña K, Van der Keur P, Kidmose J, Gnonlonfin A, Douai A, Piton G, Munir MB, Mas A, Arnaud P, Tacnet JM (2019) DELIVERABLE 6.3 DEMO Insurance value assessment report – Part 8: The Netherlands – Rotterdam DEMO. EU Horizon 2020 NAIAD Project, Grant Agreement N°730497

Le Coent P, Graveline N, Altamarino M, Arfaoui N, Benitez-Avila C, Biffin T et al (2020) Title: Is it worth investing in NBS aiming at mitigating water risks? Insights from the economic assessment of three European case studies. Manuscript submitted for publication

Levkovich O, Rouwendal J, Brugman L (2018) Spatial planning and segmentation of the land market: the case of the Netherlands. Land Econ 94(1):137–154. https://doi.org/10.3368/le.94.1.137

Pengal P et al (2017): DELIVERABLE 6.1 catchment characterization report. EU Horizon 2020 NAIAD project, Grant Agreement N°730497

# Chapter 17
# Urban River Restoration, a Scenario for Copenhagen

Morten Ejsing Jørgensen, Jacob Kidmose, Peter van der Keur,
Eulalia Gómez, Raffaele Giordano, and Hans Jørgen Henriksen

**Highlights**

- Avoided damage to subsurface housing from varying shallow groundwater levels is simulated using a hydrologic model and simplified damage functions.
- Hydrological simulations of restoring an urban river as NBS scenario to mitigate damage resulting from groundwater flooding show promising results with 35% avoided damage
- Social Network Analysis shows that the implementation of the considered restored urban river NBS needs a thorough stakeholder dialogue to obtain high NBS effectiveness by detecting collaborative and institutional barriers and include trade-offs between co-benefits

## 17.1 Introduction and Characterization

Urban flooding is an increasing hazard and is expected to aggravate with climate change. To mitigate the risk for the urban surroundings, this chapter explores a nature-based solution (NBS) focusing on a potential restoration of an urban river

M. E. Jørgensen (✉)
City of Copenhagen, Copenhagen, Denmark
e-mail: zi0m@kk.dk

J. Kidmose · P. van der Keur · H. J. Henriksen
Geological Survey of Denmark and Greenland (GEUS), Copenhagen, Denmark

E. Gómez
Climate Service Center Germany (GERICS), Helmholtz Center Geesthacht,
Hamburg, Germany

R. Giordano
Istituto di Ricerca Sulle Acque del Consiglio Nazionale delle Ricerche (IRSA-CNR),
Bari, Italy

© The Author(s) 2023
E. López Gunn et al. (eds.), *Greening Water Risks*, Water Security in a New
World, https://doi.org/10.1007/978-3-031-25308-9_17

(RUR). The restoration of a currently piped river back to its natural environment, is not only explored to mitigate urban flooding itself, but to create synergies with urban planning including climate change adaptation plans as well. In doing so, we look at quantitative and qualitative co-benefits.

The City of Copenhagen initiated its climate adaptive measures in earnest after hosting the COP 15 climate summit in December 2009. On 25 August 2011, the City Council adopted the Copenhagen Climate Adaptation Plan (City of Copenhagen 2011). This plan sets the framework for the implementation of climate adaptive measures in the City Administration area. Before disastrous flooding events in 2011 caused by a cloudburst, climate adaptation mainly focused on flooding related to storm surge, but then accelerated measures to mitigate urban flooding as well. This was translated in the adoption in 2012 of the Cloudburst Management Plan (CMP) (City of Copenhagen 2012) as an offshoot of the Copenhagen Climate Adaptation Plan from 2011. The CMP outlines the methods, priorities, and measures recommended for the area of climate adaptation including extreme rainfall. With this CMP, decisive steps forward have been taken to protect Copenhagen against high-intensity rain like the ones witnessed in August 2010 and especially in July 2011. Initiatives might include reopening streams, constructing new canals or establishing lakes and more green spaces, and using roads with high curbstones to lead the pluvial stormwater into confined spaces, notably surface water channels following a topographic gradient towards open waters.

Different nature-based solutions (NBS) are already implemented as part of the City's cloudburst management plan, which will contribute to the RUR and are therefore included in this chapter. Such NBS relevant for the RUR NBS scenario include green areas, retention areas, decoupled rainwater from roofs and blue surface water routing from street level flooding to the RUR recipient and are implemented in the hydrological model described later in this chapter.

The complex decision-making processes, for making such NBS design and implementation, require the collaboration of a wide set of decision-makers and stakeholders. Lack of cooperation could result in barriers to the NBS implementation and/or affect their effectiveness. To detect and analyze these collaborative barriers Social Network Analysis was combined with Group Model Building, in order to identify barriers hampering the implementation of the selected NBS.

## 17.2   Case Study Area and Methodology

The Copenhagen demonstration case involves a test study of the benefits of implementation of the RUR based on a currently piped river or stream running through central parts of the city (See Figs. 17.1 and 17.2). The area along the piped river is 2–4 km long and the catchment covers an area of 80.000–120.000 m$^2$ depending on the solution chosen. The primary focus regarding analysis of the benefits related to restoring the river are mitigation of damages avoided from a rising groundwater table. Specifically, the shallow groundwater table in the Copenhagen area are

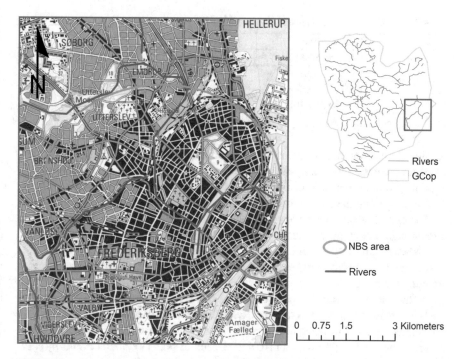

**Fig. 17.1** Focus area, area where NBS are implemented (green circle). In the NBS area the streams are presently in pipes in the subsurface. In the NBS scenario, the streams are restored to a blue/green water infrastructure (blue line). On the right, the case study is depicted as part of the Greater Copenhagen catchment

**Fig. 17.2** Bispeengbuen motorway ramp, suggested to be partially removed to create place for residential and green areas, road tunnels and climate adaptation solutions

**Table 17.1** The scenarios conducted in this study

| Vision | NBS scenario | Climate scenario | Period |
|--------|--------------|------------------|--------|
| 1 | BAU (no NBS) | Reference | 1981–2010 |
| | RUR (with NBS) | Reference | 1981–2010 |
| 2 | BAU (no NBS) | RCP8.5 | 2071–2100 |
| | RUR (with NBS) | RCP8.5 | 2071–2100 |

BAU indicates Business As Usual, RUR indicates Restoration Urban River

located only a few meters below the groundwater surface. Therefore, it is tested, using an integrated hydrological model, how future climatic changes will affect the shallow groundwater table in a part of the Copenhagen area, both by doing nothing and by reopening the river to make it function as a drainage channel.

The hydrologically based physical methodology used in this case study builds on simulating shallow groundwater with a hydrological model, which shows that the groundwater table under the current conditions and future conditions (see Table 17.1) may rise above surface and become surface inundation and adds to the stormwater flooding from e.g. cloudburst events.

Furthermore, economic analyses have been carried out to assess the avoided costs by means of the Restored Urban River (RUR) NBS scenario. Avoided costs are considered as the difference between the benefits of the RUR and the costs caused by groundwater flooding without RUR as a result of both cloudburst events by spilling over of excess water from street level and pushed up water from sewage systems to subterranean building infrastructure (e.g. cellars) and a rising groundwater level.

A simple damage function has subsequently been set up to calculate the avoided costs for the river basin.

### 17.2.1  Hydrological Model

The hydrological effects of the restored urban river in terms of its ability to divert not only rainwater but also groundwater in contact with buildings and critical infrastructure are simulated with a hydrological model, MIKE SHE/HYDRO. MIKE SHE/HYDRO is a physically based fully integrated groundwater surface water model covering the primary aquifer for the Greater Copenhagen area and large parts of Northern Zealand. The set-up of the physical environment and geology hydrological model is described in detail in Pengal et al. (2017), van der Keur et al. (2018) and LeCoent et al. (2019).

Groundwater levels are calculated for a number of deeper layers but only the most upper groundwater table is relevant for the estimation of damage level to urban buildings.

The spatial extension of the model area is in the order of magnitude 5 × 5 km for the entire urban area which is shared by the municipalities of Copenhagen and Frederiksberg (shown in Fig. 17.1) and includes the Bispeengen motorway ramp which in the considered scenario is replaced by road tunnels, residential and green areas and space for climate adaptation, Fig. 17.2.

## 17.2.2   Assessment of Management Alternatives Using Model Scenarios

To assess the effectiveness of NBS in reducing groundwater levels, we compared the NBS with the business-as-usual (BAU) alternative under different climate scenarios. We defined BAU as the present situation with a piped urban river and no cloudburst management plan projects implemented. The NBS alternative is the one of the Restored Urban River scenario (RUR): the piped urban river is restored and interaction with groundwater enabled. A flow rate of 300 l/sec is targeted to sustain a stable and ecological flow. The restored urban river is fed through the three water sources shown in Fig. 17.3 (hydraulic contact, groundwater pumping and decoupled water).

The performance of both alternatives is compared under two climate scenarios: the reference climate, 1981–2010 and the future climate 2071–2100 using the RCP8.5 projection (Collins et al. 2013). The scenarios are also shown in Table 17.1.

The Restored Urban River (RUR) scenario receives water from several sources, shown in Fig. 17.3:

1. groundwater through urban river interaction (hydraulic contact) with (shallow) groundwater resources
2. pumped groundwater from areas within the urban river catchment that experience rising groundwater, including from areas where water is infiltrated locally (Sustainable Drainage Systems, SuDS)
3. decoupled water from roofs and paved areas which otherwise would have been routed to the combined drainage-sewer system (combined sewer system)

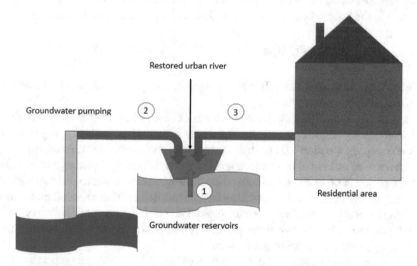

**Fig. 17.3**  Water sources to the restored urban river NBS: (1) from groundwater through hydraulic contact with the river; (2) groundwater pumping, shallow and deep; (3) decoupled water from residential areas (e.g. roofs) and other paved areas (e.g. roads)

### 17.2.3   Assessment of Avoided Damages

The scope of the economic analysis for the NAIAD Copenhagen NBS is to assess the avoided costs by means of the RUR NBS scenario under two climate scenarios with reference periods 1981–2010 and a future period 2071–2100. In Table 17.1 the scenarios are listed.

The damage levels are calculated as 30-year maximum groundwater level for the periods 1981–2010 (reference period) and 2071–2100 (future period). The generated 30-year maximum values are spatially distributed in a 100 × 100 m raster dataset relative to the surface. Then, the dataset is analysed for simulated values above the defined threshold of 2 m below surface, e.g. a number of raster grid cells with a 30-year maximum level above 2 m below surface. This dataset is then applied in GIS analysis to find areas above the defined groundwater thresholds for building damage with (RUR) and without (BAU) implementation of the NBS and with and without the hydrological effect of climate change (as prognosed by an 8.5 RCP scenario (Collins et al. 2013)).

Avoided costs are considered as costs caused by groundwater flooding as a result of both cloudburst events by spilling over of excess water from street level and pushed up water from sewerage systems to below ground building infrastructure (e.g. cellars) and rising groundwater level. Other costs as life cycle costs and opportunity costs as well as an assessment of co-benefits of the RUR – NBS scenario have not been included in the analysis.

The RUR scenario is compared to the BAU scenario with respect to shallow groundwater level and avoided damage is then calculated as the difference in affected area in which the shallow groundwater level causes damage (see Fig. 17.4). The damage caused (total damage) is estimated to € 68/m² (COWI 2014) and implemented in the adopted damage function. The damage function is expressed as:

$$\left[BAU - RUR\right]\left(Area\ affected\left[m^2\right]\right) \times unit\ cost\left[\in /m^2\right]$$

where Area affected is the cellar affected by flooding and unit cost is a constant (estimated unit damage, [€/m²]).

In this case study the total damage is set to € 68/m², which is a value derived from an analysis by COWI (2014) for the Danish assurance umbrella organisation "Forsikring og Pension" (Danish Insurance and Pension). The unit damage key numbers were estimated from reimbursed compensation from insurance companies for the period 2006 to 2012, including the catastrophic cloudburst event in July 2011.

Damages to cellars and other subsurface infrastructure from elevated groundwater levels caused by cloudburst events, prolonged rainfall, or a combination of both, can be distinguished in two main damage types, which in combination can show cascading effects, reinforcing each other:

(A) water damage due to (i) fast increased street level inundation and spilling over to lower areas, like cellars, via doorsteps, staircases to cellars or other low thresholds, as well as (ii) water pushed up from floor drains and toilets (in case no

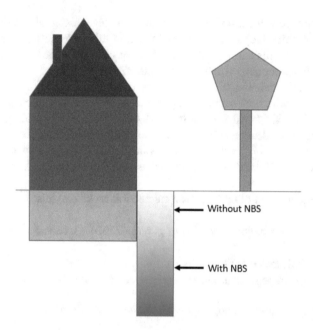

**Fig. 17.4** Avoided damage is calculated from the difference in area in which dwellings and businesses are affected by groundwater levels with and without NBS multiplied by a unit damage cost per m²

backflow blocker has been installed). Both types are the result of fast 'catastrophic' rain events. (B) slowly rising groundwater as a result of prolonged rainfall, changed (decreased) water abstraction patterns (water savings), and repaired sewage and drainage channels that also functioned as shallow groundwater drainage, all of which exacerbate the effects under (A) as drainage capacity to remove surface water is decreased.

The hydrologic model simulates the effect of (B) whereas mainly (A), and to an unknown extent (B) are previously only considered in the damage functions. This means that the calculated 'avoided damage' is considered to be a minimum avoided cost.

The avoided damage is calculated by analysing the area in which shallow groundwater level exceeds a certain threshold value where damage to cellars is probable and for which we have estimates depending on the type of built environment for that particular location. The threshold value of 2.0 m below surface has been determined from averaging known cellar depths (data provided by HOFOR, 2008) and expert judgement on expected shallow groundwater level in the RUR catchment. Groundwater may rise to this level where it interferes with the drainage and sewer system, primarily combined, and where it may cause water being pushed up to cellar or street level. In that case it adds to the stormwater flooding from e.g. cloudburst events. The applied threshold of 2.0 m below surface above which unit damage per square meter does not change. This assumption is in reality an underestimation, due to that the damage value probably will change with level higher above the threshold value, therefore the avoided damage calculated in this way must be regarded with caution and represents a minimum value (best case). Hence, the rule is damage = 0 if depth <2 m and damage = constant if depth <2 m.

Cellar depth is calculated as an overall average depth for all 100 m by 100 m model grids for which an occupation percentage of 7.5% is applied. For example, for a 100 m by 100 m grid amounting to 10,000 m$^2$ there is a building which is at risk for being flooded as a result of rising groundwater level, or spill over water from the street level. This building is assumed to have a cellar surface of 0.075*10,000 m$^2$ = 750 m$^2$, which is the value used for calculating damage for cellar flooding.

To estimate the cost of damages the calculation looks like this:

$$Damage = (number\ of\ pixel) \times 750\,m^2 \times \in 68\,/\,m^2$$

for both the RUR and BAU scenarios.

The avoided damage is the difference between the two scenarios:

$$Avoided\ damage = BAU\_damage - RUR\_damage.$$

Other avoided costs with respect to the implementation of the restored urban river RUR scenario are costs associated with climate adaptation alternatives and grey solutions. Alternatives to the opening of the urban river and its tributaries include cloudburst tunnels and retention basins, the costs of which have been estimated (Ernst and Young and Moe 2019). Before the release of the cloudburst management plan there also was the option of upgrading the traditional combined sewer system in Copenhagen which appeared to be a very costly solution (Ernst and Young and Moe 2019) as compared to alternative climate adaptation solutions. Therefore, avoided costs associated with upgrading and operating the traditional sewer solutions are not considered in this analysis for the urban river RUR NBS scenario.

## 17.3    Integrating Stakeholder's Knowledge in the NBS Designing Process

To successfully implement NBS designs for controlling the flooding from groundwater and cloudburst, the collaboration of a wide set of decision-makers and stakeholders is required. Lack of cooperation could result in barriers to the NBS implementation and/or affect their effectiveness (see Chaps. 7 and 19 for a broader discussion).

In the Copenhagen case study, stakeholders were engaged from the first stages of the project. The integration of stakeholder's knowledge in the co-design process of NBS had multiple elements of relevance. Firstly, potential barriers hampering NBS implementation were identified. Secondly, participatory modelling activities were carried out to elicit and structure stakeholder's risk perception. Finally, NBS effectiveness was assessed from a shared stakeholder perspective. To this aim, Fuzzy Cognitive Maps (FCM) (Kosko 1986) and Social Network Analysis were carried out starting from a Group Model Building (GMB) phase.

### 17.3.1   Fuzzy Cognitive Maps as a Tool for Trade-Offs Identification

Group Model Building (GMB) was used to conceptualize and understand, from a shared stakeholder perspective, different NBS strategies as well as the system where they would be applied. During a GMB Workshop key stakeholders participated in the identification of key factors and issues relevant for the modelling of Nature Based Solutions (NBS). Special attention was paid to the identification of NBS co-benefits and the main relationships between them. The main aim of the activity was to elicit and structure stakeholder's risk perception and to assess the trade-offs and synergies of the co-benefits associated to NBS implementation. For this, a conceptual qualitative model was developed in a 2 hours workshop with key stakeholders of the Copenhagen case study (Stakeholders include representatives from the City of Copenhagen (partner, but others invited to capture multiple views), the surrounding Capital administrative region of which the City of Copenhagen is part. Others represent utility companies, neighborhoods within the city area, insurance, enterprises working with innovative climate adaptation solutions, business models and a cooperation between municipalities, utility services and the Capital Region of Denmark to support efficient and sustainable climate adaptation solutions in the region. In total ca. 25 people participated). The resulting qualitative model represented key factors and variables of the system as well as the main relationships between them (See Fig. 17.5).

The qualitative model was used firstly, to identify and visualize the main causes-effects processes occurring in the system. Secondly, to increase the communication and shared vision among stakeholders in order to facilitate the consensus agreement process. Finally, the group model was used to develop a semi-quantitative Fuzzy Cognitive Map (FCM) (Kosko 1986), that was used to simulate the effect of different NBS strategies in the production of co-benefits. In Table 17.2 the simulated scenarios in the Fuzzy Cognitive Model are shown.

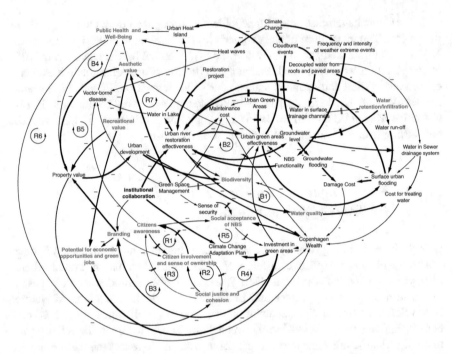

**Fig. 17.5** Showing the final qualitative group model integrating stakeholders and expert's knowledge. In green are highlighted NBS co-benefits

**Table 17.2.** Showing scenarios simulated in the Fuzzy Cognitive Map model

| Scenario | Description |
|---|---|
| BAU | Business as usual scenario without any measure applied |
| NBS1 | Restoration of the piped urban river combined with the creation of an urban green park adjacent to the river |
| NBS2 | Creation of an urban green park in the urban river area without the river restoration |
| NBS1-GSM | Restoration of the piped urban river combined with the creation of an adjacent urban green park with appropriate green space management |
| NBS1-GSM and IC | Restoration of the piped urban river combined with the creation of an adjacent urban green park with appropriate green space management and strong institutional collaboration |
| NBS2-GSM | Creation of an urban green park with appropriate green space management |
| NBS2-GSM and IC | Creation of an urban green park with appropriate green space management and strong institutional collaboration |

## 17.3.2   Fuzzy Cognitive Map Development

FCM is a methodology that allows the representation of social knowledge in order to model the decision-making process in a socio-political system (Kosko 1986; Dickerson and Kosko 1994). It has proven to be an easy-to-use method to engage stakeholders and to acquire knowledge from them. It is also useful to aggregate knowledge coming from different sources and people with different degrees of expertise (Park and Kim 1995). FCM consists of nodes representing key concepts of the simulated system and fuzzy links connecting various nodes. The links represent positive or negative causal relationships among concepts and the fuzzy part permits the description of degrees of causality (Sokar et al. 2011). The strength of a link between two concepts indicates the intensity of the relationship between them, that is to say, how strong is the influence of one concept on the other. The weights range in a normalized interval of $\{-1, 0, 1\}$. The value 1 and $-1$ represents a positive and negative causality respectively and the strongest relationship. The closer the values approach to 0, the weaker the relationships are. The relationships between variables can be represented in an adjacency matrix. In the FCM, this matrix allows the overall effects of an action on the elements in the map to be inferred qualitatively. The activation of a concept occurs when the value of a concept changes from 0 to 1. See Chap. 5 for more details.

The qualitative model developed during the group model building (GMB) exercise was post-processed and analysed and validated by a group of experts who were involved in the GMB and with experience in the system and in the FCM development. The role of the experts was assigning a weight to the causal relationship (weak, medium, strong). The strength of a relationship was represented by changing the thickness of the links between concepts composing the FCM (See Fig. 17.5). The experts also indicated potential 'delays' on the system, this means, causal processes that require time to occur. In the graphical representation of the FCM this was indicated with a delay mark (\\). Important feedback loops were also identified in order to facilitate the visualization of the causes-effect of the processes occurring in the system. Following the work developed by Giordano and colleagues in 2020 the FCM was simulated.

Identifying barriers hampering the NBS implementation using a collaborative social Network analysis approach.

The second phase of the stakeholders' engagement process in the CPH case study concerned the mapping and the complex network of interactions taking place among different decision-makers. This phase aimed at detecting potential barriers to cooperation that could hamper the actual and effective NBS implementation. Starting from the results of the Group Model Building exercise, Social Network Analysis was adopted (See Chap. 5 for more details).

To this aim, a participatory mapping exercise was organized with a group of selected stakeholders. At the beginning, participants were required to identify the key issues that need to be addressed for enabling the NBS implementation. Therefore, participants agreed upon the following list of issues: (i) water quality (discharge); (ii) traffic management and parking areas; (iii) Cost affordability and finance; (iv) quality of the urban environment; (v) control of the water flow; (vi) private property right issues; (vii) regulatory and legal issues. The discussion was organized around these issues. The main scope of the participatory mapping exercise was to model the way decision-makers interact with each other, sharing and exchanging information, and cooperating for carrying out specific tasks for addressing the above-mentioned issues.

Participants were requested to introduce in the map the actors that need to be involved in order to address the above-mentioned issues, the tasks that need to be carried out and the information needed/used by each actor in order to carry out the requested tasks. Please, refer to Chap. 5 of this book for a more detailed description of the methodology for the participatory mapping. The following Tables 17.3, 17.4 and 17.5 show the results of this first phase of the discussion.

**Table 17.3** Selected tasks within the Social Network Analysis for the Copenhagen study case group work

| Task | Acronym |
| --- | --- |
| Controlling the quality of water discharge | WDC |
| Traffic management | TRM |
| Parking areas management | PAM |
| Funding the intervention | FUND |
| Monitoring the quality of the urban environment | URBENV |
| Controlling the water flow | WFLOW |
| Adapting urban regulation | REG |
| Controlling private property rights | PPR |

**Table 17.4** Selected actors within the Social Network Analysis for the Copenhagen case study group work

| Actors | Acronym |
| --- | --- |
| City Council | CCOUN |
| European Union | EU |
| Local water utility | HOFOR |
| Citizens | CIT |
| Environmental NGO advocacy | ENVNGO |
| State authority for traffic and road | TRAUTH |
| Traffic department of Copenhagen municipality | TRDEPT |
| Consultancy | CONS |
| Climate adaptation office MIMIC | CADEPT |
| State health department | HEALTHDEPT |
| Frederiksberg municipality | FMUNC |
| Copenhagen municipality | CPH |
| Building permit office of Copenhagen municipality | BPO |
| Department municipal water regulation | WRDEPT |
| Economic department of Copenhagen municipality | ECDEPT |

**Table 17.5** Selected information within the Social Network Analysis for the Copenhagen study case group work

| Information | Acronym |
|---|---|
| Political guidelines | GUID |
| National/regional/local water planning | WPLAN |
| EU directives | EUDIR |
| Local pressure | LPRESS |
| Community awareness | CAWAR |
| Alternative water sources | ALTWAT |
| Building permits | BUILDPER |
| Traffic plan | TRPLAN |
| Cost/benefits analysis | COSBEN |
| Water needs | WNEED |
| Climate scenarios risks | CLSCEN |
| Water quality discharge technologies | WQUAL |
| Urban dynamics | URBDYN |
| Water quality limits | WQUALLIM |
| Health issue assessment | HEALTH |
| Discharge permit guidelines | DISCPER |
| Technologies for water remediation | WATREM |
| Best practices of NBS | BMPNBS |

Then, participants were requested to define the links between actors - that is, what are the actors that cooperate in order to address the issues? - between actors and tasks - i.e. who is supposed to do what? - between knowledge and tasks - i.e. what pieces of knowledge are needed to perform a certain task? - and between different pieces of knowledge - i.e. What piece of knowledge is needed in order to access another needed knowledge? Fig. 17.6 shows the Actors X Actors map of interactions among the different actors cited in Table 17.4.

## 17.4   Results and Discussion

### 17.4.1   Hydrological Modelling and Damage Functions

Figure 17.7 shows the simulated change between the future groundwater levels with and without implementation of RUR. As expected, and intended, the shallow groundwater level rise is reduced in case a RUR-NBS is implemented (the blue pixels have lowest groundwater level).

Within the catchment area of RUR NBS (see Fig. 17.7) for the future climate, the area with GW damage is reduced from $1,620,000$ m$^2$ ($162$ cells $\times$ $10,000$ m$^2$) to $1,060,000$ m$^2$ ($106$ cells $\times$ $10,000$ m$^2$), which equals a reduction of $35\%$.

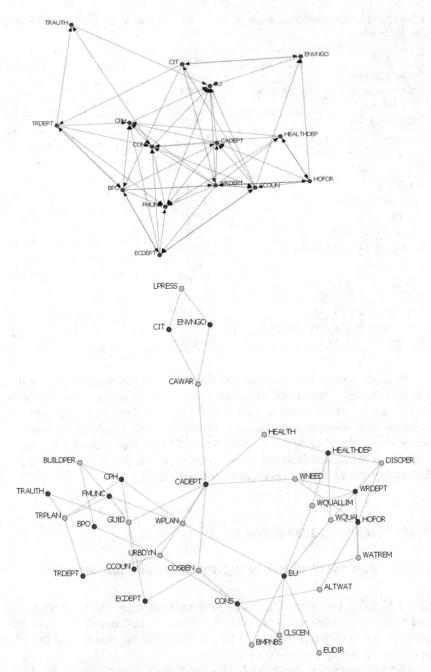

**Fig. 17.6** Maps of interactions between the different actors and the information

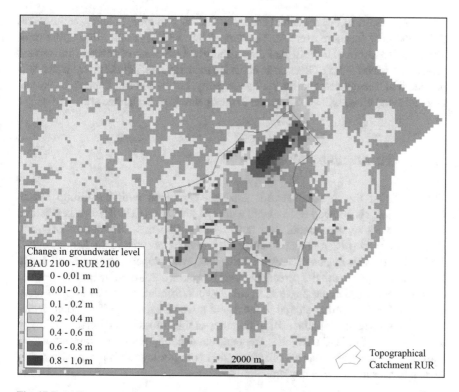

**Fig. 17.7** Difference in groundwater max levels with and without implementation of NBS (RUR) for the period 2070–2100 simulated the RCP 8.5 emission scenario (mean ensemble member)

**Table 17.6** Avoided cost of the two overall scenarios (see Table 17.1)

| Vision | NBS scenario | Climate scenario | Period | Avoided cost [€] |
|---|---|---|---|---|
| 1 | BAU (no NBS) | Reference | 1981–2010 | 2.715.700 |
| | RUR (with NBS) | Reference | 1981–2010 | |
| 2 | BAU (no NBS) | RCP8.5 | 2071–2100 | 2.856.000 |
| | RUR (with NBS) | RCP8.5 | 2071–2100 | |

This means that the avoided damage under the conditions of future climate can be calculated from the reduced number of cells for which the cellar area is 750 m² per cell. Therefore, the avoided damage is $(162–106) \times 750$ m² $\times$ € $68/m^2$ is equal to € 2.856.000.

Likewise, for the reference period, the area is reduced from 1,460,000 to 930,000 m² which results in a somewhat lower avoided cost of € 2.715.700 (Table 17.6).

Besides the direct benefits of lowering the risk of groundwater flooding by restoring the urban river, there are also many other additional benefits or co-benefits associated with the RUR scenario. The conversion of the area from an area with a heavy traffic to a greener and quieter and more ambient place, will expect to increase the liveability of residents in terms of better health because of less air pollution, outdoor facilities and the opportunity for more exercise. The temperature will be reduced during the summertime, due to the cooling effect of the greenery and reduction of concrete and asphalt (urban heat island effect). Other co-benefits are increase in real estate prices which will provide a higher tax revenue for the municipality; businesses can flourish and create more jobs and crime will go down, due to the openness of the altered area.

### 17.4.2  Participatory Modelling

The results of the Social Network Analysis combined with the Group Model Building analysis allowed us to identify the EU directive (EUDIR) as a key barrier in the network of interaction. The model shows that this information is poorly shared among the different actors, although it plays a key role in the Knowledge X Task network. This means that several tasks cannot be properly implemented due to the limited access to the EU directives. This barrier is confirmed by the low level of centrality of the EU in the Agent X Agent network. Similarly, the cost/benefits analysis (COSBEN) is poorly shared in the Agent x Knowledge map. This could represent a limit, given the high level of importance of this information. Finally, although the City Council (CCOUNS) has frequent connections with the other actors, its capability to access key pieces of information is rather low. This potentially represents a barrier in the NBS implementation.

### 17.4.3  Uncertainties and Usefulness of the Approach in Assessing NAS Effectiveness

The methodology to assess avoided damage is conducted under several sources of uncertainties.

The hydrologic model has been calibrated against discharge of the piped river to an open water recipient, but since the urban river NBS is not implemented no calibration for the anticipated urban river has been possible. However, the applied model is physically based, i.e. physical processes in the integrated groundwater and surface water environment are state-of-the-art represented. Nevertheless, assumptions have been made including the exact position and depth of the restored river, groundwater-surface water interactions, groundwater well extractions and how cloudburst management climate adaptation measures are connected to the urban river, all of them add to assessment uncertainty.

Another source of uncertainty is the missing information on cellar depths for the municipality of Frederiksberg for which similarity with data from Copenhagen has been assumed, i.e. the average depth for the Copenhagen part of 1.78 m has been assumed for the Frederiksberg part, which is probable. The model setup and overall approach however represents a credible tool which can be updated when new information becomes available.

To calculate the avoided damage, data on insured damage reimbursements for the period 2006 to 2012 were crucial. Reported damage from citizens to insurance companies has been analyzed by a consultant (COWI 2014) for which available damage data from the insurance companies to the built environment has been aggregated as 'water damage'. As a result of this, damage to cellars from both surface water overflow and from the subsurface sewerage system could not be distinguished from each other and the damage function was aggregated at the scale of residential buildings (€ 68/m²). This value does not consider whether flooded areas and buildings include businesses for which the damage costs may be higher. Also, the unit damage cost of € 68/m² does not vary with depth over the threshold value (2.0 m below surface), which represents an uncertainty and means the estimated avoided damage is a minimum value.

In addition, groundwater flooding damage as a result of sewer overflow probably also causes damage to the structure of buildings, not taken into consideration in the simple damage function used, which is why the estimated value of damage, a bit more than € 2.7 million, alone for this reason among others, probably is much higher.

When calculating the cost of avoided damages, a value of 7.5% for the affected area (e.g. cellars) is used. This value is a premise based on the best qualified assessment and is also subject to uncertainty. The affected area is likely higher, which means the calculated cost of avoided cost is higher than shown.

All this adds substantially to the overall uncertainty, but again the conceptual approach can be improved with new and improved data. Therefore, improved and more data, especially for the development of accurate damage functions are needed to reduce the overall uncertainty.

## 17.5   Conclusion

The RUR NBS scenario clearly shows that restoring (opening) a piped urban river to its natural position influences the groundwater level. The simulations show that the restored river will act as a drainage channel for both cloudbursts induced surface water flooding and a rising groundwater level and thereby avoid groundwater flooding and subsequently damage to subterranean structures.

Calculations show that by restoring the urban river to its natural position, 35% of the catchment area for the RUR NBS scenario will remain below the applied threshold of 2 m, where no damage will occur, compared to the BAU scenario.

Economic analyses show that the avoided costs by restoring the urban river will be at least € 2.7 million for the catchment area of the RUR scenario, and probably higher considering the overall uncertainty.

Besides the direct benefits of lowering the risk of groundwater flooding by restoring the urban river, there are also many other additional benefits or co-benefits associated with the RUR scenario. These co-benefits have not been converted into direct economic values in this study, but they will certainly mean a lot in the overall assessment of RUR NBS.

The restoration of the urban river (RUR) combined with an adjacent urban green area is perceived by the stakeholder as an effective solution capable to deliver a wide range of co-benefits. The results of the simulation show that the integration of both measures is likely to be more effective in the reduction of groundwater and surface flooding. The effectiveness of NBS increases considerably when good strong institutional collaboration and good green management follows. The FCM simulation has also revealed potential trade-offs between co-benefits. For example, an increase of the recreational and aesthetic value of the area where the NBS are applied could lead to a decrease of the social justice and cohesion co-benefit. This is because green spaces tend to increase property value. The results of the study have emphasized the need of adopting an integrated view of the system to assess and compare different NBS and adaptation strategies.

However, NBS implementation requires effective cooperation among different decision-makers. Therefore, efforts are needed to overcome the existing collaborative barriers, in order to (i) enhance the flow of information among the different decision-makers, (ii) reduce the level of conflicts and (iii) to facilitate collaborative decision-making. The implemented SNA helped to detect those barriers. A dialogue is needed among the different decision-makers for defining potential interventions – i.e. networking interventions – whose main scope is to enhance the collaboration over NBS design and implementation.

# References

City of Copenhagen (2011) Climate Adaptation Plan. https://international.kk.dk/artikel/climate-adaptation

City of Copenhagen (2012) Cloudburst Management Plan 2012. https://international.kk.dk/artikel/climate-adaptation

Collins M, Knutti R, Arblaster J, Dufresne J-L, Fichefet T, Friedlingstein P, Gao X, Gutowski WJ, Johns T, Krinner G, Shongwe M, Tebaldi C, Weaver AJ, Wehner M (2013) Long-term climate change: projections, commitments and irreversibility. In: Stocker TF, Qin D, Plattner G-K, Tignor M, Allen SK, Boschung J, Nauels A, Xia Y, Bex V, Midgley PM (eds) Climate change 2013: the physical science basis. Contribution of Working Group I to the Fifth Assessment Report of the Intergovernmental Panel on Climate Change, Cambridge University Press, Cambridge

COWI (2014) Prepared for: Insurance and Pension, City of Copenhagen & FRB, HOFOR, FRB utilities (2014). Unit-costs for flooding (35 pp., in Danish)

Dickerson JA, Kosko B (1994) Virtual worlds as fuzzy cognitive maps. Presence Teleop Virt Environ 3:173–189. https://doi.org/10.1162/pres.1994.3.2.173

Ernest & Young (EY.com) and MOE (moe.dk). Prepared for: City of Copenhagen & Frederiksberg (2019) Analyses on alternatives to Bispeengbuen (53 pp., in Danish) and building on HOFOR (2018) VEL 46–47 report, including various cost scenarios and business cases.

Giordano R, Pluchinotta I, Pagano A, Scrieciu A, Nanu F (2020) Enhancing naturebased solutions acceptance through stakeholders' engagement in co-benefits identification and trade-offs analysis. Sci Total Environ 713(136):552

HOFOR Utility (2008) Spread sheet containing results of a mapping of cellars and drainage levels in Copenhagen. Results goes up to year 2008.

Kosko B (1986) Fuzzy cognitive maps. Int J Man Mach Stud 24:65–75. https://doi.org/10.1016/S0020-7373(86)80040-2

LeCoent et al (2019) Deliverable D6.3. NAIAD. http://naiad2020.eu/media-center/project-public-deliverables/

Park KS, Kim SH (1995) Fuzzy cognitive maps considering time relationships. Int J Hum Comput Stud 42:157–168. https://doi.org/10.1006/ijhc.1995.1007

Pengal et al (2017) Deliverable D6.1 NAIAD. http://naiad2020.eu/media-center/project-public-deliverables/

Sokar IY, Jamaluddin M, Abdullah M, Khalifa ZA (2011) KPIs target adjustment based on trade-off evaluation using fuzzy cognitive maps. Aust J Basic Appl Sci 5:2048–2053

van der Keur et al (2018) Deliverable D6.2 NAIAD. http://naiad2020.eu/media-center/project-public-deliverables/

# Chapter 18
# Enabling Effective Engagement, Investment and Implementation of Natural Assurance Systems for Water and Climate Security

**Josh Weinberg, Kanika Thakar, Roxane Marchal, Florentina Nanu, Beatriz Mayor, Elena López Gunn, Guillaume Piton, Polona Pengal, and David Moncoulon**

## Highlights

- Key issues to improve enabling conditions to support the uptake of NBS and NAS range from connecting an evidence-base to an experience gap through to creating an enabling regulatory environment.
- Opportunity areas to promote the uptake of NBS and NAS arise by facilitating their financing and implementation, which include finding solutions to de-risk private sector investment in NBS.
- Further opportunity areas to effectively engage the insurance sector include increased scope for scientific exchange and cooperation, awareness raising on climate risks and policy dialogue on risk reduction and environmental regulation.

J. Weinberg (✉) · K. Thakar
Stockholm International Water Institute (SIWI), Stockholm, Sweden
e-mail: josh.weinberg@siwi.org

R. Marchal · D. Moncoulon
Caisse Centrale de Réassurance (CCR), Paris, France

F. Nanu
Business Development Group (BDG), Bucharest, Romania

B. Mayor · E. López Gunn
ICATALIST S.L., Las Rozas, Madrid, Spain

G. Piton
Université Grenoble Alpes, INRAE, CNRS, IRD, Grenoble INP, IGE, Grenoble, France

P. Pengal
Institute for Ichthyological and Ecological Research (REVIVO), Dob, Slovenia

## 18.1 Introduction

While there is increasing enthusiasm and support at a global level to promote NBS; scaling investment still requires enhanced coordination, capacity, and confidence among public authorities that would be primarily responsible for accessing their financing and overseeing their implementation. The demonstration cases in the NAIAD project detailed in the previous chapters show that the implementation landscape is very diverse across European countries, and even more so when comparing contexts across different continents.

This chapter investigates the enabling conditions and policy settings that are more conducive to the uptake of NAS and discusses how to effectively engage with the insurance sector as part of that process. Building on learnings and resources from the NAIAD project, it highlights opportunities and challenges to support the mobilization of green infrastructure as part of NAS schemes.

## 18.2 Overview of Key Challenges and Enablers for NBS and NAS

The recent *Review of progress on implementation of the EU green infrastructure strategy* (EU 2019) concluded that "experience illustrates that ecosystem-based approaches such as GI, nature-based solutions, ecosystem-based adaptation, natural water retention measures and ecosystem-based disaster risk reduction measures are cost-efficient policy tools; but they are not used to their full extent and their potential should be further strengthened at EU level." This conclusion leads to further questions: if NBS are cost-effective, what impedes their implementation? And how can these challenges be overcome? This chapter examines four frequently cited issues that limit NBS implementation that also directly relate to NAS (several of which have been covered across multiple chapters in this book): (1) connecting an evidence base to an experience gap; (2) capturing full value on cost-benefit assessment; (3) capitalizing on investor demand; and (4) creating an enabling regulatory environment.

### 18.2.1 Connecting an Evidence-Base to an Experience Gap

An obstacle historically mentioned that can prevent investment in NBS is a perceived lack of evidence of the performance of NBS relative to traditional infrastructure assets (EU 2019, Nesshöver et al. 2017). This implies that potential projects are effectively stopped by the engineers and technical project development staff that are not comfortable with NBS before they reach the stage of arranging financing. This can be particularly important when considering NBS to provide a service such as flood risk mitigation. While there is a growing number of cases that demonstrate NBS service delivery at a global level, a local evidence or experience gap can still exist. Thus, a key enabler is to mainstream effective performance assessment methodologies for NBS and NAS so practitioners have confidence in using them.

### 18.2.2   Capturing Full Value in Cost-Benefit Assessment

A second commonly noted issue is a perception that it is difficult to assign economic value to ecosystem services and perform adequate cost-benefit analyses (CBA) of NBS, which are normally a critical aspect of finance preparation and qualification. In this case, it may often be the case that the benefits in terms of water and climate security provided (e.g. flood/drought risks reduction, water supply) are in fact more straightforward to calculate than other benefits provided by ecosystems or a specific NBS. However, learning from the NAIAD cases (see case study chapter Medina) has shown that CBA analysis of an NBS that only considers a single benefit (such as flood mitigation) may not be sufficient to show justification for investment, and it may also fail to include additional benefits of even greater value that would transform the CBA proposition. Thus, the key enabler is the development of integrated CBA methodologies to capture multiple values provided by NBS and NAS (Le Coent et al. 2020). NAIAD has developed tools that support evaluation of risk reduction potential provided by a proposed NBS or GI, as well as tested of integrated Cost-Benefit Analysis methodologies in several cases. These have shown NBS outperforming grey infrastructure alternatives, but also found that the DRR benefits provided were not able to be conclusively assessed as outweighing the cost of the intervention on their own. Thus, NBS aiming at solely reducing water risks cannot be assumed to be economically efficient. Indeed, NBS often appear to be economically efficient particularly when all the benefits generated are considered. This requires strategies to translate multiple co-benefits into revenue flows that can be used for as an argument for project funding. In many cases, it will also involve adapting regulations and procedures to apply public funding to NBS, as different financing bodies may be willing to pay for different types of co-benefits provided.

### 18.2.3   Capitalizing on Existing and Potential Investor Demand

Another issue often raised is insufficient access to capital for NBS investments (Weinberg et al. 2018). This lack of investor demand from financing institutions is explained by them being either averse to NBS, not aware of them, and/or not able to find appropriate finance instruments to fund them. In recent years, however, very significant developments have occurred in producing financing mechanisms in this area, such as the creation of bonds that explicitly signal to investors that the project has environmental benefits and/or contains NBS components (e.g., green and climate bonds), the emergence of private-public partnerships that explicitly look at ecosystem functions and interventions for investment (e.g., Flood Re in the UK), and the EU sustainable finance initiative's approach to develop a new regulatory framework that can encompass new and emerging issues (e.g., climate adaptation, NBS). Explosive growth has occurred in the green and climate bonds market in recent years, expanding from upper-income countries (USA, western Europe, Australia) to middle-income countries (China, India, Brazil), including lower-middle income countries (e.g., Nigeria).

A better diagnosis of the finance challenge may be described as a "market gap". Under investment in NBS is seen a failure, at large, of those looking to access financing to produce viable projects, to sustain and pay back investment. It also, however, signals a failure in the market where the value of services provided by NBS may be undervalued, or the values provided are not monetized sufficiently or made possible through other KPIs to enable investment. This can also be a problem where familiar finance instruments are ill-fitted to non-traditional investments. Each of these scenarios may require public sector interventions and be better considered within NBS/GI strategies. The shortage of bankable projects therefore requires not only capacitation to enable proposals with improved business models, but also can require shifts in financing processes that can better understand and value what these projects offer. Now in many places it may be theoretically more possible to get a loan to implement an NBS but the lack of awareness of the investor (those taking the loans to finance a project) of the advantages of NBS relative to more traditional infrastructure investments and a lack of political will by regulators - or other public policy groups to encourage or favor NBS- may remain as important obstacles. Procedural issues come into play as well. Concern over transaction costs and requirements on staff resources are also likely slowing adoption of more effective and progressive approaches to both climate risk assessment and mitigation options. For plans and projects to access funding and financing it is necessary to prepare a full business case for the entire investment program and each of the projects that make part of this investment program.

To meet this opportunity, there are a number of emerging tools and innovations to support improved proposals for NBS as viable investment projects. This includes those featured in this book, such as the *Handbook for the implementation of nature-based solutions for water security* (Altamirano et al. 2021) can be used to support proponents of NBS to create a more structured project plans and implementation strategies. *The NAS Business Canvas* is another tool that can be applied to any NBS project or strategy in order to identify the most suitable business model for the case. Each can make it easier to engage with public authorities, collectivities and private investors into further project preparation. These tools can be applied, refined and promoted to accelerate uptake of NBS in Europe and globally.

## 18.2.4    Creating an Enabling Regulatory Environment

Broadly speaking, there are few direct directives, policies or governing institutions with exact mandate over supporting or implementing NBS or NAS. However, there are numerous policies, institutions and governance instrument that are potentially relevant and can impact capacity for investing and implementing them. This can add complexity and challenges for coordination as well as clear processes to facilitate their uptake.

In many countries prioritized actions may still be required to ensure water-risks are recognized appropriately by governments and citizens. One key point of analysis within a national context is clarifying who pays for DRR and ecosystem protection and who benefits from it, particularly when developing dialogue with the insurance sector in this field.

Citizens are often unaware of the natural hazard coverage and terms in their insurance contracts. Here it is important to consider that worldwide it is relatively rare that insurers themselves make direct investments in risk mitigation measures (Atreya et al. 2015). Rather insurers may make insurance coverage conditional upon the uptake of risk reducing actions taken through warranties or 'must-do' clauses; provide premium/deductible discounts or client awareness raising (Kunreuther 2019). Instead of investing directly, it is more common for insurance groups to advocate public investment or stricter regulations for DRR (Kousky 2019). The primary actors investing in NBS or NAS schemes will be from public authorities and public financing unless there are reforms or structural changes to the way insurance is provided in most places.

That NBS are considered as viable options for risk reduction is an important step beyond this to ensure the best measures are taken. There are several specific policy mechanisms developed in EU member states that can be considered for application in other countries or even at EU level that have been highlighted in this book (see Chap. 3). The Barnier Fund, for example, which supports implementation activities for DRR by the national insurance scheme, has proven an effective example in France. Together with GEMAPI Law in France, which allows river basin authorities to authorize a tax to finance actions for risk reduction and environmental protection/restoration, there is a strong enabling environment for viable NBS to be implemented.

Policy frameworks to mainstream NAS into the insurance industry need to evaluate the DRR policy processes at the national level, their implementation at local scale, as well as the current and potential financing arrangements and mechanisms to integrate it into the insurance system. Current adaptation policy frameworks will often need to prioritize short term actions to reduce long-term vulnerabilities and impacts. This may further benefit from actions that can facilitate cross-sectoral cooperation, multi-stakeholder involvement, knowledge sharing, bridging local gaps and international cooperation. Importantly, this must effectively connect to the broad ambitions to advance uptake and implementation of NBS to support a wide range of green growth, low-carbon development and sustainability strategies developed at multiple levels. Challenges and opportunities in this endeavor are discussed in more detail in the following section.

## 18.3  Enabling Conditions for NAS – Learning from Case Studies

One of the key strengths of the NAIAD project was its ability to test the assessment methodologies and integration tools it has developed in real life scenarios through demonstration projects located across Europe (see case study chapters in this book). Here we first we provide a short overview of the strikingly different conditions for implementation faced and learnings that can be drawn to develop responses that make sense for different contexts. In Slovenia, the institutional implementation environment made it extremely challenging for a NBS to even be considered as option. In France, there are multiple pathways to implementation created by with new and existing measures taken by the government that enable investments in NBS. A third case in the Lower Danube in Romania highlighted a different challenge: public perception of flood risks are low and not in line with actual risk levels faced. This requires investment in risk reduction and finding ways to enable public support for NBS or measures taken to achieve this.

### 18.3.1  Lessons from Glinščica, Slovenia – Overcoming Political Challenges to Considering Nature's Solutions

Through a series of interviews and stakeholder engagement processes in the Glinščica catchment looking specifically at the adoption of NBS for flood risk management, the NAIAD team unearthed a broader issue of failed transition to adaptive integrated water management in Slovenia. Ershad Sarabi et al. (2019) point out that barriers to implementation of new technologies, such as NBS, are often socio-institutional rather than technical. Common challenges include a lack of coordination between institutions, unclear roles and responsibilities between parties, low levels of community engagement, and little or no monitoring and evaluation. For the Glinščica catchment specifically, barriers to the adoption of NBS for flood risk management included: a high degree of scepticism from decision makers to engage with local research projects; fragmented practices in water management; institutional knowledge gaps; low interagency cooperation; weak participatory processes and, an acute lack of enforcement and accountability. While legislation has greatly advanced quickly in recent years it is not being fully implemented.

Once the participating stakeholders developed and agreed on the best future strategy for the Glinščica catchment that included an NBS, another barrier was identified: land ownership. While this is a broader issue connected to the struggles of the EU Common Agricultural Policy, the implementing institutions faced the opposition from land owners (mostly farmers) as the main barrier preventing specifically the implementation of NBS measures (which usually require more space than grey solutions). The Ministry for agriculture, forestry and food, expressed their opposition to using agricultural land for water risk management. On the other hand, the opposition from farmers was not absolute and was impacted by past bad experience with governmental institutions that usually fail to understand and accommodate their requirements. One farmer, for example, expressed his concern with the governmental institutions that prohibited him from building small retention reservoirs on several of his fields, a practice that he learned from German colleagues was supported by the German government. Building a more enabling environment for NBS uptake in Slovenia will take time, effort and resources. Several participating stakeholders noted that top-down guidance and pressure from the EU through the changes in its own policy, regulations and enforcement is a critical lever needed to push more integrated and adaptive water management at large, and considerations of NBS as a result of such circumstances.

### 18.3.2   Lessons from Brague, France – A Strong Enabling Environment Still Requires Political Will

The Brague DEMO, as Glinščica, is a catchment of intermediate size (5–200 km$^2$) located in south France along the Mediterranean coast. Similar flash flood problems are experienced but the institutional context is very different. Several laws created national funds funded by taxes on insurance premiums to finance flood protection measures and by water bill finance river restoration measures. A recent law, GEMAPI, closed the loop by making it mandatory for any works performed in a river to address risk reduction and environmental restoration. It also allows river managers to raise a local tax, up to 40€/person/year, to finance it. NBS are *per se* measures consistent with this policy context but are not yet mainstreamed.

Lack of confidence regarding the flood reduction capacity of NBS is a key barrier to their implementation. River managers are advised at a daily basis by civil engineers working from consulting companies. Their expertise on computing the physical effects of large civil engineering structures on flood processes are ancient and solid,. This experience is, however, much lower regarding NBS effects. Evidence of significant cumulated flood reduction effect of NBS as well as capacity building for design engineers to compute and design NBS with targeted risk reduction objective are needed.

Another barrier is related to lack of decision maker commitment for ambitious NBS projects. In the Brague River, urban sprawling during the 1980s and 1990s confined the river channel in an excessively narrow stripe unable to convey extreme flows. The proposed NBS strategy, namely small natural retention areas in the upper basin and giving room to the river in the lowlands, requires demolishing about 50 houses. Even though these houses have experienced repeated flood events over the last decade and most people are willing to leave, decision makers are reluctant to launch large-scale and ambitious expropriations operations to achieve a consistent vision of the river corridor for the next 50 years. Bringing significant improvements in both river quality and flood risk is feasible but require strong political willingness to launch a long-term integrated plan that will start by a difficult phase of real estate acquisition.

### 18.3.3 Lessons from the Lower Danube in Romania: How to Confront Overly Optimistic Risk Perception?

Research performed in the Lower Danube case study aimed at understanding the role of natural assurance schemes in complex natural, economic and social contexts. The vulnerability to water-related hazards in the region was mapped and two scenarios were analyzed focused on the total flooding (optimistic planning scenario) or partial flooding (realistic planning scenario) of an enclosure. The NBS risk mitigation solution (a restored pond) would protect communities up- and downstream in the identified location while supporting the diversification of economic activities and sustainable development for the local community (fish farming and ecotourism). The application of a model developed by CCR provided estimates of the amount of damages related to destroyed and affected dwellings and an average cost per claim. To increase the relevance and the potential for replicable results, the large scale demo approach was complemented with a focus on the valuation of a specific NBS along with other soft-institutional measures to allow assessment of effectiveness, based on a combined bottom-up interest of communities for diversification of economic activities and a top-down concern for reducing the pressure on the grey infrastructure for flood protection by means of a cascade system of green solutions. The approach was synthesized in a business model and a proposal for a financing framework for water security that, with further elaboration, could be used for upscaling the approach within a strategy for recreating the green corridor along the Danube.

The research in the Lower Danube case study included an open dialogue with the local insurance sector. The increased frequency and intensity of natural disasters and the increasing value of both private and public property will result in higher risk exposure. The sector is recognizing the urgency for action (including for prevention) but remains client oriented. It prioritizes ensuring accuracy of forecasting and risk quantification methods, but not prevention or mitigation mechanisms. The Romanian insurance sector is additionally coping with the relatively recent incorporation of insurance concepts in citizens' financial education requiring focus towards raising awareness and building trust. At the same time different regulatory obligations (e.g. GDPR, professional accreditations, etc.) often result in prioritizing the core business and leaving little capacity for actively seeking new business models.

The low level of public risk perception is a recognized barrier towards sector involvement in strategic actions for mitigation of water related risks at the national level. A study conducted by UNSAR[1] (2018) showed that 90% of Romanians surveyed perceived that the climate is changing and 71% are interested in property insurance. Yet, most do not consider floods and droughts as major risks. Partnership with scientists and knowledge providers along with functionality of public-private-partnerships at local level are essential for the identification and implementation of resilience measures in which the sector can get involved or can contribute to. There is a clear acceptance on the advantages for sharing actuarial data and mutualisation of risk levels, however a successful approach needs the support from the financial supervisory authority regulating the market to create and manage the needed instruments at national level. With a concerted approach the sector can contribute to prioritization of green solutions by adapted premiums and underwriting criteria. Incorporating climate change in different sector policies (such as agriculture, territorial development, water, health, education, etc.) can impact in the same direction. The government is the key player here as they are the ones that ultimately have responsibilities to reduce hazards, exposure and vulnerability, influencing policies in different sectors and utilize legislative power via central and local authorities to implement resilience building measures.

## 18.4   Priorities to Promote Uptake of NBS and NAS

Public institutions responsible for water resources and disaster risk management, such as utilities and local governments, are the main actors currently investing and implementing NBS (Browder et al. 2019). A key driver is often to address risks and comply with national and supra-national environmental regulations, such as the EU Flood Directive, etc. (Mayor et al. 2019, Somarakis et al. 2019). NBS are promoted as strategic investments that can boost the overall performance and climate resilience of infrastructure. Numerous reports and studies have discussed enablers and barriers to the investment and uptake of NBS projects (Weinberg et al. 2018; Stagakis et al. 2019). Common observations highlight perspectives ranging from financial sector professionals stating a lack of access to viable projects ready for investment (e.g. a pipeline of projects), to developers claiming that the key barrier is to access capital investments to implement projects at the scale required (Browder et al. 2019). The truth is that these paradoxical conditions co-exist. Mobilizing investments require both the development of new and improved application of existing financing models, as well as interventions to improve the enabling conditions, like regulations, access to good data and management capacity of NBS projects at all levels. The following sections gives an overview of multiple strategic areas to focus on to enable NBS and NAS to be funded and effectively implemented, and

---

[1] The National Association of Insurance and Reinsurance Companies in Romania.

highlights opportunities to effectively engage with the insurance sector in this process.

The financing of NBS is often hindered by perceived risks on key performance indicators and level of service that can provide, which can lead to reduced confidence by governments, utilities and local authorities to invest (Browder et al. 2019). These perceived risks frequently center around either the institutional structures and capabilities to manage them; or questions around accessing evidence that the NBS will provide the specific service stated (Ershad Sarabi et al. 2019).

### 18.4.1  Solutions to De-risk Private Sector Investment in NBS

Unlocking private finance can involve overcoming multiple hurdles. Mayor et al. (2019) points out that financers and institutional investors still consider NBS as risky. There is a high perception of risk of preconstruction and construction phases, and first years of operation. Bankability of projects often depends on appropriate risk mitigation for investors. Private financing can still play a role using refinancing vehicles, allowing entry options to institutional investors in less risky periods (e.g., post-completion of construction). In these cases, additional parties that can guarantee the financing (and have a higher ability to take on the risk), can be important.

Strategies for de-risking investments in green infrastructure for private investors is needed to level the playing field and make NBS easier to develop. The financing for de-risking may come from the public sector, investment banks or other sources, but the basis for the assessment may include the insurance sector. For example, a model where the insurance investor could act as the warrantor for the project, provided the project reaches agreed upon benchmarks is being tested in other regions like in the US and may be considered for replication. This can be done within a green or environmental bond, making a portion of payments to a project developer contingent upon reaching specific performance targets that can provide assurances to the financer (Hindlian et al. 2019). This should keep in mind that more complex financing mechanisms can increase transaction costs and make them only suitable for larger-scale interventions (Mayor et al., 2019). Specific guidance for investors and developers on how to access and develop these mechanisms may therefore also be needed.

### 18.4.2  Quotas for Financing Natural Infrastructure Projects in Initiatives and Funds

A key lever to explore is to promote policies that mandate investment in risk reduction to receive eligibility for certain forms of financing. This could begin with reviews of viable funding sources where criteria can be added to ensure options to provide risk reduction, including through green infrastructure/ NBS, are included or at least considered.

### 18.4.3   Placing Explicit Criteria for NBS, DRR and Adaptation in Green Finance

The EU Sustainability Taxonomy (and the EU Sustainable Finance Action Plan) represents an important opportunity, as it will both widen the total amount of green capital (by gathering more financers and capital deployed) and deepen its impact (by specifying more criteria for investments that are green in specific areas). This means the explicit criteria for nature-based solutions, green infrastructure for climate adaptation (disaster risk mitigation) will be critically important. As will be the setting of criteria and guidance for the measurement and operation of these projects. This will enable more capital to be invested in NBS, and more of those investments to explicitly target the use of NBS to reduce water related risks.

## 18.5   Enabling Effective Engagement with the Insurance Sector

Chapter 3 explained the multiple roles that insurers can play to support uptake of NBS and wider understanding of NAS as a strategy to promote resilience and sustainability. Multiple roles apply to both the industry itself, as well as to society at large. This section outlines several ways to effectively leverage support from the insurance sector as a partner, provider, innovator and investor for greater effect, and also points out areas that require increased consideration. These are presented as a sequence of key opportunities for effective engagement between the insurance sector and relevant stakeholders. These are grouped as 'easy wins' as well as critical (but more difficult) chances to increase water and climate security.

### 18.5.1   Scientific Exchange and Joint Action to Raise Awareness on Climate Risks

Insurance companies commonly view themselves as social actors that serve as institutional partners to support resilience in societies (Marchal et al. 2019). There is a clear concern from insurance professionals that investments related to climate change and disaster risk mitigation need to be increased, and investors (both generally and specifically those from the insurance sector) see this as an area the sector will expand in future. The sector welcomes increasing exchanges with scientists, private companies, governments and new partnerships with stakeholders involved in eco-DRR. There is high willingness and on-going research within the insurance sector to engage directly in the assessment of green infrastructure and NBS. This includes both valuation of the avoided damages and possible valuation of green infrastructure as an insurable asset. The insurance industry can also play a key role in financing studies on nature-based solutions in risk reduction and for the longer-term monitoring of natural infrastructures.

Marchal et al. (2019) further indicated through their survey that there is high willingness to increase the role of insurance companies in awareness raising activities related to the risks posed by climate change and loss prevention to respond to their customers. This role can both be a viewed as a form of good corporate governance and social responsibility, as well as a market opportunity for service provision. In some cases, the insurance sector will drive action where the sector sees risks to the sustainability of their business; in others the sector will need to be persuaded more actively to engage as knowledge provider and investor. There are cases where insurance companies act as partners, taking a lead to push for more practical actions and investments to reduce unsustainable (and un-insurable) disaster risk. Denmark Pensions and Insurance, for example, regularly engages with local authorities and has established partnerships with government and developers, so that their company can directly invest in projects to mitigate risks for properties and then recoup the costs for those investments. This form of leadership may be more of an exceptional case than common practice but underscores the importance of the regulatory environment and policy priorities set for the insurance sector and their interactions with other actors like e.g. local authorities, to function.

### 18.5.2    Policy Dialogue on Risk Reduction and Environmental Regulation

Several reports have highlighted that environmental regulation is also one of the important drivers promoting the implementation of NBS in the EU (Stagakis et al. 2019). Previous studies also identified comprehensive lists of policies and regulations relevant to investment, promotion and implementation of NBS overall (e.g. Stagakis et al. 2019; Ershad Sarabi et al. 2019), though these studies have mostly missed including specificities of national insurance schemes and corresponding legislation. At the EU level, directives specifically relevant to driving investment into NBS effective to flood and drought risks were identified by NAIAD (Joyce et al. 2018).[2] These studies also commonly note the lack of a single method or regulatory framework suited to this process. Instead the analysis recommends focusing on

---

[2] This included the following EU policy, directives and strategies: Cohesion Policy; Biodiversity Strategy; Environmental Liability Directive; Environmental Impact Assessment Directive; Strategy Environmental Assessment Directive; Adaptation Strategy; Mitigation Strategy; Water Framework Directive; Floods Directive; Habitats Directive; and Green Infrastructure Strategy.

finding ways to streamline relevant policies, plans and strategies to support NBS at each level (Somarakis et al. 2019). When initiating such a process, the insurance sector should be considered and included. Here the inputs from insurance companies as risk management service providers can be leveraged, notably through their role in assessing risk and potential avoided damages. This could serve as an enabler for investment improve knowledge on prevention.

### 18.5.3 Guidance for Insurance Companies to Contribute to Resilience Planning and Investment

This guidance may differ due to specific local, national, and project contexts. However, there are many common, large processes (e.g. in issuing of bonds, city-level adaptation investment planning, land-value capture strategies, etc.) where guidance and road-mapping can be refined and used. The NAS canvas (Mayor et al. 2021) provides a potentially useful framework that can be applied to do this.

### 18.5.4 Capitalize on the Insurance Sector as Investors

The Global Commission on Adaptation estimates that an additional 1.8 trillion USD of investment is needed over the next decade to increase resilience worldwide (GCA 2019). The GCA claims that targeting this investment to specific areas like disaster risk reduction, water management, the natural environment and more resilient infrastructure would provide over 7 trillion USD in benefits and savings. The World Bank made similar assessments, estimating 1 trillion USD needed to be invested in resilient infrastructure in low- and middle-income countries to provide more than 4 trillion USD in benefits (Hallegatte et al. 2019). The general conclusion is consistent: investment in resilience now avoids higher costs long-term.

The insurance sector is one of the largest institutional investors world. In Europe, it generates annual gross written premiums of over 1.2 trillion EURO and invests more than 10 trillion EURO in the economy (Insurance Europe 2019). Given the scope of their influence and assets, engaging the insurance sector clearly represents a very important opportunity to increase investment in resilience, including in nature-based solutions.

The role of insurance companies as investors is best seen as separate from their role as an insurer and agent for disaster risk recovery or reduction. When issuing green (or sustainability) bonds, companies have interest in finding and showcasing socially responsible investment (SRI) and are keen to find good projects with operational models to fund them and include in their sustainable investment portfolio. This requires a clear demonstration of the benefits (e.g. DRR impacts) and a viable business model from the borrower to pay back the bond. Like all investors, opportunities that offer a good combination of ROI, stability, and lowest level of risk are sought. While there is increasing interest and activity in increasing green and sustainable investment, a recent report by Share Action (Uhlenbruch 2019) noted there is greater focus of among insurance company investors on the climate change mitigation side than adaptation. The report also stated climate-data was better available for the investment branches of insurance companies than for the underwriting branches for many of the interviewed companies (ibid). These are still obviously important steps being taken. In 2017, the top 15 European insurance and re-insurance companies had total investments in fossil fuel sector of over 130 billion USD; and underwriting of fossil fuel projects and operation were of high importance (in terms the size of their total business) for a majority of those companies (Bosshard 2017). Actions taken to invest in green energy, and divest in activities as coal production and mining, are more straightforward to communicate externally and reduce reputational risks. They can also be viewed as less complex investments with lower perceived uncertainties and transaction costs than (for example, to invest in a solar panel installation project versus an NBS) those taken for adaptation and risk reduction. Incentives are needed to make financing resilience (through NBS) equally attractive as other areas of green investment.

The propensity of insurers to purchase bonds makes investments in NBS projects through green bonds a natural fit (Filkova et al. 2018). Marchal et al. (2019) found that increased investment in sustainability and resilience building actions (sustainable and responsible investments) to decrease their risks and costs (particularly under climate change) was a frequently articulated objective of insurance and re-insurance companies surveyed. Many respondents stated a strong willingness to issue green bonds, as well as to participate in sustainable finance and the circular economy. At the same time, many others indicated that it can be challenging for the insurance sector to directly invest in loss prevention (ibid). Continued efforts are therefore needed to instill confidence and comfort with investments from these institutional investors in NBS as part of NAS.

### 18.5.5    Leverage Loss Data for More Resilient Municipalities

One area with potentially high value for engagement with insurers is on the provision of loss data. The insurance sector holds information on the historical impacts of weather events, with the most detailed data set on the location and level of damages incurred and how this has evolved over time. Loss data collected by insurance companies can be an invaluable resource for municipalities looking to better plan resilient communities and mitigate their risk from natural hazards such as floods, storm surge, cloud bursts and/or drought, etc. For example, experiences in Norway and other countries in Europe show use how the use of this data could greatly improve the capacity of public authorities to invest in risk mitigation measures (see e.g. Klima 2050). The different aspects surrounding this issue can be explored at the local, national and EU levels. Insurers require a combination of positive incentives and clear regulation to ensure a level playing field for all companies to be equally required to provide data, as well as the guarantee that the privacy of customer data is duly protected.

The greatest concern from insurance companies stems from how the data can be protected. European privacy laws, for example, make it complex to share address level data. Sharing data at the address level risks the identification of an individual person. In order to avoid breaching the EU General Data Protection Regulation (GDPR), data needs to be aggregated to a level where no person could risk being identified. Otherwise, it holds the potential to disadvantage or discriminate against consumers whose properties have experienced loss previously. Importantly, this should still be done in a way that the data does not lose its value. The aggregation of data at a larger scale, and the sharing of hazard or damage maps is considered as part of the expertise that could be provided by the insurance sector in the frame of contractual agreements with e.g. municipalities. Another important concern for insurance representatives is that data can become available to competitors. This can influence the competitive edge that many insurance companies gain from the ability to out predict their competitors. The public sharing of insurance loss data could also present risks that smaller or foreign insurers will be able to infiltrate the market with little risk or investment, creating an unfair advantage for the different insurance companies. The experience from the KLIMA 2050 project (Hauge 2019) in Norway indicated that companies became more positive to sharing their loss data following multiple rounds of dialogue but stipulated that this kind of data sharing needs to be done at the request of a higher authority. This is to ensure that all companies share equally - or are equally obliged - and guarantee insurance data will be handled appropriately.

### 18.5.6 Ensure Institutional Investors Underwriting Risks Fully Consider Climate

Currently, we face an environment of increasing levels of financial risk (due to climate change) taken on by the insurance sector; but also increasing cost-competition for underwriting those risks created by investors banking on turning over investments within a shorter time frame and selling their investments for quicker profit return. This strategy enables profit for some investors, but also places them at risk to go bankrupt (at the public expense if they are not solvent). This also undermines the ability of re-insurance to accurately price risk, in line with increasing climate uncertainty. This in turn undermines our ability to guide public investment for protection in line with actual risk levels faced and maintain solvency of insurance and re-insurance markets. It also undercuts the recognition of risks and exposure that may be reduced through NAS, thereby lowering interest in investing in them. Resolving this challenge is both complex and challenging and goes beyond questions on NBS investment.

## 18.6   Conclusion

The chapter provided an overview of key issues to consider to improve enabling conditions to support the uptake of NBS and NAS designed to enhance climate and water security. This includes taking action to connect an evidence-base to an experience gap, capture full value in cost-benefit assessment, capitalize on existing and potential investor demand and create an enabling regulatory environment.

It pointed out a range of opportunity areas to promote the uptake of NBS and NAS and facilitate their financing and implementation. This includes finding solutions to de-risk private sector investment in NBS, as well as the use of explicit criteria (as well as quotas for allocated financing where relevant) for NBS, DRR and climate adaptation within green financing mechanisms.

Further opportunity areas to effectively engage the insurance sector to promote investment in NBS were also highlighted. This includes broadening partnerships for scientific exchange and cooperation, joint actions to raise awareness on climate risks, as well as policy dialogue on risk reduction and environmental regulation. More challenging but critical areas for collaboration were also elaborated, such as finding ways to better leverage the insurance sector as institutional investors, both to allocate finance to NBS projects and also to ensure that the underwriting of investments adequately consider climate change impacts on risk. Collaboration between public authorities, insurance sector actors and civil society will be critical for effective resilience planning and investment at all levels.

## References

Altamirano MA, de Rijke H, Basco-Carrera L, Arellano B (2021) Handbook for the implementation of nature-based solutions for water security: guidelines for designing an implementation and financing arrangement, DELIVERABLE 7.3: EU Horizon 2020 NAIAD Project, Grant Agreement N°730497 Dissemination

Atreya A, Hanger S, Kunreuther H, Linnerooth-Bayer J, Michel-Kerjan E (2015) A comparison of residential flood insurance in 25 countries. Working Paper of the Zurich Alliance on community flood insurance

Bosshard P (2017) Underwriting climate chaos: insurance companies, the coal industry and climate change. Published by Friends of the Earth France, Greenpeace Switzerland, Market Forces, Re:Common, Sierra Club, The Sunrise Project and Urgewald. https://unfriendcoal.com/wp-content/uploads/2017/06/Coal-insurance-briefing-paper-0417.pdf

Browder G, Ozment S, Rehberger Bescos I, Gartner T, Lange GL (2019) Integrating green and gray: creating next generation infrastructure. Washington, DC: World Bank and World Resources Institute.© World Bank and World Resources Institute. https://openknowledge.worldbank.org/handle/10986/31430. License: CC BY 4.0

EC (2019) Commission Staff Working Document. Additional information on the review of implementation of the green infrastructure strategy. Accompanying the document report from the Commission to the European Parliament, the Council, the European an Economic and Social Committee and the Committee of Regions. Review of progress on implementation of the EU green infrastructure strategy. SWD/2019/184 final

Ershad Sarabi S, Han Q, Romme L, Georges A, de Vries B, Wendling L (2019) Key enablers of and barriers to the uptake and implementation of nature-based solutions in urban settings: a review. Resources 8(3):121

Filkova M, Frandon-Martinez C, Meng A, Rado G (2018) The green bond market in Europe. Climate bonds initiative. Available at https://www.climatebonds.net/files/reports/the_green_bond_market_in_europe.pdf

Global Commission on Adaptation (2019) Adapt now: A global call for leadership on climate resilience. Available at https://cdn.gca.org/assets/2019-09/GlobalCommission_Report_FINAL.pdf

Hallegatte S, Rentschler J, Rozenberg J (2019) Lifelines: the resilient infrastructure opportunity. Sustainable infrastructure Washington, DC: World Bank. © World Bank. https://openknowledge.worldbank.org/handle/10986/31805. License: CC BY 3.0 IGO

Hauge ÅL (2019) Attitude in Norwegian insurance companies towards sharing loss data. SIWI World Water Week, Stockholm, Sverige, 25–30.08.2019. Downloaded at https://static1.squarespace.com/static/54ff1c6be4b0331c79072679/t/5d778390fe618371434635f3/1568113557410/WWW_Attitudes+to+sharing+loss+data_25.08.2019_%C3%85LH.pdf

Hindlian A, Lawson S, Banerjee S, Duggan D, Hinds M (2019) Taking the heat: making cities resilient to climate change. Published by Goldman Sachs Global Markets Institute. Downloaded at https://www.goldmansachs.com/insights/pages/gs-research/taking-the-heat/report.pdf

Insurance Europe (2019) Annual Report 2018–2019. Downloaded at https://insuranceeurope.eu/sites/default/files/attachments/Annual%20Report%202018-2019_WEB.pdf

Joycé J, Marchal R, Graveline N, Thakar K, Altamirano M, Tacnet J-M, Zorrilla P Weinberg J, López-Gunn E (2018) DELIVERABLE 8.2; first roundtable report and policy brief. EU Horizon 2020 NAIAD Project, Grant Agreement N°730497

Kousky C (2019) The role of natural disaster insurance in recovery and risk reduction. Ann Rev Resour Econ 11:399–418

Kunreuther H (2019) The role of insurance in risk management for natural disasters. In: The future of risk management. University of Pennsylvania Press, pp 267–285

Le Coent P, Herivaux C, Calatrava J, Marchal R, Mouncoulon D, Benitez-Avila C, Altamirano M, Gnonlonfin A, Graveline N, Piton G, Daartee K (2020) Is-it worth investing in NBS aiming at mitigating water risks? Insights from the economic assessment of NAIAD case studies, EGU General Assembly 2020, Online, 4–8 May 2020, EGU2020-22537, 2020. https://doi.org/10.5194/egusphere-egu2020-22537

Marchal R, Piton G, Lopez-Gunn E, Zorrilla-Miras P, Van der Keur P, Dartée K, Pengal P et al (2019) The (re)insurance Industry's roles in the integration of nature-based solutions for prevention in disaster risk reduction—insights from a European survey. Sustainability 11(22):6212. https://doi.org/10.3390/su11226212

Mayor B, Zorrilla-Miras P, Coent PL, Biffin T, Dartée K, Peña K, Graveline N, Marchal R, Nanu F, Scrieu A et al (2021) Natural assurance schemes canvas: a framework to develop business models for nature-based solutions aimed at disaster risk reduction. Sustainability 13:1291. https://doi.org/10.3390/su13031291

Nesshöver C, Assmuth T, Irvine KN, Rusch GM, Waylen KA, Delbaere B, Haase D, Jones-Walters L, Keune H, Kovacs E, Krauze K (2017) The science, policy and practice of nature-based solutions: an interdisciplinary perspective. Sci Total Environ 579:1215–1227

Somarakis G, Stagakis S, Chrysoulakis N (eds) (2019) Thinknature: nature-based solutions handbook. ThinkNature. EU Horizon 2020 programme grant agreement No 730338

Stagakis S, Somarakis G, Poursanidis D, Chrysoulakis N, Enzi (CNR) S, Bernardi A, Mortimer N, Nikolaidis N, Lilli M, Goni E, Van Rompaey S, Jurik J (2019) Policy proposals and decision-making mechanisms for risk management and resilience. Deliverable 5.4. ThinkNature. EU Horizon 2020 programme grant agreement No 730338

Uhlenbruch P (2019) Insuring a low-carbon future: a practical guide for insurers on managing climate-related risks and opportunities. Published by Shareaction. Downloaded at https://aodproject.net/wp-content/uploads/2019/09/AODP-Insuring-a-Low-Carbon-Future-Full-Report.pdf

Weinberg J, Kjellen M, Coates D (2018) Enabling uptake of nature based solutions. In: Coates D, Connors R (eds) World Water Development Report 2018: Nature based solutions for water. UN World Water Assessment Programme, Geneva

# Chapter 19
# The Natural Assurance Schemes Methodological Approach – From Assessment to Implementation

**Nora Van Cauwenbergh, Raffaele Giordano, Philippe Le Coent, Elena López Gunn, Beatriz Mayor, and Peter van der Keur**

**Highlights**

- We demonstrate that most case studies achieved high levels of technology readiness, given the large amount of data-driven and physical modelling driven approaches combining engineering and natural sciences expertise.
- A transdisciplinary approach to NBS planning and design further increased technology readiness, by generating understanding of NBS performance across stakeholders.
- Most multifaceted tailoring was needed to assess and generate institutional readiness and investment readiness.
- To cope with the inherent uncertainty of NBS and their implementation, we propose an adaptive planning and management approach to provide sufficient flexibility on the risk-benefit transfers while providing needed investment security.

This chapter is based on Van Cauwenbergh, N., Dourojeanni, P., Mayor, B., Altamirano, M., Dartee, K., Basco-Carrera, L., Piton, G., Tacnet, JM., Manez, M., Lopez-Gunn, E. (2020). *Guidelines for the definition of implementation and investment plans for adaptation*. EU Horizon 2020 NAIAD Project, Grant Agreement N°730497.

N. Van Cauwenbergh (✉)
IHE Delft Institute for Water Education, Delft, the Netherlands
e-mail: n.vancauwenbergh@un-ihe.org

R. Giordano
CNR-IRA, Bari, Italy

P. Le Coent
G-Eau, BRGM, Université de Montpellier, Montpellier, France

E. López Gunn · B. Mayor
ICATALIST S.L., Las Rozas, Madrid, Spain

P. van der Keur
Geological Survey of Denmark and Greenland (GEUS), Copenhagen, Denmark

## 19.1  Introduction: NBS and NAS Implementation Readiness

The previous chapters of this book have provided an overview of the concept and role of nature-based solutions (NBS) aimed at disaster risk reduction in view of the limitations of grey infrastructure. They illustrate how NBS deal with societal challenges and mitigate water related natural hazards while at the same time being cost effective and providing environmental, social and economic benefits to society.

Lopez-Gunn et al. (2020) developed the novel concept of Natural Assurance Schemes (NAS), defined as "ecosystem-based risk reduction measures that reduce the level of risk in one area". The central idea of NAS is that nature can ensure some assets in real monetary terms while also assuring (restoring or protecting) the ecosystems in a context of anthropogenic pressure. The question then is, how can we build NAS?

In this chapter, we aim to answer that question by discussing how NAS are set up to operationalize the assurance value of NBS, i.e. their ability to reduce the flood and drought risk while generating a series of co-benefits. Operationalizing NBS and NAS requires a context-specific understanding of drivers and barriers that exist.

To manage uncertainty, overcome barriers and capitalize on existing drivers, we propose an improved planning process for NBS and NAS that explicitly leads to implementation and investment planning. Rather than framing the process as one of overcoming barriers, we present it as a process to increase readiness for the implementation of NBS as described in detail by van der Keur et al. (2022, Chap. 1 – this volume). Following this, we further divide the readiness into three types:

- Technology readiness (TR) – linked to barriers on knowledge and absence of clear evaluation of NBS performance and uncertainties in the natural and technical system (generation of evidence) + inclusion of certain benefits such as aesthetic appeal in the design– related to setting up an appropriate level of experimentation in a context of trust. Levels run from 1 to 9 and an NBS is considered to be ready for implementation at large scale (or aggregated smaller scale) at TRL 8–9.
- Institutional readiness (IR) – linked to barriers on acceptance, trust, handling uncertainty and ambiguity, multi-functional solutions and coordination, as well as innovative regulatory frameworks to deal with the inherent uncertainty of NBS and potential liabilities. IR is positioned on the crossroads of the natural/ technical and social system and is constituted by 8 categories (e.g demand for NBS, sustainability) that exist in parallel and each have to achieve sufficient maturity for the overall institutional readiness to be achieved.
- Investment readiness (IvR) – linked to capturing multiple values and valorising the multiple benefits in public-private-people partnerships and related to funding/finance barriers and economic/financial uncertainties in the social system. IVR is related to the building of innovative business models such as the NAS canvas. In analogy to TRL, IVRL consists of 9 levels and an NBS is considered to be ready for funding/financing at IVRL 8–9.

This chapter aims to document the experiences of increasing readiness in different contexts, scales and starting conditions. We present an ex-post analysis of readiness levels before and after the application of NAS methods and tools and discuss the key lessons learned.

The insights from case studies aim to help practitioners from different disciplines to design NAS with methods and tools appropriate for the context they encounter themselves in. Key messages are (1) importance of self-check to choose the right tools/methods, (2) guidance to tailoring tools and methods to specific context. Our findings also contribute to further developments in the science-policy arena on NAS/NBS approaches and methods that are explicitly considering investment and institutional readiness, in addition to the already widespread TRL.

## 19.2   NAS Approach: From Assessment to Implementation

The NAS approach consists of the participatory step-wise creation of NAS with NBS targeting flood and drought risks. It is based on a quantitative and qualitative assessment of NBS and their implementation schemes. The approach includes a detailed assessment of risks, costs and co-benefits and formulates adaptive implementation plans that provide a blueprint for the fair distribution of investment, risks and benefits of NBS in a context of uncertainty.

### 19.2.1   Participatory Adaptive Planning Framework and Readiness Levels

To increase readiness for NAS and the integration of NBS in climate adaptation and water security plans, this handbook proposes a participatory and adaptive planning (PAP) and implementation process. The framework outlining this process is shown in Fig. 19.1 and discussed in detail in by Basco & Van Cauwenbergh et al. 2022 (Chap. 7, this volume).

At the core of the PAP approach is the recognition that for NBS to be integrated in water security and climate adaptation plans, the planning and implementation process needs to address not only technology readiness, but institutional and investment readiness as well. To increase the readiness level of innovative technology such as NBS, uncertainty needs to be managed. A number of methods and tools are used at different stages of the planning process to increase knowledge and thereby reduce uncertainty related to the process and address variability. Given that the uncertainty is not only related to variability (irreducible) and incomplete knowledge (reducible), but also to ambiguity reflected by diverse stakeholders involved, management of an agreement on information transfer between parties is key.

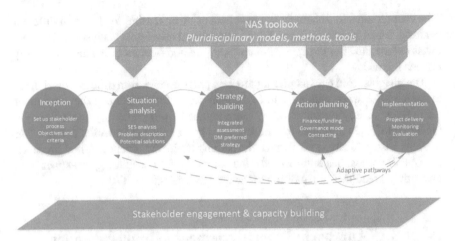

**Fig. 19.1** Participatory adaptive planning process. (Adapted from Van Cauwenbergh et al. 2020)

Those elements are structurally addressed in all the different steps of the process as to assure the level of readiness is high enough to formulate implementation and investment plans. Uncertainty management is not limited to the steps of adaptive action planning and implementation using adaptive pathways, it starts in the early phases of the planning by recognizing ambiguity in problem framing, design of scenarios and potential measures, but also in the interpretation of evidence (either from models or empirical evidence) in the integrated assessment.

### 19.2.2 NAS Framework and Selection of Methods/Tools

To generate the needed readiness to implement NBS, a suite of methods and tools have been developed and optimised for the different case studies discussed in previous chapters. As introduced in Chaps. 4, 5, 6, 7, 8 and 9, the design of NAS involves a myriad of assessments, methods and tools in an interdisciplinary and transdisciplinary approach. At the basis of NAS design, lies a structured analysis of the system, aiming to (1) identify and formulate feasible management actions; and (2) generate and present quantitative information to enable better decisions on proposed actions targeting natural assurance. The NAS toolbox combines a number of disciplines and approaches to deal with the complexity of NBS and its multiple benefits in a risk-based context. Indeed, to address this complexity, pluri-disciplinary methods and tools are needed. Apart from methods, models and tools facilitating quantitative assessment of biophysical system behaviour, economic impacts and social risk perceptions (see Chaps. 4, 5, 6 and 7), a number of semi-quantitative and qualitative methods are used to incorporate the less tangible values of stakeholders.

As for the quantitative models, the effect of NBS in the case studies (Chaps. 10, 11, 12, 13, 14, 15, 16 and 17) has been assessed using integrated hydrological modelling, including surface- and groundwater models as well as hydraulic models. Each of these models are associated with uncertainty with respect to the availability and quality of data to set up the model, run, calibrate and validate it, but also to structural uncertainty, i.e. incomplete understanding of the representation of physical processes (Refsgaard et al. 2006) and uncertainty guidelines have been developed in e.g. van der Keur et al. (2010, 2016). Policy and decision makers within disaster risk reduction and climate adaptation need transparency and guidance on the, often long-term (deep) (e.g. Herman et al. 2020), uncertainties to make informed decisions and to consider measures that could reduce uncertainty where it matters most.

Whereas some of the methods follow a more technology heavy and data demanding top-down approach, using remote sensed data or data and process intensive hydrological modelling, several of the methods are grounded in the stakeholder reality in often local environments and where results heavily depend on the quality of the stakeholder engagement process that is described below.

The use of models, methods and tools is channeled through the planning process. Understanding of complex issues is mainly aided through the use of data and modelling in the situation analysis and strategy building steps. Different models are used in this step to cover the socioeconomic, political and bio-physical dynamics of the water system to be managed and planned. These dynamics are captured in indicators that provide comparable metrics between alternative options. As different indicators are expressed in different units, and not all indicators can be monetized toward a cost-benefit assessment, multi-criteria analysis is proposed to generate integrated assessment of alternative strategies.

## 19.2.3   Stakeholder Engagement at the Core of the NAS Approach

For the models and tools to be useful in the process and generate the readiness that is needed, participatory approaches with sufficient attention to capacity building and fostering social learning are needed. Throughout the entire planning process, involvement of stakeholders is key to a number of issues. First of all it helps to assure a good understanding of the often complex issues and to handle trade-offs in a societal acceptable way. But stakeholder involvement is also necessary to anticipate and adapt to a number of implementation issues to avoid producing results that those potentially impacted will not support. Indeed, choices about managing water-related risks and other natural resources trade-offs involve more than hydrology and economics. They involve people's values, ethics, and priorities that have evolved and been embedded in societies over thousands of years (Priscoli et al. 2004). Finally, as mentioned earlier, uncertainty is not only related to variability and incomplete knowledge, but also to ambiguity in the diverse stakeholders involved.

Therefore the transfer of information between stakeholders needs to be well understood, managed and agreed upon. Stakeholder involvement brings both knowledge and preferences to the planning process—a process that typically will need to find suitable compromises among all decision-makers and stakeholders if a consensus is to be reached.

By hypothesising NBS design and implementation as a collaborative decision-making process, we assume three premises: (i) NBS design and implementation need to be based on inclusive and equitable participatory processes, capable to ensure the active involvement of all different categories of stakeholders and decision-makers; (ii) collaborative decision-making for NBS implementation requires a clear understanding of the ambiguity among different decision-makers in perceiving and valuing NBS co-benefits (Giordano et al. 2020); (iii) decision-makers do not take decisions in a vacuum, but social interactions can alter preferences, choices and hence decisions (Kolleck 2013; Siegel 2009; Sueur et al. 2012).

Nevertheless, divergences in values, beliefs and problem frames may lead to collaboration structures that encourage stakeholders and decision-makers to avoid each other, turning the participatory process into a controversial and futile process (Brugnach and Ingram 2012; Giordano et al. 2017; Howe et al. 2014; Jacobs 2016; Shrestha and Dhakal 2019), resulting in a barrier to NBS (Eisenack et al. 2014; Therrien et al. 2019).

Most of the approaches described in the literature concerning conflicts analysis and resolution assume that conflicts among decision-actors derive from ambiguity in problem framing and non-conformity in their individual objectives and preferences towards alternatives. However, through effective interaction mechanisms, different decision-actors tend to align their problem frames, overcoming the barriers caused by ambiguity in problem framing. Conflicts may not occur between decision-makers with a rather different problem frame and good relationships (Liu et al. 2019).

## 19.3 Methods: Ex-post Analysis of NAS Using an Integrated Readiness Framework

We tested the above described NAS approach for readiness creation in a number of European case studies, ranging from small scale NBS for flood and drought management (e.g. hybrid UBW, the Netherlands), to large scale projects with focus on either drought (e.g. Medina, Spain and Danube, Romania) or floods (e.g. Lez, France and Glinščica, Slovenia, Copenhagen, Denmark). Details of these case studies are described in Chaps. 10, 11, 12, 13, 14, 15, 16 and 17 of this book.

For a number of selected case studies, we performed an ex-post self-assessment of NBS readiness both before and after implementing a series of methods and tools to support NAS design. The self-assessment uses the definition of readiness levels provided in Sect. 19.1 and attributes LOW, MEDIUM or HIGH readiness for each of the levels.

We then discuss the changes in readiness achieved in relation to the case studies' (1) varying biophysical conditions, spatial scale and vulnerability to water related natural hazards that require diverse NAS approaches, but also to (2) varying starting levels of technology (TRL), institutional (IRL) and business (IVRL) readiness for implementation of nature based solutions in NAS.

### 19.3.1   Selected Case Studies

Below, we briefly describe the context and projected NBS for the case studies that were selected for an in-depth analysis of readiness generation.

- Urban water buffer in the Spangen area, Rotterdam, the Netherlands: this case study is discussed in detail by Dartee et al., 2022 (Chap. 16 – this volume) and concerns a hybrid NBS at urban neighbourhood scale targeting flood and drought risk. It consists of an innovative grey underground water storage for buffering storm-water runoff with controlled release into a natural filtration system which creates a green space for the benefit of the community. The treated water is then stored subsurface in an aquifer bubble and pumped up when needed for use in the neighboring football stadium. The local water operator, municipality and regional water authority all manage part of the NBS.
- Copenhagen restored urban river scenario, Denmark: this case study is discussed in detail by Jørgensen et al., 2022 (Chap. 17 – this volume) and concerns a restored urban river stretch scenario to lessen the risk for urban groundwater flooding which result from high and rising groundwater levels due to changed patterns in water use, sewage system management and climate change. Key stakeholders involved are the city of Copenhagen (and adjacent Frederiksberg), Copenhagen Water Utility (HOFOR), an insurance and pension umbrella organization, national and regional authorities, environmental NGOs, legal advisors and urban planners.
- Lez watershed (including the city of Montpellier), France: this case study at medium basin and (peri-) urban scale discussed by LeCoent et al. 2022 (Chap. 14 – this volume) explored whether different scenarios of green infrastructure (water retention basins, bioswales, green roofs…) and conservation of peri-urban natural and agricultural land, considered as NBS scenarios may reduce runoff flood risks and address climate adaptation challenges. Key stakeholders involved are Montpellier city, the Lez river basin authority, CCR (French reinsurer), local communities, environmental associations, and local and national government.
- Glinščica, Slovenia: this case study at medium scale discussed by Pengal et al., 2022 (Chap. 15 – this volume) explores NBS river restoration and management to reduce flood risk in the Glinščica Stream, upstream of Slovenia's capital, Ljubljana. The torrential character of this river, together with advancing urbanization, climate change (less frequent, but higher intensity rainfall) and hard regulations, results in regular flooding of the Vič and Rožna dolina districts of Ljubljana.

- Medina del Campo aquifer recharge, Spain: in this large scale case study discussed by Mayor et al., 2022 (Chap. 11 – this volume) a number of different NBS were considered to deal with increased flood and drought risk; Managed Aquifer Recharge (MAR), change of crops, agricultural soil conservation, and water reuse. Stakeholders involved are the Duero river basin authority, regional and provincial government as well as associations in the environmental and agricultural sectors, local cultural associations, municipal councils, universities, private companies and civil protection.
- Lower Danube basin, Romania: in this large scale case study discussed by Scrieciu et al., 2022 (Chap. 10 – this volume), the focus was on identifying NBS to reduce natural hazards, mainly focusing on flood management, but also on reducing drought and desertification aggravated by climate change. Involved stakeholders are the ministries of environment, water, agriculture and rural development as well as the national administration of Romanian waters, General Inspectorate for Emergency Situations, the National Association of Insurance and Reinsurance Companies, the Lower Danube River Administration and Local authorities & NGOs.

### 19.3.2   Checklist of Questions

To assess how readiness was created (or not) in the selected case studies, we performed a readiness assessment before and after the NAS approach and analysed how the NAS toolbox and the larger (changes in) context contributed in the creation of readiness. Findings are qualitative and based on a self-assessment by the leading researcher of each case study, using the checklist of questions in Table 19.1.

## 19.4   Results: Assessment of Readiness and Its Increase Using the NAS Approach

Using the checklist above, key experts of each of the case studies assessed the technology, institutional and investment readiness before and after the interventions of the NAS approach. Table 19.2 summarizes the assessment and lists the key methods and tools used. In continuation, we discuss how the NAS approach in general and the specific methods/tools mobilized have contributed to the increase in readiness.

### 19.4.1   Urban Water Buffer, The Netherlands

This case study started at high readiness levels. The technology of water storage and bioremediation – infiltration had been tested at lab scale. Institutionally, there had been prior experience in the municipality with building green infrastructure as part

**Table 19.1** Practitioners' checklist for participatory NBS implementation/investment planning and relation to readiness (*darker colors = higher contribution to readiness*)

| Phases | Actions/Information to be gathered | Check list questions | Readiness | | |
|---|---|---|---|---|---|
| | | | T | IV | I |
| **Inception phase** | | | | | |
| Set up the stakeholder process | Identify and characterize (key) stakeholders and their interlinkages<br>Define core working group<br>Agree level of involvement and modes of communication<br>Assure continuity in the process | Support for adaptation guaranteed at high level<br>A core team on adaptation in place<br>Institutional cooperation set up<br>All affected stakeholders involved<br>Human and financial resources secured in the long term<br>Target group-specific formats for awareness raising carried out | | | |
| Define Objectives, criteria | Based on prior assessment of risks and existing policies, what are key objectives of NBS in the hydro-social system (reduced risk as primary objective/benefit and increased socio-economic opportunities as co-benefits)<br>Identification of criteria/yardsticks that allow you to evaluate whether the objectives are reached? | Appropriate indicators developed | | | |
| **Situation analysis** | | | | | |
| Assess baseline<br>Situation analysis | Description of natural, socio-economic and institutional system<br>Identification of problems and controversies<br>Stakeholder analysis, DPSIR<br>Assessment of ambiguity and risk perception | A first overview on climate-related impacts gained<br>Ongoing activities with relevance for adaptation identified<br>Common understanding on climate change adaptation gained<br>A systematic overview on past weather events, their consequences and response actions in place<br>Understanding of future climate change gained<br>Non-climatic stress factors identified and considered<br>Main concerns are identified that require an adaptation response<br>Transboundary issues considered<br>Knowledge gaps and uncertainties in climate change summarized and made explicit | | | |

(continued)

**Table 19.1** (continued)

| Phases | Actions/Information to be gathered | Check list questions | Readiness | | |
|---|---|---|---|---|---|
| | | | T | IV | I |
| **Inception phase** | | | | | |
| Identify indicators, methods and tools | Identification of quantitative and qualitative assessment needed to provide relevant information (related to policy objectives and stakeholder preferences)<br>Checklist of available methods/tools | An approach on how to deal with uncertainties developed | | | |
| Scenarios | Storylines of possible futures (drivers of change, external to the system)<br>List of drivers and translation into quantitative scenarios to be modelled | | | | |
| **Strategy building** | | | | | |
| Formulate solutions (measures, alternatives) | Identification of interventions/measures/alternatives in collaboration with key stakeholders<br>(Participatory) prioritization of promising measures that will be modelled | Gaps and barriers that hindered an adequate response in the past identified and understood<br>A full portfolio of adaptation options considered<br>Measures considered are aligned with existing regulations and policy priorities (SBC) | | | |
| Combine priority measures into strategies | Separation of interventions in short/medium/long term | Suitable adaptation options were described in detail (SBC)<br>Clear scope of preferred solution (SBC) | | | |
| Assess strategies under different scenarios, including capacity building | Quantification of indicators (including life cycle costs)<br>Identification of no-regret interventions (based on scenarios)<br>Definition of no-regret strategies | A prioritisation system of adaptation options developed in cooperation with stakeholders<br>Reliable information on hydro-meteorological risk reduction by measures (SBC)<br>Possible synergies and conflicts identified and taken into account | | | |
| Decision making (negotiation, choice and compensation): select preferred strategy | Communication of indicator results for different strategies<br>Use of different decision-making methods (e.g. MCA, CBA, scorecard, voting, …)<br>And (group) decision making aided by the assessment scores provided in previous step | Cost-benefits of options assessed<br>Reliable evidence of positive cost-benefit considering direct and indirect costs (EBC)<br>Preferred adaptation options selected for implementation (EBC)<br>Adaptation strategy developed and politically adopted (EBC)<br>Direct beneficiaries and willingness to pay identified (EBC)<br>Compensation for losers identified (EBC) | | | |

| Action planning | | |
|---|---|---|
| Enabling environment, develop business plan and distribute responsibilities | Discussion of role and responsibilities for the implementation of the preferred strategies<br>Timeline of actions<br>Identification of funding/finance<br>Development of business plan | Action plan developed<br>Steps for implementation set<br>Timeline of actions and capital investment clearly defined (CBC)<br>Risk and liabilities are clarified (CBC)<br>Identification of life-cycle cost (FBC) |
| Mobilize financing and funding | Definition of service levels, performance indicators and contract negotiation<br>Risks and liabilities<br>Funding/Finance strategy | Quantification of revenue streams considering tariffs, transfers and taxes (FBC)<br>Access to funding or financial instruments to close funding or financing gaps (FBC)<br>Responsibilities of organizations in procurement, construction and operation are clarified (MBC) |
| Implementation | | |
| Implement | Construction of green-grey infrastructure and enforcing regulatory arrangements | Contracts needed and procurement strategy are clear (MBC) |
| Monitor and evaluate | Data collection on performance indicators<br>Communication of monitoring results<br>Check of evolution monitoring levels with initially established service levels and NBS objectives | Appropriate M & E provisions for both your adaptation policy's objectives and selected adaptation options developed<br>Appropriate indicators from step 1 checked and operationalized in M & E plan |

Own elaboration after European Commission 2013; HMTreasury 2018; Altamirano et al. 2021

*SBC* Strategic business case, *EBC* economic business case, *CBC* Commercial business case, *FBC* financial business case, *MBC* management business case, *TRL* Technology readiness level, *IVRL* investment readiness level, *ISRL* intitutional readiness level

**Table 19.2** Readiness level before and after the NAS approach

| Readiness level | T | Iv | I | Methods/tools used | T | Iv | I |
|---|---|---|---|---|---|---|---|
| Urban water buffer, NL | 6 | 5 | 6 | Hydrological modelling, stakeholder workshops, willingness to pay surveys | 8 | 9 | 8 |
| Copenhagen city plan, DK | 6 | 5 | 6 | Hydro-geological modelling, economic assessment, system dynamics modelling and social risk perception | 8 | 7 | 8 |
| Lez basin, FR | 6 | 5 | 5 | Runoff flooding risk modelling, economic assessment suite of methodologies, economic valuation and perception of co-benefits | 7 | 8 | 7 |
| Glinščica, SL | 5 | 4 | 4 | Hydrologic and hydraulic modelling, implementation of Free Station monitoring, stakeholder involvement | 8 | 7 | 8 |
| Medina aquifer recharge, ES | 4 | 4 | 4 | Hydro-geological modelling, ecosystem services assessment, economic modelling, stakeholder workshops, integrated modelling (meta-model) and scenario analysis, social risk perception and system dynamics modelling, NAS canvas generation | 5 | 7 | 7 |
| Danube delta, RO | 5 | 4 | 4 | Hydro-geological modelling, ecosystem services assessment, economic assessment, NAS canvas generation and framework for funding and finance, | 7 | 6 | 6 |

*T* Technology, *Iv* Investment, *I* Institutional

of climate adaptation plans and there was interest from the neighborhood organizations who wanted to increase green spaces in the area. Finally, the investment readiness started at a high level as well, with support from a technology and innovation fund (TKI) toward the design of the system and interest by the neighboring football stadium to buy the water once the system would be built. As a large water user, the football stadium was interested in reduced costs for irrigating its field. Through the NAS approach, the readiness was further increased through a series of workshops, interviews and co-design sessions that looped a number of key stakeholders into the conversation on detailed design and operation of the system and by doing so, furthered the confidence of the stakeholders that the proposed scheme would work. The conversation was championed by a local actor at the municipality, who contributed largely to overcoming the disconnects between municipal silos and highlighted the potential co-benefits of the project as it would be contributing to government programs around resilience and climate adaptation, while reducing the flood risk in a non-privileged area of the city. Investment in the building and implementation of the scheme was further secured by the connection between the football stadium as large water user and the water utility company Evides that joined the project triggered by the support of the regional water authority and supporting the incorporation in their network of the bioremediated water stored underground.

## 19.4.2   Copenhagen City Plan, Denmark

The case study identified the potential for river restoration, including an estimation of avoided costs of groundwater flooding induced damage for insurance companies, the city of Copenhagen and citizens. Also, barriers for NBS implementation were identified by stakeholder involvement and subsequently analysed by participatory modelling. The Copenhagen case study (Jørgensen et al., Chap. 17, this volume) contributed to increasing the technological readiness level (TRL) linked to advancing knowledge and performance with respect to developing a hydrological modelling approach for exploring the effect on rising groundwater level by reestablishing an urban river from a currently piped stream in Copenhagen. The developed integrated surface- and groundwater hydrological model can subsequently be (re) applied to evaluate additional and new scenarios including climate change and decisions by the municipality for climate adaptation measures. As the modelling tool is physical based it can be applied in other (urban) environments as well. The investment readiness level (IVRL) has been addressed by considering a simple damage function approach by combining hydrologically modelled effects on groundwater level as a result of the draining effect of the restored urban river NBS scenario with reported insured damage. Assumptions on how shallow groundwater levels relate to incurred damage makes it possible to value the avoided damage which affects the IVRL. The results show that by reopening the river, an economic benefit is obtained because the river now functions as a drainage channel which prevents flooding by groundwater of subterranean structures, notably housing cellars, and potentially as a recipient of stormwater events by connecting to cloudburst management measures. Valuation of co-benefits in addition to avoided damage is anticipated to contribute substantially to the IVRL. Finally, the institutional readiness (IRL) is explored by the integration of stakeholder's knowledge in the co-design and implementation process of NBS to support complex decision-making processes and was carried out by (1) participatory modelling activities to elicit and structure stakeholder's risk perception, (2) mapping the interaction among decision-makers and stakeholders, and by (3) deriving Fuzzy Cognitive Maps (FCM) from Group Model Building. The FCM simulation showed that NBS implementation requires effective cooperation among different decision-makers to define potential interventions and to reduce the level of conflicts and to facilitate collaborative decision-making.

## 19.4.3   Lez Basin, France

In the Lez study (Le Coënt et al., 2022 – Chap. 14 this volume), the Lez basin demonstration site at the at the watershed/city scale showed the potential of scenarios of NBS (green infrastructure and urban sprawl control) at the watershed scale to reduce urban flooding risks and address territorial challenged. The TRL was increased by designing spatially-explicit green infrastructure development scenarios and modeling their impact on urban flood hazard. The economic assessment revealed that NBS could reduce flood damage cost by 14–20%. In addition, a survey with 400 citizens demonstrated the large value granted by residents to NBS co-benefits, notably climate change mitigation, landscape conservation and air quality improvement. Overall the cost-benefit analysis revealed the economic interest of a large NBS programme as well as the magnitude of revenue streams that should be mobilized to finance NBS (increased IVRL). The institutional readiness (IRL) was addressed and strengthened through the involvement of stakeholders to increase the knowledge on the potential of NBS to mitigate flood risk and other challenges and to help identifying potential strength and barriers for implementation. To increase IRL, the results of this study will need to be translated into strategies/programs led by municipalities of the watershed in which smaller scale projects at the neighborhood scale may be developed for concrete implementation. The leadership of municipalities is key to increase IRL and reach that new step of implementation.

### 19.4.4 Glinciska River Basin, Slovenia

The Slovenian demonstration site (Pengal et al., 2022 – Chap. 15, this volume) considered as NBS measures to mitigate flooding hazards: retention areas, re-meandering of the river and wetland restoration in the Glinščica catchment area. While none of NBS technologies are new or unproven, implementation and evaluation of NBS strategies were enhanced and supported by integrated HEC-HMS - HEC-RAS and FEV based hydrological/hydraulic rainfall-runoff modelling. The increased know-how for achieving this in combination with implementing FreeStation multifunctional monitoring of the effects of implemented NBS increased the TRL. In order to assess the investment readiness, an economic analysis was performed to compare business as usual (BAU) with NBS strategies over a 30-year timeframe. The cost of NBS strategies were approximately 60% lower that BAU, although large barriers for implementation remain. Institutional barriers include poorly coordinated institutions at several levels, in-effective regulatory and legislative frameworks, but stakeholder based consultation and demonstration of co-benefits increased awareness and may decrease uncertainty and ambiguity on NAS and NBS and contribute to increased institutional readiness (IRL) on the longer term.

### 19.4.5 Medina Aquifer, Spain

Technological readiness level has been increased substantially through a geo-hydrological and geophysical assessment of managed aquifer recharge (MAR) based ecosystem services as well as the role of groundwater sustained ecosystem services. Stakeholder workshops were then organized to co-create viable business options for the public/private financing of management measures that increase the Investment readiness level (IVRL) of groundwater related ecosystem services.

The demonstration site identified a number of feasible technical and institutionally supported NBS strategies to positively contribute to the institutional readiness level (IRL) and evaluated the acceptability of the NBS solutions. Apart from a series of structural NBS, stakeholders in this case study ranked "increase awareness and environmental education" as well as "Regulatory fees and improving users' organization" as most appropriate to deal with the increasing climate variability in this area.

### 19.4.6  Danube Floodplain, Romania

In the Danube case study, the NAIAD project aimed to create an efficient network of stakeholders trained to apply the methods and scenarios identified in the project, in order to promote sustainable development for extreme events mitigation by using the ecosystems services, ecological (re)construction and green solutions. The technical readiness level for Danube floodplain restoration NBS planning scenarios was substantially supported by integrating local stakeholder knowledge in hydraulic modelling (HEC-RAS) for assessing river flooding vulnerability. Stakeholder knowledge was incorporated by means of two workshops from which a causal loop diagram (Vensim model) was derived to explain and support expected impacts, benefits and co-benefits of planned NBS. This also supported the increase in institutional readiness. Finally, the investment readiness was supported by assessing the economic parameters related to damage as a consequence of flooding with and without the implementation of floodplain restoration NBS. The economic assessment was based on a GIS aided analysis and collected information from various sources on flood damage.

## 19.5  Discussion

In this section we discuss how the different elements of the NAS approach facilitate the increase in readiness for NBS/NAS to reduce flood and drought risk. We divide the discussion in reflections on the methods/models and tools used and on the participatory process. We then discuss some lessons learned and provide recommendations for the use of the approach in different contexts.

### 19.5.1  The NAS Toolbox and Contribution of Methods and Tools to Technology and Investment Readiness

Technology readiness is the first necessary step to ensure consideration by local decision makers of the relevance of NBS for water risk management. In the case of flood risk, the civil engineer culture remains dominant and the demonstration of the effectiveness of NBS for flood risk management remains a challenge. In the case studies above, the modeling of the effectiveness of NBS as compared to grey solutions for the reduction of flood risk has been key in the pathway towards implementation, especially in those case studies initially strongly biased towards grey solutions, as for example in the Brague case (Chap. 13 – this volume). The assessment of the effectiveness of NBS also provides the basis for the economic evaluation of NBS.

The economic assessment compares elements to evaluate the magnitude of the costs and benefits generated by NAS. It is built on the preliminary assessment of the effectiveness of NBS using key indicators, whose monetary value is subsequently evaluated. The proposed Cost-Benefit analysis method (Chap. 6), helps (i) identifying whether a given NAS presents positive net benefits, (ii) determining among different NAS which one is preferable from an economic standpoint. The economic assessment also helps identifying the magnitude of the different benefits of the NAS, which is the basis to identify revenue flows and a viable business model, necessary to achieve investment readiness. Some indicators such as non-monetary impacts on water risks and co-benefits that can not or only partially be valued monetarily such as social and environmental indicators are fundamental in NBS assessment and the decision making process for the development of NBS. Economic assessments of NBS should therefore be complemented with other integrative approaches such as Multi-Criteria Decision Analysis (MCDA) described in Chap. 7.

Investment readiness can be pursued through the generation of the NAS business canvas (Mayor et al. 2021) and can be translated in investment plans built around the 5 business cases for water security proposed by (Altamirano et al., 2020), which are further discussed by Mayor et al. (2022 – Chap. 7, this volume). When analysing the different boxes of the business canvas, it becomes clear how investment readiness is generated in the planning process; from the start of the inception phase, throughout situation analysis, strategy building and action planning. A clear understanding of the monitoring and evaluation as well as how different parties of the public, private, and communities are related to it, is further increasing investment readiness.

## 19.5.2   Importance of Capacity Building and Stakeholder Engagement for Institutional Readiness

Experiences in the different case studies show that capacity and readiness building is key for the creation of an implementable NBS. Given the multitude of stakeholders involved project and their multiple objectives and interests as well as knowledge frameworks. Different levels of capacitation will relate to different targets for the readiness (individual, group and institutional). With the participatory process in PAP we are mainly aiming at capacity building at group level (the multi-stakeholder platform) and at the institutional level (which directly relates to institutional readiness).

The *ex-post* assessment of the activities in the selected case studies demonstrate to what extend the PAP process allowed pursuing the three key elements of a stakeholders' engagement process for NBS/NAS effective implementation, i.e. (i) equitable engagement of different stakeholders; (ii) based on a clear understanding of ambiguity in problem framing and risk perception; and (iii) enabling cooperation among different institutional actors. Pros and cons of the adopted approaches are discussed further in the text.

The elicitation and analysis of the different risk perceptions and problem understanding (for more details on the implemented methodologies, please, refer to Chap. 5 of this book) were at the basis of the stakeholders' activities in several demo sites. These activities contributed in enhancing the institutional readiness level. Specifically, the implementation of the PAP process contributed in making clear that different stakeholders' needs and concerns need to be accounted for during the NBS design phase. Contrary to most of the works mentioned in the scientific literature, in which NBS are mainly described as solutions for addressing different risks, the experiences carried out in the case studies demonstrated that the co-benefits are, in many cases, as important as the risk reduction itself. Therefore, accounting for the stakeholders' co-benefits perceptions and valuation since the NBS design phase is of utmost importance. NAS activities demonstrated the suitability of disciplined methods and tools that facilitate stakeholders' dialogue and help reflecting on the different sources of ambiguity in co-benefits definition and valuation.

Among the different enabling elements supporting institutions in dealing with a complex issue such as NBS implementation, the institutional cooperation demonstrated to play a key role in different case studies. NBS implementation requires effective flow of information and knowledge among the different institutional and non-institutional actors. Lack of trust or limited understanding of the role played by the others, could hamper the cooperative implementation of important tasks required for the NBS implementation. It is worth noting that the improvement of institutional cooperation has been defined as one of the most important steps for the institutional readiness by several stakeholders during NBS implementation. Specifically, we learned that, in order to be effective in reducing water-related risks and in producing the expected co-benefits, NBS implementation needs to be supported by several socio-institutional measures, claiming the involvement and cooperation of other institutional actors. Finally, the experiences carried out in the Copenhagen demo demonstrated that, in urban areas, NBS need be thought as a part of an urban systemic interventions' strategy, whose implementation requires the cooperative intervention of different decision-actors.

### 19.5.3   Lessons Learned for NAS Building in Europe and Other Contexts

Our results also point to some important implications for NBS uptake. For one, our detailed case study analysis showed that decision support models and tools were only marginally used during the planning and implementation process.

Government actors did not rely on the extensive cost-benefit and multi-criteria assessments that were available, focusing on political and institutional issues instead. This is somehow contradicting (Droste et al., 2017), who emphasizes the importance of a comprehensive assessment of the multi-functionality of NBS through elaborated cost-benefit or multi-criteria assessment methods. Findings suggest that for NBS uptake it is far more important to have willingness and commitment from the key stakeholders. Nevertheless the need for evidence on cost-benefit ratios of the NBS in the case studies was highlighted during a mock funding pitch at a January 2020 stakeholder meeting in Copenhagen. The repetitive feedback of experts from the private and public funding and financing community (such as TNC, EIB, and private investors) here was that costs and benefits of the proposed projects should be better evidenced before investors. This indicates that importance of evidence might arise at later stages of the NBS planning process and also toward upscaling, calling for support by above mentioned methods and tools.

Secondly, we found that co-benefits can be a driver for success when the funding is available, a clear owner of the NBS project exists and there is a concretized level of service. In the case of Rotterdam, the NBS' ability to generate cheaper water supply for the sport arena nearby, leveraged the needed support for TKI funding and ownership, with flood reduction and recreational value as co-benefits functioning as leverage for the willingness and acceptability of the project by other stakeholders. In cases where the added value of the NBS is not clearly linked to an existing operator, co-benefits have to play a stronger role. This was for example the case of the Lez and Braque demo, where public-good co-benefits (air quality improvement, biodiversity, climate regulation) represent the largest value given by residents to NBS scenarios but may be more challenging to turn into revenue streams for project funding as potential mobilizer of institutional support. However more in-depth analysis is needed in all demos to see whether co-benefits can play this role in general.

Thirdly, we made a number of observations on the aspect of integration that underlies successful planning and implementation of NBS. Case study analysis shows a reality where objectives and related indicators are driven by sectoral interests. This makes that what is defined as a benefit or co-benefit depends on the viewpoint of the stakeholders involved. In the Rotterdam case, the decision making on the NBS was defined by the leading organization (related to mandate and funding) and the clear risk/benefit cycle (involving Evides and Stadium) proved crucial to facilitate that decision making (see point above). The case shows that institutional coordination is a key barrier to implementation (and that this is happening even within the municipality). Finally we observed that in order to mainstream the NBS, evidence of performance across (co-) benefits is needed. However, little to no monitoring incentives or interest exists.

Our findings show that the investment and institutional readiness are an important factor to consider in the mainstreaming of NBS and NAS. While TRL are generally higher at the start of the projects, large differences existed in the IRL and IvRL and tools and methods need to be adapted to address this appropriately.

This has implications for the implementation of NBS in non-EU contexts. A study of the relation between NBS and policy support by Van Cauwenbergh et al. (2021) highlight that while international policies such as Sendai, the Paris agreement and SDGs are generally favorable for the integration of NBS into NAS, policy support at national and regional level are equally important. Indeed, for NBS to be integrated into operational management plans at different scales, they need to be linked to the practices and policy frameworks at lower institutional levels. Likewise, the presence of funding and financing opportunities is a fundamental condition for the implementation of NBS and NAS. While nature restoration and ecosystem-based investment is starting to become accepted in more developed nations, earmarking funds in less developed nations is challenging. It remains to be seen to what extent global and international finance players and investment funds such as the Green Climate Fund and the Natural Capital Financing Facility are able to promote mainstreaming of NBS in context with low national and regional investment capacity.

## 19.6   Conclusion and Recommendations

This chapter set out to discuss the methodological approach for natural assurance schemes (NAS) in a broad range of case studies. Incorporating inter- and transdisciplinary approaches in a structured participatory adaptive planning process, we discuss and assess how the step-wise use of multiple methods and tools in combination with stakeholder engagement and capacity building, is able to increase readiness for NBS and NAS. To structure and support this process, (i) technological, (ii) investment and (iii) institutional readiness levels are considered to assess the potential of NBS operationalization in different physical, socio-economic and institutional settings. This is demonstrated for contrasting cases to facilitate upscaling and replication.

Results of selected case studies show the assessment of investment, institutional and technology readiness before and after the participatory adaptive planning (PAP) approach. The PAP approach is endorsed to address the inherent uncertainty in the NBS implementation process and in turn, increase readiness. It has been demonstrated that most case studies have achieved substantial technology readiness, given the large amount of data-driven and physical modelling driven approaches combining engineering and natural sciences expertise. Hydrogeological and hydraulic modelling techniques were applied from urban scale to large floodplain scale and physically based assessments obtained of NBS effects to mitigate water related hazards. In addition, system dynamic modelling mapped stakeholder risk perception and the interaction among stakeholders and decision maker in the planning process, to support the assessment of institutional readiness.

Obtained knowledge and experience from the included case studies showed that most multifaceted tailoring was needed to assess and generate institutional readiness and investment readiness. Institutional readiness is generated throughout the entire planning and design process, through a combination of joint assessment of risk perceptions, crafting of institutional set-up and facilitation of awareness and agreement on responsibilities in the NBS planning process. Investment readiness is supported through the generation of the NAS business canvas to highlight the value proposition and opportunities for risk-benefit transfers in a regulated environment. The NAS canvas can be translated in investment plans built around the 5 business cases for water security proposed in the Financing Framework for water security.

Recommendations from the work presented in this book take point of departure in the developed stepwise approach to assist in generating the natural assurance schemes, demonstrated in case studies at contrasting scales as a guideline for NBS planning and using the concept of technological, investment and institutional readiness in the participatory and adaptive planning process. Considering the inherent uncertainty of NBS and their implementation in the future (related to the multitude of actors involved and the dynamic nature of NBS performance), the proposed adaptive planning and management approach aims to provide sufficient flexibility on the risk-benefit transfers while providing needed investment security. These findings provide operational guidelines for practitioners and researchers to facilitate the creation of NAS.

# References

Altamirano MA, de Rijke H, Basco-Carrera L, Arellano B (2021) Handbook for the implementation of nature-based solutions for water security: guidelines for designing an implementation and financing arrangement, DELIVERABLE 7.3: EU Horizon 2020 NAIAD Project, Grant Agreement N°730497 Dissemination (1st edn)

Brugnach M, Ingram H (2012) Ambiguity: the challenge of knowing and deciding together. Environ Sci Pol 15(1):60–71

Eisenack K, Moser SC, Hoffmann E, Klein RJ, Oberlack C, Pechan A et al (2014) Explaining and overcoming barriers to climate change adaptation. Nat Clim Chang 4(10):867–872

European Commission (2013) Guidelines on developing adaptation strategies. https://ec.europa.eu/clima/sites/clima/files/adaptation/what/docs/swd_2013_134_en.pdf

Giordano R, Brugnach M, Pluchinotta I (2017) Ambiguity in problem framing as a barrier to collective actions: some hints from groundwater protection policy in the Apulia region. Group Decis Negot 26(5):911–932. https://doi.org/10.1007/s10726-016-9519-1

Giordano R et al (2020) Enhancing nature-based solutions acceptance through stakeholders' engagement in co-benefits identification and trade-offs analysis. Sci Total Environ 713:136552. https://doi.org/10.1016/j.scitotenv.2020.136552

Herman JD, Quinn JD, Steinschneider S, Giuliani M, Fletcher S (2020) Climate adaptation as a control problem: review and perspectives on dynamic water resources planning under uncertainty. Water Resour Manag 56. https://doi.org/10.1029/2019WR025502

HMTreasury. (2018) In: Lowe J (ed) Guide to developing the project business case – better business cases for better outcomes. HM Treasury

Howe C, Suich H, Vira B, Mace GM (2014) Creating win-wins from trade-offs? Ecosystem services for human well-being: a meta-analysis of ecosystem service trade-offs and synergies in the real world. Glob Environ Chang 28:263-275

Jacobs S (2016) The use of participatory action research within education – benefits to stakeholders. World J Educ 6(3):48–55

Kolleck N (2013) Social network analysis in innovation research: using a mixed methods approach to analyze social innovations. Eur Journal Futur Res 1(1):1–9

Liu B, Zhou Q, Ding RX, Palomares I, Herrera F (2019) Large-scale group decision making model based on social network analysis: trust relationship-based conflict detection and elimination. Eur J Oper Res 275(2):737–754

Mayor B, Zorrilla-Miras P, Coent PL, Biffin T, Dartée K, Peña K, Graveline N, Marchal R, Nanu F, Scrieu A, Calatrava J, Manzano M, López Gunn E (2021) Natural assurance schemes canvas: a framework to develop business models for nature-based solutions aimed at disaster risk reduction. Sustainability 13:1291. https://doi.org/10.3390/su13031291

Priscoli JD, Dooge J, Llamas R (2004) Water and ethics. United Nations Educational, Scientific and Cultural Organization, Paris, p 33pp

Refsgaard JC, van der Sluijs JP, Brown J, van der Keur P (2006) A framework for dealing with uncertainty due to model structure error. Adv Water Resour 29(11):1586–1597. https://doi.org/10.1016/j.advwatres.2005.11.013

Shrestha S, Dhakal S (2019) An assessment of potential synergies and trade-offs between climate mitigation and adaptation policies of Nepal. J Environ Manag 235:535–545

Siegel DA (2009) Social networks and collective action. Am J Polit Sci 53(1):122–138

Sueur C, Deneubourg JL, Petit O (2012) From social network (centralized vs. decentralized) to collective decision-making (unshared vs. shared consensus). PLoS One 7(2):e32566

Therrien MC, Jutras M, Usher S (2019) Including quality in Social network analysis to foster dialogue in urban resilience and adaptation policies. Environ Sci Pol 93:1–10

Van der Keur P, Brugnach M, DeWulf A, Refsgaard J-C, Zorrilla P, Poolman M, Isendahl N, Raadgever GT, Henriksen HJ, Warmink JJ, Lamers M, Mysiak J (2010) Identifying uncertainty guidelines for supporting policy making in water management illustrated for Upper Guadiana and Rhine sub-basins. Water Resour Manag 24:3901–3938. https://doi.org/10.1007/s11269-010-9640-x

van der Keur P, van Bers C, Henriksen HJ, Nibanupudi HK, Yadav S, Wijaya R, Subiyono A, Mukerjee N, Hausmann H-J, Hare M, Terwisscha van Scheltinga C, Pearn G, Jaspers F (2016) Identification and analysis of uncertainty in disaster risk reduction and climate change adaptation in South and Southeast Asia. Int J Disaster Risk Reduct 16(2016):208–214. https://doi.org/10.1016/j.ijdrr.2016.03.002

# Chapter 20
# Looking into the Future: Natural Assurance Schemes for Resilience

Elena López Gunn, Nina Graveline, Raffaele Giordano, Nora Van
Cauwenbergh, Philippe Le Coent, Peter van der Keur, Roxane Marchal,
Beatriz Mayor, and Laura Vay

**Highlights**

Main lessons learned include the new knowledge acquired, its integration and application in real environments presenting different geographical conditions and scales, with very diverse socio-economic arenas and very different institutional and regulatory settings.

## 20.1 Introduction

In Greek mythology, the Naiads (Ναϊαδες) were the spirits of small brooks, fountains, wells, springs, and other freshwater bodies. Different to the river gods, the naiads were smaller, more adaptable with different shapes and forms. The EU H2020 NAIAD project that provided the background for the work presented in

E. López Gunn (✉) · B. Mayor · L. Vay
ICATALIST S.L., Las Rozas, Madrid, Spain
e-mail: elopezgunn@icatalist.eu

N. Graveline
University of Montpellier, UMR Innovation, INRAE, Montpellier, France

R. Giordano
CNR-IRA, Bari, Italy

N. Van Cauwenbergh
IHE Delft Institute for Water Education, Delft, Netherlands

P. Le Coent
G-Eau, BRGM, Université de Montpellier, Montpellier, France

P. van der Keur
Geological Survey of Denmark and Greenland (GEUS), Copenhagen, Denmark

R. Marchal
Caisse Centrale de Réassurance (CCR), Paris, France

© The Author(s) 2023
E. López Gunn et al. (eds.), *Greening Water Risks*, Water Security in a New
World, https://doi.org/10.1007/978-3-031-25308-9_20

this book takes inspiration from this freshwater ancient wisdom to look at disasters and the role of nature in risk prevention and management.

Particularly, by considering the prevention and reduction part of the disaster risk management cycle looking at nature, not just as part of the problem but as part of the solution. The ancient Greeks thought of the world's waters as all one system, which percolated in from the sea in deep cavernous spaces within the earth, to the sea. This systemic view on risks is very much at the heart of NAIAD. The approach is also focused on this versatility afforded by nature and the interest in understanding the protective role of nature-based solutions (NBS) in buffering risks posed by natural hazards through the development of natural assurance schemes (or NAS from now on).

Flood events have huge impacts worldwide. In Europe, numerous examples can be found from the past decade, that cause extensive damages (e.g., a cloudburst in Copenhagen in 2011, the Elbe floods in 2002, 2013, Danube floods in 2006, Alpes Maritimes floods in 2015, Lez floods in 2014, Seine floods in 2016 and 2018, Germany and Belgium floods in 2021, etc.). Around 90% of natural hazards are water-related and these are likely to become more frequent and more severe due to climate change. For example, climate change is projected to increase damages up to 50% by 2050 in France (Marchal et al., 2022), Chap. 3 this volume has presented a conceptual framework, a series of methods to implement this conceptual framework and their validation through examples to apply and test these in nine case studies, providing critical insights from practice.

This chapter reflects on what we have learnt from our conceptual frame, and the methods and tools to understand, assess and implement NAS. Our aim is that this conceptual frame, the tools and methods to develop NAS perdure in time, are adopted, improved and adapted to suit other contexts or challenges. This will lead over time to a better alignment of both the conceptual frame and methodologies to different contexts, institutional settings, risk types, scales, etc. to create a baseline for future actions to build on ecosystem services to mitigate water risks. One of the *reason d'étre* of the book is to show how others could develop their own natural assurance schemes, based on these tried and tested methods and tools, as well as others that might emerge to complement and strengthen the methods and tools presented here.

The book thus provides a comprehensive guide to select, assess and implement NBS considering the effectiveness of the implementation of these NBS specially for risk reduction, and thus the potential for investment based on a risk prevention and mitigation frame. An important lesson is that the identification and assessment of the co-benefits in the economic analysis of NBS considering the assurance value is built on the combination of the quantitative benefits (calculated avoided damages) and its qualitative co-benefits. The importance for context specific indicators to monitor the effectiveness and performance of NBS in DRR, for instance, is especially important. The recently published Handbook on NBS Impact will also help in this area of impact assessment and evaluation.

This final chapter will now revisit the main modules or blocks from the book, as well as the original questions we posed initially to summarise the main takeaways for the design, implementation, and evaluation of natural assurance schemes.

## 20.2   Conceptual Framing: What Added Value Does a Natural Assurance Scheme Bring into the Picture?

One of the main advances has been the conceptual framing developed that under-pins the concept of a "Natural Assurance Scheme". This is an area where future further work could go deeper and wider, to address the central role that nature-based solutions can play to address the most frequent and costly natural hazards: water risks, through the development of natural assurance schemes. That is, building on the potential to avoid damages and generate co-benefits through the use of nature-based solutions for water related hazards.

We have argued in this book that there is a subtle but important difference between the "assurance" value and the "insurance value" of nature based solutions. The "assurance value" is the protective value of nature and the regulation functions that can both mitigate and help prevent water risks. This value oftentimes comes accompanied by the additional co-benefits from other ecosystem services that are also generated at the same time, like biodiversity, carbon capture etc., that add addi-tional layers of value as multifunctional nature-based solutions. The insurance value we argue is the monetisation of this value as a risk transfer mechanism.

Assurance is a guarantee, a promise of something (Cambridge Dictionary). For the purpose of this book, we have defined Natural Assurance Schemes (Lopez Gunn et al., 2023 – Chap. 2 this volume) as "ecosystem-based risk reduction measures that reduce the level of risk in one area". The assurance value of nature, or of an ecosys-tem is the role that nature plays to help mitigate natural risks while also providing long term guarantees of the resilience of the ecosystem itself and in delivering flows of the full range of ecosystem services, like e.g. avoiding or minimising the risk of biodiversity loss, that will enable long term resilience of the socio-ecological system. When properly accounted and internalised through e.g., new SEEAW accounting or new sustainability company accounts, accrued benefits might be larger than short term risk reduction benefits. The Natural Assurance Scheme is underpinned by this central idea that nature both insures some assets in real monetary terms and that these schemes also assure (restores, protect) the ecosystem from anthropocentric threats. There is a double "dividend" from the Natural Assurance Scheme and double mate-riality in both financial and pure environmental returns.

The insurance value of nature or ecosystems is a money value and share of the insurance value that represents only the money value provided by the short-term protection provided by the NBS. It mimics the classical financial insurance instru-ment: an insurance service is a transfer mechanism between two parties- the insur-ance company and the subscriber that pays regular risk premiums in exchange for financial risk coverage, but there is no significant reduction in the global risk, it is just a reduction in the individual risk level (the damage will still occur). The eco-nomic value of the insurance value is conceptualized by Baumgärtner & Strunz (2014) as the value of one specific function of resilience: to reduce an ecosystem user's income risk from using ecosystem services under uncertainty. The nature assurance value is wider than just the financial insurance value: the compensation of the timely loss of a good or service is limited compared to the existence of an ecosystem and the provision of continued protection derived from healthy natural ecosystems that are fully functional (e.g. for the regulation of floods).

The following models can illustrate these different concepts:

- basic insurance scheme: asset owners pay a fee to have a certain level of risk, if they face more risk, these asset owners are compensated (not totally and in fact only for the monetary value). the insurance would pay out in case of an insured hazard occurring. Insurers and asset owners can decide or not to take action in terms of risk reduction investment (and as we demonstrated e.g. in the case of Lez, go for a grey strategy, a green strategy, a hybrid strategy or opt to do nothing).
- natural insurance scheme: an asset owner pays a fee to a fund; the fund maintains the basic and short-term functioning of the ecosystem regulation services, but could not substitute this by monetary value (ie. there is a limit in terms of substitution).
- natural assurance scheme: an asset owner and other stakeholder pay a fee or contribute with other values to maintain the ecosystem (the NBS) that ensures both short term risk reduction and long-term resilience of the SES.

We hope that other cases will emerge to deepen our understanding of these concepts as well as their quantification, and their limitations.

## 20.3   Physical Assessment

Nature-based solutions (NBS) have become a valid alternative to grey infrastructures for coping with climate-related risks in urban and rural areas alike. NBS are increasingly recognized for their capacity to foster the functioning of ecosystems and to generate additional environmental, economic and social benefits that are considered as essential backbones of actions for climate-change mitigation and adaptation.

Our overall aim is to better understand how to operationalise the assurance and/ or the insurance value of ecosystems, i.e. a better knowledge and methods to help both prevent and mitigate risks associated with water (floods and droughts), while helping to generate valuable co-benefits like biodiversity, health, recreation, etc.

The baseline to provide evidence of the role of NBS on risk management associated with water (e.g. floods and droughts), was to gather biophysical information at the nine case studies and characterise the biophysical hazards present at each location (see Sect. 2.1). The ecosystem services delivered in each of the case studies were assessed applying several tools, methods and approaches at different levels depending on the readiness of the case studies. These approaches were further developed or adapted to the case studies, e.g. with decision support tools like

Eco:actuary, which assesses the role of natural capital in hazard mitigation (applied at the large scale case studies of the Thames and the Danube), and a set of tailored approaches suitable for the different conditions present in the rest of the case studies (with different geographical settings and scales). To assess the impact of the different NBS, monitoring stations that gather environmental data in real time (named Freestations) were set out in two of the case studies (Thames and Lower Danube) as an early warning system for near real time forecasting (see Sect. 20.4). Plausible climate, land and ecosystem scenarios were used to assess the role of ecosystems in providing ecosystem services in different conditions and a series intervention scenarios were explored for each case study.

## 20.4   Codesign and Stakeholder Participation: Lessons Learnt and Next Steps

The engagement of different stakeholders in the co-design and assessing the NBS is key for enhancing the social acceptance of these NBS solutions. Specifically, case study results showed that addressing the socio-institutional barriers – e.g. the lack of community's engagement, low level of institutional cooperation, lack of community risk awareness, etc. – could be even more important that overcoming the physical barriers. The experiences carried out in the different case studies demonstrated the suitability of two approaches. Firstly, ambiguity analysis allowed us to account for the diversity of risk perception, addressing potential conflicts among different groups of stakeholders in the early phases of the NBS co-design. Secondly, the participatory modelling approach enabled a collective learning process enhancing participants' understanding about the need to adopt an integrated approach in NBS design and implementation. In building the model for the design of the NBS, stakeholders become aware of the wide range of potential barriers hampering the NBS implementation, and of the need to define socio-institutional measures capable of transforming the barriers into enabling factors.

Co-benefits (reduced air pollution, reduction of heat in cities, improved landscape, climate change mitigation…) represent the largest share of the value generated by NBS strategies. They are needed for these solutions to be economically beneficial, despite the fact that these measures were initially designed to reduce water risks.

When all costs and benefits are considered in cost benefit analysis (CBAs), the overall cost-efficiency of NBS strategies appears to be context-specific, with positive and negative CBA results. The cost-benefit ratio nevertheless remains superior to alternative grey solutions for all of our case studies. In order to improve the economic balance of NBS projects, developers should consider closely the impact of land cost and the choice of NBS that maximize the production of co-benefits in addition to their risk reduction function.

- The assessment of NBS co-benefits represents a challenge considering the diversity of values associated with NBS: physical, socio-cultural and monetary values. The use of combined approaches is required to fully grasp the value of NBS co-benefits. Among these, the monetary estimation of co-benefits are key to convince decision makers and investors of the economic advantages of NBS. Economic methods relying on the involvement of the public such as stated choice approaches may be particularly useful to reveal peoples' preference for NBS.
- It is important to consider that NBS can also be associated with negative effects/disbenefits that should not be overlooked, since they may have a strong impact on residents' acceptance and the value local residents put on NBS. In addition, the heterogeneity of perception of NBS co-benefits may generate equity issues like for examples difficulties in the social acceptance of NBS when these harm some people and benefit others.
- Our results suggest that co-benefits represent a large share of the overall benefits of NBS aimed at reducing water risks. Cobenefits are therefore as important as risk-reduction benefits and should therefore be systematically considered in project design and assessment.

From a non-economic assessment of co-benefits, two main lessons were learned concerning the production or generation of co-benefits. Firstly, the stakeholders' engagement showed that, in several cases, the generation of ancillary environmental, economic and social benefits can be considered as the actual drivers of NBS implementation. Secondly, we learned that diverse beneficiaries might have diverse perception of – and preferences – over the co-benefits. The production of one co-benefit might hamper the production of others – e.g. the creation of wetlands could reduce agricultural productivity – causing potential conflicts between stakeholders with different preferences. The work done demonstrated the importance of detecting potential trade-offs among different co-benefits in the early phases of NBS design. Therefore, the co-benefits definition was used as the basis for designing the most suitable NBS, accounting for the potential conflicts due to the trade-offs.

## 20.5 Economic Valuation of NBS for Risk Reduction and Co-benefits

We developed an economic assessment framework, with detailed guidelines aimed at comparing the main costs and benefits generated by NBS for water related risks (Chap. 6). We particularly described and implemented methods for the monetary assessment of different costs and benefits:

- Costs of implementation are those that are necessary for the implementation and maintenance of the NBS included in the NBS strategies

- Opportunity costs are related to the loss of benefits of areas that are taken out of production, or land that is used for NBS and that cannot be used for other profitable purposes such as the construction of buildings. These are the indirect costs of the NBS strategies
- Avoided damages are the damages avoided due to the reduction of water risks generated by NBS strategies. Avoided costs are the primary benefit generated by NBS strategies aiming at reducing water risks.
- Co-benefits are the additional environmental, economic, and social benefits generated by NBS.

The economic assessment subsequently compares these costs and benefits over the life-time of alternative projects, grey, hybrid and NBS, with a Cost-Benefit Analysis. This assessment methodology was partially implemented in four case studies (Lower Danube, Thames, Medina and Copenhagen) and fully implemented in three (Lez, Brague and Rotterdam) with the following conclusions.

The cost of implementation of NBS appears to be lower than the cost of grey solutions for the same level of reduction of water risks (Braque and Rotterdam). This reinforces claims about the cost-effectiveness advantage of NBS and would urge decision makers to consider more systematically these solutions to address water risks. However, in urban areas, taking into account the opportunity costs of NBS can change the appreciation of their cost advantage. NBS may indeed take space that may not be available for real estate development. Considering that NBS can require a large area to be implemented as compared to traditional grey solutions, the inclusion of opportunity costs has a strong weight in the overall cost estimation, especially in urban areas where land cost is high. In terms of benefits, NBS have a significant impact on the reduction of water risks that translate into the monetary benefits of damage reduction. In our cases, the monetary benefits related to the reduction of flood damages are however not sufficient to fully cover capital expenses, operation and maintenance costs. This problem is however even more serious for grey solutions. Meanwhile, the economic value of co-benefits (reduced air pollution, reduction of heat in cities, improved landscape, climate change mitigation...) is very significant and may be the stronger argument for the development of NBS for water risks. Finally, there are no clear-cut conclusions on the results of the Cost-Benefit Analysis of NBS in our assessments. Indeed, NBS strategies have a Benefit-Cost-Ratio close to 1 or slightly superior in Lez and Brague and below 1 in Rotterdam. The picture is however more positive if we exclude opportunity costs from the economic analysis. Importantly, for Brague and Rotterdam, the economic efficiency of NBS strategies is much higher than the economic efficiency of grey strategies.

Our conclusion of that the large share of co-benefits in the overall value of NBS aimed at reducing water risks, combined with the limited avoided damages, has strong implications for NBS funding and business models. Indeed, support from sectoral policies is generally conditioned to a positive cost-benefit analysis on the specific benefit they target, such as for example flood risk reduction. However, NBS appear to be economically efficient only when all the benefits they generate are considered. Implications for project set up and financing are very significant. Rules applying for the public funding of NBS should therefore be adapted in order to take into account cross-sectoral benefits of NBS. This requires modifications of the silo approach currently still prevailing in the application of public water risk policies.

## 20.6   Decision Making Processes

We present a methodology developed to fully consider the biophysical, social and economic assessment of these potential natural assurance schemes, in a co-designed approach with stakeholders. The interactions with stakeholders are fundamental to engage the local community and decision makers to work together on risk perception and the potential NBS offer to address water related hazards in specific regions and locations from the city level to transboundary basins.

Several levels of integration exist; from science-society-policy integration, through disciplinary integration, through multiple objectives and multiple-method integration over time, space, resource and sectors. Analyzing decision processes ranging from technical decisions in a disaster risk reduction context to strategic planning of climate proof development can be a versatile methodological approach, as it combines a combination of data, information, stakeholders and procedures that are to be integrated in – for example – multi-criteria decision making, adaptive planning etc. As NBS address both hydrological risk reduction and the generation of a number of co-benefits, decisions will involve some level of multi-criteria analysis, where multiple objectives or benefits can be assessed with indicators.

This book showcases a series of tools and methods to help integrate these assessments, like adaptive planning, the natural assurance business canvas, or the financing framework for water security. This integration of knowledge and experience aims to come up with viable projects that can be implemented on the ground. The role of insurance is also incorporated into this analysis, i.e. the roles the sector can play in the effective implementation of NBS and in disaster risk reduction and prevention. The different roles of the insurance sector are discussed in relation to investment, new insurance models, data or through new modes of public private partnerships that ensure the insurability of the system under climate change scenarios.

A conscious use of information and the involvement of stakeholders in the framing of problems, the range of possible solutions and the metrics to assess e.g. the effectiveness of these solutions, will facilitate learning throughout the different steps of the planning process. This means that local stakeholders, consultants or experts involved in the evidence generation and communication around NBS have to pay specific attention to the needs of their audience when communicating information. This is also an essential part of managing uncertainty in the form of ambiguity and can help to avoid potential conflicts in the development of NBS-based adaptation plans.

The importance of capacity building in NBS across organizations and throughout the entire planning process is illustrated by Droste et al. (2017). It is however important to note that stakeholders are not only the recipients of capacity building, some stakeholders also build capacity either of other stakeholders or the very experts/ managers mandating the planning process. Different actors in the process can improve the overall understanding of the system, the enabling environments, the definition of the problem, the identification of potential solutions, find ways on how to assess preferences and impacts, decide implementation arrangements are feasible and how to monitor and adapt NBS when changes in performance are assessed.

## 20.7    Business Models, Enabling Frameworks and Investments for Risk Prevention and Reduction Through Nature Based Solutions

One of the main challenges in ecosystem services supply, e.g. in insurance/regulation, is how to internalize the positive contribution that ecosystem services can bring to society or the economy, which normally are unaccounted for, and which if perceived and valued, could help to change behaviours. We have aimed to reflect on how to create value chains associated to risk reduction and other ecosystem services, as well as on how to set up the right incentives. Innovative business models – which tell the story of how value is created - have a financial and an institutional function. The cash profile of a project can be improved for instance by community in kind contributions for maintenance and/or to safeguard of ecosystem health. Changes towards NBS can be achieved in three ways, (i) the insurance as commercial proposition, (ii) government interventions (e.g. through economic instruments), (iii) markets for externalities[1] (which could also be seen as economic instruments). A number of chapters in the book have provided insights on how to effectively mobilize money for investments required, how to enable the economic transfer of values between investors, beneficiaries and "polluters" e.g. what are the business models and economic instruments that could help making this internalization of values happen (see Sect. 20.7).

The innovative business models, economic and financial approaches developed have focused on gathering and generating the required knowledge and tools to work out and operationalise the financing aspects required to take NBS or NAS projects from design to implementation.

This goal has been achieved through four stages: description of existing funding and financing instruments for NBS, the development of Natural Assurance Schemes

---

[1] Externalities are costs (negative externalities) or benefits (positive externalities), which are not reflected in free market prices. Externalities are sometimes referred to as 'by-products', 'spillover effects', 'neighbourhood effects' 'third-party effects' or 'side-effects'.

canvas, the elaboration of a Financing Framework for Water Security and documenting successful business models (Box 20.1).

**Box 20.1 The Value of Prevention**

Expertise in catastrophe risk modelling has been applied to the two French case studies (Brague and Lez) to assess the potentially avoided damages related to the implementation of preventive measures. For the hazard part, the overflow and runoff hazards have been modelled and mapped integrating locally based information at a 25 m-resolution. It produced maps of overflow and runoff for the studied flood events that occurred in these catchments for the October 2015 for Brague and the 2014 Cevenol events for Lez. Secondly, CCR provided information about the insured damage for the studied events at the catchment scale. The four 2014-flood events on the Lez case study represented 65 M€ of losses. The 2015 flood event on the Brague case study represented 200 M€ of losses.

*Developing damage curves*

Specific damage curves have been developed for the events and these two case studies, focusing only on residential homeowners. Damage curves are the correlation between hazard characteristics (height or flows) and observed damages defined by the destruction rate (the ratio of total claims divided by the insured value).

For the Brague, the damage curves for the overflow show a destruction rate of 40% for 1.5 to 3.5 m, with a 20 cm of water threshold on damages, which is clearly visible. This is the threshold when the electricity network is damaged. For the Lez, the damage curves for runoff hazard show a destruction rate of 12% for 4 to 5 m3/s. This high level of sinistrality is related to the high intensity of the September 2014 event.

*Implementing damage curves to assess the effectiveness of fictive preventive measures*

To then fulfil the objective of assessing NBS effectiveness on avoided damages for both case studies, a solution was developed by CCR. Once the damage curves were calibrated with the two events mentioned earlier and for residential homeowners only, it was possible to assess the effect of a potential percentage of runoff hazard reduction on insured damages.

- For the Brague: simulated and calibrated results show that with a reduction of 50% runoff hazard, the damages would be reduced by 45%. The damage could be lower at 2.2 M€ (as the current damage are 4 M€ for residential homeowners)

(continued)

- For the Lez: simulated and calibrated results show a reduction of 50% runoff hazard, which would reduce the damages to ~ 1.9 M€ (or −40.45%) (as the current damage are 3.3 M€ for residential homeowners)

***Considering the future consequences of climate change, how important should prevention be?***

Finally, Marchal et al. (this volume) introduced the climate change scenario of RCP8.5 to assess future exposure of the two case studies based on the potential climate change impacts on insured losses for the year 2050 (CCR, 2018). The evolution of the damage between the current climate and future climate was developed using calibrated damage curves on residential homeowners.

- For the Brague case study: the result highlights that climate change at horizon 2050 will increase losses by 25.5% for individual homeowners only. In order to maintain the losses to the current business as usual, the runoff hazard has to be reduced by 40% to limit the effect of climate change.
- For the Lez case study the losses are estimated to increase by 30% for the year 2050 according to the RCP 8.5. Then, CCR calculated the estimated percentage of potential hazard reduction, using NBS, to reduce the impacts from climate change compared to business-as-usual scenario estimated losses. CCR estimated that the runoff hazard must be reduced by 35%, using NBS, to limit the impact of climate change compared to the current business-as-usual losses.

The results summarized demonstrate the potential to use catastrophe risk models to assess the effectiveness of loss prevention and to the importance of sharing this information with stakeholders. It also demonstrates that preventive measures have to be ambitious to reduce the losses toward a bearable threshold today and in the future.

## 20.8   Capacity Building and Additional Resources – Do Your Own NAS

Training activities are a central element for the uptake of Natural Assurance Schemes. As part of the NAIAD project that supported the conceptual, methodological and case studies presented in this book a Massive Online Open Course "Greening Risk Reduction with Nature Based Solutions" on how to integrate the insurance value of ecosystem in environmental planning and infrastructure investment was designed and launched. The Materials to support the course introducing and explaining the work developed under the framework of the project. In addition

an Electronic Guide on all the materials, publications and resources available was
also developed and is freely available (see Box 20.2 Below).

---

**Box 20.2 The H2020 NAIAD Eguide Online Resource**

The E-GUIDE is a tool whose objective is to support external users to navigate through the extensive H2020 NAIAD results, so that interested parties can take advantage of these tools and examples. It also aims to make it easier for users to explore its possibilities, including those at the scientific level that are relevant to study local problems and formulate solutions.

The E-GUIDE is aimed at a wider audience of potential users of NAIAD's models, tools and methods, and all those who are interested in acquiring knowledge about its results. The models, tools and methods generated are presented schematically in compact web pages and in a language understandable to a non-expert audience. Particular attention is paid to the following questions that potential users of NAIAD's products might ask regarding each product:

What are the categories of users (technical, political, interested observer,…) for whom the product can be most useful?

What are the applications of the product for decision making/management in the water sector?

What is the added value of the product?

What management decisions/processes can be strengthened or improved thanks to this product?

For those readers interested in obtaining additional information on the products presented, references are provided to the relevant reports and publications generated within the framework of NAIAD, hyperlinks to the corresponding web pages, and contact details of the persons responsible for the product within the team of the draft.

The E-GUIDE is structured in a series of categories that collect the different types of knowledge and tools generated as a result of the project. These categories include the following:

(a) Information, reports and knowledge acquired in the project's DEMOs
(b) Models for evaluating the costs and benefits of SBNs and the costs of their implementation (including their insurance value)
(c) Publications, including scientific texts, and the hyperlinks to access them
(d) Access to the MOOC (Massive Online Open Course) on the implementation of SBNs and other training activities generated within the framework of NAIAD.
(e) Tools for use in various contexts

(continued)

These categories can be accessed through the web from the following entries:

(f) NAIAD Strategic Objectives
(g) Specific questions
(h) Specific use cases
(i) Guide to Nature-Based Solutions
(j) Type of result or product
(k) Demos or case studies

Therefore, this guide allows you to create a customized and ordered view of the knowledge and products generated by the project, allowing the user to navigate and access those that are of greatest interest, with a lower or higher level of detail.

## 20.9 Lessons Learnt and Main Conclusions

Main lessons learned are presented on the role of Natural Assurance Schemes and implemented nature-based solutions and demonstrated for a broad range of case studies across Europe. This books summarises the knowledge acquired, its integration and application to real environments presenting different geographical conditions and scales, from the neighbourhood scale in the Rotterdam case study to the very large scale of the Danube basin, with very diverse socio-economic conditions and very different institutional and regulatory settings. As a result of this not only have new knowledge, methods and tools emerged from the work presented in this book, but also lesson learnt can be shared and recommendations made for an effective implementation of NBS for risk reduction. NAS as a specific type of NBS schemes become a strategy in the reduction of risks associated with water-related climate events which are anticipated to increase in frequency and intensity under climate change. Biophysical assessments in most cases were done by physically based models, e.g. a groundwater model in Copenhagen and a hydraulic model in the Brague catchment, as well as ecosystem services DSS (Eco-Actuary) for the Thames catchment.

The main lessons learned from the biophysical assessments were (i) that evidence based decision making is key in the effective implementation of NBS. Monitoring and modelling are essential for evaluating the effect of Natural Flood Management (NFM) and NBS interventions; (ii) NBS strategies implemented as area-based interventions (e.g. conservation agriculture) have greater potential to reduce flood risk and provide more valuable co-benefits than point based interventions (e.g. leaky dams or retention ponds, as demonstrated for the Thames catchment); (iii) Operational and Maintenance costs of NBS interventions must be accounted for and kept low to be an attractive alternative for grey or hybrid solutions. Maintenance of NBS are not always well known and may have high costs, but without maintenance, interventions can fail, as for traditional grey solutions, and worsen flood risk.

The main lessons learnt from the social assessments were: (i) NBS effective implementation requires effective cooperation among different decision-makers (e.g. municipalities, river Basin authorities, regional government) and between them and the main stakeholders (e.g. local citizens, end users, NGOs); (ii) Eliciting stakeholders' perception about co-benefits plays a key role in NBS co-design, and not only in the NBS assessment; (iii) Ambiguity in risk perception should be considered as an enabling factor for NBS design and implementation rather than a barrier.

The main lessons learned from the economic assessment were: (i) Co-benefits represent the largest share of the value of NBS, despite the fact that these were initially designed to address water-related risks; (ii) The cost of implementation of NBS is lower than the cost of grey solutions for the same level of water risk management, confirming the cost-effectiveness advantage of NBS. Benefits in terms of avoided damages alone are however generally not sufficient to fully cover investment and maintenance costs; (iii) NBS cannot automatically be assumed to present benefits larger than their cost of implementation and opportunity costs. The evaluation of several combinations of NBS strategies aiming at maximizing the benefit cost ratio is therefore needed.

The main lessons learned for deploying an integrative framework were: (i) An integrated, multidisciplinary, co-designed framework is key to analyze the NBS multifactorial effects. For risk assessment, checking the physical effectiveness is a first key step before addressing other co-benefits. And in parallel communicating and making the information understandable are both necessary and crucial to assess NBS effectiveness; (ii) A new hybrid design methodology has emerged to combine classical and eco-engineering approaches with decision-aiding frameworks (e.g. multi-criteria, economic methods) to support decision-making process; (iii) a Climate Change perspective is necessary for understanding the impact of NBSs and also how climate change itself could affect the effectiveness of NBS under different climate change scenarios.

Main lessons learned from innovative business models and financial instruments: (i) One of the main difficulties in building business models was to engage indirect beneficiaries of co-benefits within the pool of payers and funders. This is critical because co-benefits often have an even higher value than the risk reduction itself; (ii) Legislation can become both a critical enabler and barrier for the development and implementation of business models for NAS schemes. In the case of the EU, the environmental legislation plays a critical driver pushing the interest in NBS and opening opportunities for its implementation; (iii) the Importance of flexible tools that enable replication, e.g. NAS canvas is a flexible and replicable tool applicable to any NAS scheme or NBS strategy regardless of the stage or the context; (iv) Normally, the proponents of NBS are organizations with an advocacy and/or scientific background with limited experience in public and private investment planning processes. As a result, often NBS pilots and demonstration projects are shaped more as awareness raising projects than as "investment projects" that could attract funds from either public authorities aiming at reducing a risk or private impact investors willing to accept lower returns in exchange for social and environmental impacts; (v) There is an Information gap between evidence proposed by NBS proponents and the required by public and private investors or implementers. The criteria and level of detailing regarding implementation costs and risks differ greatly between the project descriptions of NBS proponents and the requirements for allocation of public funding or granting of loans by impact investors; (vi) In order to move towards the implementation at scale of NBS it is of extremely important to move from pilots and monitoring systems that are designed to raise awareness towards the real monitoring of systems that develop the evidence base and the baseline required to move towards performance-based contracts and payment mechanisms.

Main lessons learned from policy uptake: (i) Scaling investment still requires enhanced coordination, capacity, and confidence among public authorities that would be primarily responsible for accessing their financing and overseeing their implementation. The demonstration cases also show that the implementation landscape across the EU is diverse. In many countries prioritized actions from the EU may still be required to ensure water-risks are appropriately recognised by governments and citizens, and that NBS are viable options for risk reduction; (ii) The use of NBS as part of strategies to maintain the insurability of assets under changing climate scenarios should be more strongly promoted. This may be a more urgent political motivator to local populations and governments than their potential to reduce the impacts from catastrophic hazards; (iii) Challenges to finance NBS show a "market gap", not only a funding gap. Under investment in NBS is seen a failure, at large, of those looking to access financing to produce viable projects to sustain and pay back investment. It also, however, signals a failure in the market where the value of services provided by NBS may be undervalued, or the values provided are not monetized sufficiently or through other KPIs to enable investment, or familiar finance instruments seem ill-fitted to non-traditional investments. Each of these scenarios may require public sector interventions and should be better analysed.

Our results provide valuable evidence that co-benefits represent the largest share of the NBS benefits in the design and implementation of Natural Assurance Schemes. Therefore co-benefits should be defined at an early stage to support the NBS co-design, and the structuring of their funding and financing. However, policy makers and decision makers have to be realistic: NBS alone will not be able to completely reduce the impacts from large events. However, NAS can however play an important role to help increase overall system resilience and help to reduce the effects of less frequent extreme events, reducing running or operational costs, and overall pressure on the system. This is the case for example of the urban water buffer in Rotterdam, which has allowed a football stadium to cope better with both drought and flood events through its continuity of service with water guaranteed through rainwater harvesting for the football fields.

This frees capacity and additional human and financial resources and thus allows for a higher response capacity during large events. Another lesson learnt is the critical importance of local knowledge and capacity, the tacit knowledge of stakeholders is critical not only for the problem definition itself, but also to fully document the range of social, environmental and economic benefits derived from NBS which by their very nature are multifunctional. Thus end user co-design is essential to fully characterize the problem and viable implementation contexts, options and key potential barriers and opportunities early in the design. It also raises awareness but most importantly, it increases the legitimacy and acceptance of the process and the chosen outcomes. Finally, we have also learnt how the regulatory framework and policy framing can help provide the right incentives to the different actors to reach collective natural assurance schemes at these locations.

In terms of future work, the conceptual frame and methodology provided seeks to provide a venue for practitioners, researchers, and others to help advance the entry points for NAS projects for risk reduction, advance a better quantification of the benefits of NAS, and finally, provide additional evidence of what works and what does not work. In short to develop a deeper understanding of the NAS concept and its potential implementation.

# References

Baumgärtner S, Strunz S (2014) The economic insurance value of ecosystem resilience. Ecol Econ 101:21–32

CCR (2018) Département Analyses et Modélisation Cat Conséquences du Changement Climatique sur le Coût des Catastrophes Naturelles en France à Horizon 2050. CCR, Paris, France

Droste N, Schröter-Schlaack C, Hansjürgens B, Zimmermann H (2017) Implementing nature-based solutions in urban areas: financing and governance aspects. In: Nature-based solutions to climate change adaptation in urban areas. Springer, Cham, pp 307–321

Printed in the United States
by Baker & Taylor Publisher Services